T0360539

LECTURES IN
NONLINEAR FUNCTIONAL ANALYSIS

Synopsis of Lectures Given at the Faculty of Physics
of Lomonosov Moscow State University

LECTURES IN
NONLINEAR FUNCTIONAL ANALYSIS

Synopsis of Lectures Given at the Faculty of Physics
of Lomonosov Moscow State University

Maxim O Korpusov

Lomonosov Moscow State University, Russia

Alexey V Ovchinnikov

Lomonosov Moscow State University, Russia

Alexander A Panin

Lomonosov Moscow State University, Russia

NEW JERSEY · LONDON · SINGAPORE · BEIJING · SHANGHAI · HONG KONG · TAIPEI · CHENNAI · TOKYO

Published by

World Scientific Publishing Co. Pte. Ltd.

5 Toh Tuck Link, Singapore 596224

USA office: 27 Warren Street, Suite 401-402, Hackensack, NJ 07601

UK office: 57 Shelton Street, Covent Garden, London WC2H 9HE

Library of Congress Control Number: 2021059099

British Library Cataloguing-in-Publication Data
A catalogue record for this book is available from the British Library.

LECTURES IN NONLINEAR FUNCTIONAL ANALYSIS
Synopsis of Lectures Given at the Faculty of Physics of Lomonosov Moscow State University

ISBN 978-981-124-892-4 (hardcover)
ISBN 978-981-124-893-1 (ebook for institutions)
ISBN 978-981-124-894-8 (ebook for individuals)

For any available supplementary material, please visit
https://www.worldscientific.com/worldscibooks/10.1142/12603#t=suppl

Printed in Singapore

Dedicated to the shiny memory of Professors
Il'ya Andreevich Shishmarev
and
Stanislav Ivanovich Pokhozhaev

Preface

This monograph is addressed to mathematicians and physicists who apply methods of nonlinear functional analysis to the study of various problems of mathematical physics. It is based on lectures given by the authors for graduate and undergraduate students specializing in mathematical physics at the Faculty of Physics of the M. V. Lomonosov Moscow State University. We introduce and discuss in detail the notions of Gâteaux and Fréchet derivatives and various types of continuity of nonlinear operators. We describe variational methods, including M. A. Krasnosel'sky's theory of genus of a set. Galerkin's method combined with the monotony and compactness methods, the method of lower and upper solutions, and the topological Schauder principle are presented. The monograph ends with chapters devoted to the abstract Picard theorem, which generalizes the classical theorem on the existence and uniqueness of a solution to an ordinary differential equation, and its applications to Sobolev-type equations.

The authors are deeply grateful to Professors A. G. Sveshnikov, A. N. Bogolyubov, and N. N. Nefedov for helpful discussions and to the reviewers, Professors V. Yu. Popov, S. A. Zagrebina, and M. V. Falaleev for valuable remarks that significantly improved the text.

Contents

Part 2. Nonlinear Operators: Continuity, Differentiability, and Compactness   51

Part 3. Variational Methods of the Study of Nonlinear Operator Equations 119

PART 1
Functional Analysis: Preliminaries

Lecture 1

Linear Operators

In this lecture, we introduce the important notion of the transposed operator and prove the theorem on the equality of duality brackets. Note that in this lecture, we consider only *linear* operators.

1.1 Notation

Let $\mathbb{E}$ and $\mathbb{F}$ be two Banach spaces with the norms

$$\|\cdot\|_e \quad \text{and} \quad \|\cdot\|_f$$

and the corresponding dual Banach spaces $\mathbb{E}^*$ and $\mathbb{F}^*$ with respect to the duality bracket

$$\langle e^*, e \rangle_e \quad \text{for all } e \in \mathbb{E} \text{ and } e^* \in \mathbb{E}^*$$

and

$$\langle f^*, f \rangle_f \quad \text{for all } f \in \mathbb{F} \text{ and } f^* \in \mathbb{F}^*.$$

In a standard way, we introduce the duality brackets between the pairs of Banach spaces $(\mathbb{E}^*, \mathbb{E}^{**})$ and $(\mathbb{F}^*, \mathbb{F}^{**})$:

$$\langle e^{**}, e^* \rangle_{e^*} \quad \text{for all } e^* \in \mathbb{E}^* \text{ and } e^{**} \in \mathbb{E}^{**}$$

and

$$\langle f^{**}, f^* \rangle_{f^*} \quad \text{for all } f^* \in \mathbb{F}^* \text{ and } f^{**} \in \mathbb{F}^{**}.$$

The zero element of a Banach space $\mathbb{E}$ is denoted by $\theta_{\mathbb{E}}$ or simply by θ if this does lead to confusion.

In this monograph, we will often use the following notions.

Let $\mathbb{B}$ be a separable real Banach space, $\mathbb{B}^*$ be the dual space, and $\langle \cdot, \cdot \rangle$ be the duality bracket between them. We denote by $\|\cdot\|$ and $\|\cdot\|_*$ the norms of the Banach spaces $\mathbb{B}$ and $\mathbb{B}^*$.

Definition 1.1. An operator $A : \mathbb{B} \to \mathbb{B}^*$ is said to be

    (i) *radially continuous* if for all fixed $u, v \in \mathbb{B}$, the real-valued function $s \mapsto \langle A(u + sv), v \rangle$ is continuous on the segment $[0, 1]$;

3

(ii) *demicontinuous* if the condition "$u_n \to u$ strongly in $\mathbb{B}$" implies that

$$Au_n \rightharpoonup Au \quad \text{$*$-weakly in } \mathbb{B}^* \text{ as } n \to +\infty;$$

(iii) *Lipschitz continuous* if there exists a constant $M > 0$ such that

$$\|Au - Av\|_* \leq M\|u - v\| \quad \text{for any } u, v \in \mathbb{B}.$$

Remark 1.1. Note that the demicontinuity of an operator $A : \mathbb{B} \to \mathbb{B}^*$ implies its radial continuity. Indeed, let $\{s_n\} \in [0, 1]$ and

$$s_n \to s \quad \text{as } n \to +\infty.$$

Obviously, $s \in [0, 1]$. Moreover, for any $u, v \in \mathbb{B}$,

$$u + s_n v \to u + sv \quad \text{strongly in } \mathbb{B} \text{ as } n \to +\infty.$$

Therefore, due to the demicontinuity of the operator A, we have

$$\langle A(u + s_n v), v \rangle \to \langle A(u + sv), v \rangle \quad \text{as } n \to +\infty.$$

Therefore, the function

$$\varphi(s) := \langle A(u + sv), v \rangle$$

belongs to the class $C[0, 1]$ for any fixed $u, v \in \mathbb{B}$.

Next, we present various versions of the monotonicity property for operators.

Definition 1.2. Let u and v be arbitrary elements from $\mathbb{B}$. An operator $A : \mathbb{B} \to \mathbb{B}^*$ is called

(i) *monotonic* if

$$\langle Au - Av, u - v \rangle \geq 0;$$

(ii) *strictly monotonic* if

$$\langle Au - Av, u - v \rangle > 0 \quad \text{for } u \neq v;$$

(iii) *strongly monotonic* (with the monotonicity constant m) if

$$\langle Au - Av, u - v \rangle \geq m\|u - v\|^2, \quad m > 0;$$

(iv) *locally bounded* if for any fixed $u \in \mathbb{B}$, there exist constants $\varepsilon > 0$ and $M > 0$ such that $\|Av\|_* \leq M$, where $\|u - v\| \leq \varepsilon$.

Finally, we recall the definition of coercive operators.

Definition 1.3. An operator $A : \mathbb{B} \to \mathbb{B}^*$ is said to be *coercive* if there exists a real-valued function $\gamma(s)$ defined on $[0, +\infty)$ satisfying the limit property

$$\lim_{s \to +\infty} \gamma(s) = +\infty$$

and such that

$$\langle Au, u \rangle \geq \gamma(\|u\|)\|u\|.$$

1.2 Transposed Operator and Its Norm

We denote by $\mathscr{L}(\mathbb{E}, \mathbb{F})$ the space of all linear continuous operators acting from a Banach space $\mathbb{E}$ into another Banach space $\mathbb{F}$. Let $T \in \mathscr{L}(\mathbb{E}, \mathbb{F})$. This space is a Banach space with the norm

$$\|T\|_{e \to f} \stackrel{\text{def}}{=} \sup_{\|e\|_e = 1} \|Te\|_f.$$

Lemma 1.1. *For an arbitrary operator* $T \in \mathscr{L}(\mathbb{E}, \mathbb{F})$, *the following inequality holds:*

$$\|Te\|_f \le \|T\|_{e \to f} \|e\|_e \quad \text{for all } e \in \mathbb{E}.$$

Remark 1.2. In particular, due to Lemma 1.1, we have

$$|\langle e^*, e \rangle_e| \le \|e^*\|_{e^*} \|e\|_e \quad \text{for all } e \in \mathbb{E} \text{ and } e^* \in \mathbb{E}^*.$$

Definition 1.4. For an operator T, the *transposed operator*

$$T^t : \mathbb{F}^* \to \mathbb{E}^*, \tag{1.1}$$

is defined as follows:

$$\langle T^t f^*, e \rangle_e \stackrel{\text{def}}{=} \langle f^*, Te \rangle_f \quad \forall e \in \mathbb{E}, \ \forall f^* \in \mathbb{F}^*. \tag{1.2}$$

Theorem 1.1. *If* $T \in \mathscr{L}(\mathbb{E}, \mathbb{F})$, *then* $T^t \in \mathscr{L}(\mathbb{F}^*, \mathbb{E}^*)$. *Moreover,*

$$\|T\|_{e \to f} = \|T^t\|_{f^* \to e^*}.$$

Proof. First, we define the norm in the space $\mathscr{L}(\mathbb{F}^*, \mathbb{E}^*)$, which is obviously a Banach space since $\mathbb{E}^*$ is a Banach space.

1. We have

$$\|T^t\|_{f^* \to e^*} \stackrel{\text{def}}{=} \sup_{\|f^*\|_{f^*} = 1} \|T^t f^*\|_{e^*},$$

where the norms in the dual spaces $\mathbb{E}^*$ and $\mathbb{F}^*$ are defined in a standard way:

$$\|e^*\|_{e^*} \stackrel{\text{def}}{=} \sup_{\|e\|_e = 1} |\langle e^*, e \rangle_e|, \quad \|f^*\|_{f^*} \stackrel{\text{def}}{=} \sup_{\|f\|_f = 1} |\langle f^*, f \rangle_f|.$$

First, we prove that the operator T^t is linear. We have

$$\begin{aligned}
\langle T^t(\alpha_1 f_1^* + \alpha_2 f_2^*), e \rangle_e &\stackrel{\text{def}}{=} \langle (\alpha_1 f_1^* + \alpha_2 f_2^*), Te \rangle_f \\
&= \alpha_1 \langle f_1^*, Te \rangle_f + \alpha_2 \langle f_2^*, Te \rangle_f \\
&= \alpha_1 \langle T^t f_1^*, e \rangle_e + \alpha_2 \langle T^t f_2^*, e \rangle_e \\
&= \langle \alpha_1 T^t f_1^* + \alpha_2 T^t f_2^*, e \rangle_e
\end{aligned}$$

for all $\alpha_1, \alpha_2 \in \mathbb{R}^1$ and $f_1^*, f_2^* \in \mathbb{F}$.

2. Now we prove the boundedness of the operator T^t. By Lemma 1.1 and Remark 1.2, we have

$$\left| \langle T^t f^*, e \rangle_e \right| = \left| \langle f^*, Te \rangle_f \right| \le \|f^*\|_{f^*} \|Te\|_f \le \|f^*\|_{f^*} \|T\|_{e \to f} \|e\|_e. \tag{1.3}$$

From the definition of the norm of a linear operator (in particular, a linear functional) we obtain

$$\|T^t\|_{f^* \to e^*} = \sup_{\|f^*\|_{f^*}=1} \|T^t f^*\|_{e^*} = \sup_{\|f^*\|_{f^*}=1} \sup_{\|e\|_e=1} \left| \langle T^t f^*, e \rangle_e \right|. \tag{1.4}$$

Without loss of generality, we assume that $f^* \neq \theta$ and $e \neq \theta$ in the inequalities (1.3). Then from (1.3) we obtain the inequality

$$\left| \left\langle T^t \frac{f^*}{\|f^*\|_{f^*}}, \frac{e}{\|e\|_e} \right\rangle_e \right| \leq \|T\|_{e \to f}.$$

This and (1.4) imply the inequality

$$\|T^t\|_{f^* \to e^*} \leq \|T\|_{e \to f}. \tag{1.5}$$

Therefore, the operator T^t is bounded, $T^t \in \mathscr{L}(\mathbb{F}^*, \mathbb{E}^*)$.

3. Now we prove that the inequality

$$\|T\|_{e \to f} \leq \|T^t\|_{f^* \to e^*}. \tag{1.6}$$

Since $T^t \in \mathscr{L}(\mathbb{F}^*, \mathbb{E}^*)$, due to Lemma 1.1 and Remark 1.2, we obtain

$$\left| \langle f^*, Te \rangle_f \right| = \left| \langle T^t f^*, e \rangle_e \right| \leq \|T^t f^*\|_{e^*} \|e\|_e \leq \|T^t\|_{f^* \to e^*} \|f^*\|_{f^*} \|e\|_e. \tag{1.7}$$

By the definition of the norm in the space $\mathscr{L}(\mathbb{E}, \mathbb{F})$, we have

$$\|T\|_{e \to f} \overset{\text{def}}{=} \sup_{\|e\|_e=1} \|Te\|_f = \sup_{\|e\|_e=1} \sup_{\|f^*\|_{f^*}=1} \left| \langle f^*, Te \rangle_f \right|. \tag{1.8}$$

Without loss of generality, we assume that $e \neq \theta$ and $f^* \neq \theta$.

Then from (1.7) we obtain the inequality

$$\left| \left\langle \frac{f^*}{\|f^*\|_{f^*}}, T \frac{e}{\|e\|_e} \right\rangle_f \right| \leq \|T^t\|_{f^* \to e^*}.$$

This and Eq. (1.8) imply the inequality (1.6). From (1.6) and (1.5) we obtain the second assertion of the theorem.   □

Lemma 1.2. *If*

$$\mathbb{E} \xrightarrow{T_1} \mathbb{F} \xrightarrow{T_2} \mathbb{G},$$

then the corresponding transposed operators satisfy the relation

$$(T_2 T_1)^t = T_1^t T_2^t.$$

1.3 Topological Embedding Operators

Definition 1.5. An operator $T : \mathbb{E} \to \mathbb{F}$ is said to be *injective* if the condition $Te = \vartheta_f$ implies $e = \theta_e$.

We emphasize that now we discuss only linear operators; in the general case, the notion of injectivity is introduced differently.

Remark 1.3. We denote by $\theta_e \in \mathbb{E}$, $\theta_f \in \mathbb{F}$, $\theta_e^* \in \mathbb{E}^*$, $\theta_f^* \in \mathbb{F}^*$, $\theta_e^{**} \in \mathbb{E}^{**}$, and $\theta_f^{**} \in \mathbb{F}^{**}$ the zero elements of the corresponding vector spaces.

An operator T is noninjective if there exists an element $\overline{e} \neq \theta_e$ such that $T\overline{e} = \vartheta_f$. By the definition of the operator T^t, in this case we have

$$\left\langle T^t f^*, \overline{e} \right\rangle_e = \left\langle f^*, T\overline{e} \right\rangle_f = 0 \quad \text{for all } f^* \in \mathbb{F}^*. \tag{1.9}$$

Definition 1.6. A set $A^* \subset \mathbb{E}^*$ is said to be *orthogonal* to the set $A \subset \mathbb{E}$ with respect to the duality bracket between these Banach spaces if

$$\langle a^*, a \rangle = 0 \quad \text{for all } a \in A \text{ and } a^* \in A^*.$$

Now we consider a particular case of operators from the Banach space $\mathscr{L}(\mathbb{E}, \mathbb{F})$, namely, a linear, continuous, and injective embedding operator

$$J_{ef} : \mathbb{E} \to \mathbb{F}.$$

First, this operator is linear, i.e.,

$$J_{ef}(\alpha_1 e_1 + \alpha_2 e_2) = \alpha_1 J_{ef} e_1 + \alpha_2 J_{ef} e_2 \quad \forall \alpha_1, \alpha_2 \in \mathbb{R}^1 \text{ and } \forall e_1, e_2 \in \mathbb{E}.$$

Second, this operator is continuous and hence — due to the linearity — is bounded:

$$\|J_{ef} e\|_f \leq c_1 \|e\|_e.$$

In the sequel, the notation $\mathbb{E} \overset{ds}{\subset} \mathbb{F}$ means that the Banach space $\mathbb{E}$ is dense in the Banach space $\mathbb{F}$ (i.e., $\overline{J_{ef} \mathbb{E}} = F$).

Theorem 1.2. *Let $\mathbb{E}$ and $\mathbb{F}$ be two Banach spaces and $T \in \mathscr{L}(\mathbb{E}, \mathbb{F})$. Then the following assertions are valid:*

(i) *$T(\mathbb{E}) \overset{ds}{\subset} \mathbb{F}$ if and only if T^t is injective;*

(ii) *if $T^t(\mathbb{F}^*) \overset{ds}{\subset} \mathbb{E}^*$, then T is injective. Moreover, if $\mathbb{E}$ is reflexive, then the converse assertion is also valid.*

Proof. 1. First, we prove that if $T^t f^* = \theta_e^*$, then $f^* = \vartheta_f^*$.

Let $T^t f^* = \theta_e^*$; then for all $e \in \mathbb{E}$ we have the equalities

$$0 = \left\langle T^t f^*, e \right\rangle_e = \langle f^*, Te \rangle_f \quad \text{for all} \quad e \in \mathbb{E}.$$

Since $T(\mathbb{E}) \overset{ds}{\subset} \mathbb{F}$, for any $f \in \mathbb{F}$ and any $\varepsilon > 0$, there exists $e_\varepsilon \in \mathbb{E}$ such that

$$\|f - Te_\varepsilon\|_f < \varepsilon;$$

therefore,

$$\langle f^*, Te_\varepsilon \rangle_f = 0 \quad \Rightarrow \quad \langle f^*, f \rangle_f + \langle f^*, Te_\varepsilon - f \rangle_f = 0.$$

We have

$$|\langle f^*, f \rangle_f| = |-\langle f^*, Te_\varepsilon - f \rangle_f| \le \|f^*\|_{f^*} \|Te_\varepsilon - f\|_f \le \varepsilon \|f^*\|_{f^*}.$$

The number $\varepsilon > 0$ is independent of $f \in \mathbb{F}$ and $f^* \in \mathbb{F}^*$; therefore, passing to the limit as $\varepsilon \to +0$, we conclude that

$$|\langle f^*, f \rangle_f| = 0 \quad \text{for all } f \in \mathbb{F} \quad \Rightarrow \quad f^* = \vartheta_f^*.$$

Now we prove the converse assertion. Let T^t be injective and assume that $T(\mathbb{E})$ is not dense in $\mathbb{F}$. Then there exists an element

$$f \in \mathbb{F} \backslash \overline{T(\mathbb{E})}.$$

By a consequence of the Hahn–Banach theorem on the separating functional, there exists a functional $f^* \in \mathbb{F}^*$ such that

$$\langle f^*, Te \rangle_f = 0 \quad \text{for all } e \in \mathbb{E} \text{ and } \langle f^*, f \rangle_f \ne 0.$$

Then

$$0 = \langle f^*, Te \rangle_f = \langle T^t f^*, e \rangle_e \quad \text{for all } e \in \mathbb{E}.$$

Therefore,

$$T^t f^* = \vartheta_e^* \quad \Rightarrow \quad f^* = \vartheta_f^*,$$

but the last equality contradicts the property $\langle f^*, f \rangle_f \ne 0$. This proved the second part of (i).

2. The first part of the assertion (ii) was actually proved on the first step. However, we repeat the corresponding arguments. Let

$$T^t(\mathbb{F}^*) \overset{ds}{\subset} \mathbb{E}^*. \tag{1.10}$$

We prove that the equality $Te = \theta_f$ implies that $e = \theta_e$. The following equalities hold:

$$0 = \langle f^*, Te \rangle_f = \langle T^t f^*, e \rangle_e \quad \text{for all } f^* \in \mathbb{F}^*. \tag{1.11}$$

Due to (1.10), for any $e^* \in \mathbb{E}^*$ and any $\varepsilon > 0$, there exists an element $f_\varepsilon^* \in \mathbb{F}^*$ such that

$$\|T^t f_\varepsilon^* - e^*\|_{e^*} < \varepsilon. \tag{1.12}$$

Therefore, due to (1.11), we obtain the equality

$$\langle e^*, e \rangle_e = \langle e^* - T^t f_\varepsilon^*, e \rangle_e, \tag{1.13}$$

which, together with (1.12), implies the inequality

$$|\langle e^*, e \rangle_e| = |\langle e^* - T^t f_\varepsilon^*, e \rangle_e| \le \|T^t f_\varepsilon^* - e^*\|_{e^*} \|e\|_e \le \varepsilon \|e\|_e. \tag{1.14}$$

Due to the arbitrariness of $\varepsilon > 0$, we conclude that

$$\langle e^*, e \rangle_e = 0 \quad \text{for all } e^* \in \mathbb{E}^* \quad \Rightarrow \quad e = \theta_e.$$

Now let T be injective. We prove that the functional $e^{**} \in \mathbb{E}^{**}$, which vanishes on $T^t \mathbb{F}^*$, also vanishes on the whole space $\mathbb{E}^*$; then by the Hahn–Banach theorem we obtain the required result. Let

$$\langle e^{**}, e^* \rangle_{e^*} = 0 \quad \text{for all } e^* \in T^t \mathbb{F}^*;$$

this is equivalent to

$$\langle e^{**}, T^t f^* \rangle_{e^*} = 0 \quad \text{for all } f^* \in \mathbb{F}^*.$$

The last expression is equal to

$$\langle T^{tt} e^{**}, f^* \rangle_{f^*} = 0 \quad \text{for all } f^* \in \mathbb{F}^*,$$

where T^{tt} is the operator transposed to T^t. The last equality immediately implies that

$$T^{tt} e^{**} = \theta_e^{**}.$$

Here we encounter the following difficulty: *generally, the injectivity of the operator T does not imply the injectivity of the operator T^{tt}.* Thus, we need an explicit representation of the operator T^{tt} through the operator T.

Consider the operator T^{tt} transposed to the operator T^t. By definition, we have

$$\langle T^{tt} e^{**}, f^* \rangle_{f^*} = \langle e^{**}, T^t f^* \rangle_{e^*} \quad \forall e^{**} \in \mathbb{E}^{**}, \ \forall f^* \in \mathbb{F}^*. \tag{1.15}$$

Owing to the isometric embeddings

$$J_e : \mathbb{E} \to \mathbb{E}^{**}, \quad J_f : \mathbb{F} \to \mathbb{F}^{**},$$

where $J_e \mathbb{E} = \mathbb{E}^{**}$ due to the reflexivity of $\mathbb{E}$, we can rewrite (1.15) as follows:

$$\langle T^{tt} J_e e, f^* \rangle_{f^*} = \langle J_e e, T^t f^* \rangle_{e^*} \quad \forall e \in \mathbb{E}, \ \forall f^* \in \mathbb{F}^*. \tag{1.16}$$

On the other hand, we have the equalities

$$\langle J_e e, T^t f^* \rangle_{e^*} = \langle T^t f^*, e \rangle_e = \langle f^*, Te \rangle_f = \langle J_f Te, f^* \rangle_{f^*} \quad \forall e \in \mathbb{E}, \ \forall f^* \in \mathbb{F}^*.$$

From this and (1.16), we obtain the equality

$$T^{tt} J_e = J_f T.$$

Due to the reflexivity of the space $\mathbb{E}$, there exists the inverse operator J_e^{-1}, and hence we arrive at the equality

$$T^{tt} = J_f T J_e^{-1}.$$

This and the injectivity of T imply the injectivity of the operator T^{tt}; therefore, we obtain

$$T^t(\mathbb{F}^*) \overset{ds}{\subset} \mathbb{E}^*.$$

The theorem is proved. $\qquad\qquad\qquad\qquad\qquad\qquad\qquad\qquad\qquad\qquad\qquad\quad \square$

Now we can apply the general result (Theorem 1.1) to the important particular case of the injective and continuous embedding operator

$$J_{ef} : \mathbb{E} \to \mathbb{F}$$

and its transposed operator

$$J_{ef}^t : \mathbb{F}^* \to \mathbb{E}^*.$$

Theorem 1.3. *Let $\mathbb{E}$ and $\mathbb{F}$ be two Banach spaces and $\mathbb{E} \subset \mathbb{F}$. Then the following assertions hold:*

(i) $\mathbb{E} \overset{ds}{\subset} \mathbb{F}$ *if and only if J_{ef}^t is injective;*

(ii) *if $\mathbb{F}^* \overset{ds}{\subset} \mathbb{E}^*$, then J_{ef} is injective. Moreover, if the space $\mathbb{E}$ is reflexive, then the converse assertion is also valid.*

1.4   Theorem on the Equality of Duality Brackets

Let $\mathbb{E}$ and $\mathbb{F}$ be two Banach spaces and $\mathbb{E}$ be reflexive; moreover, let $\mathbb{E} \overset{ds}{\subset} \mathbb{F}$, i.e., there exists a linear, injective, and continuous embedding operator

$$J_{ef} : \mathbb{E} \to \mathbb{F},$$

such that $J_{ef}\mathbb{E}$ is dense in $\mathbb{F}$. Thus, to each element $u \in \mathbb{E}$, we assign a certain element $v = J_{ef}u$. On the other hand, for the operator J_{ef}, the transposed operator is defined:

$$J_{ef}^t : \mathbb{F}^* \to \mathbb{E}^*;$$

by Theorem 1.1, the operator J_{ef}^t is linear, continuous, and injective and, moreover, $J_{ef}^t\mathbb{F}^*$ is dense in $\mathbb{E}^*$ due to the item (ii) of Theorem 1.3 and the reflexivity of $\mathbb{E}$.

Thus, to each element $f \in \mathbb{F}^*$, a certain element $J_{ef}^t f \in \mathbb{E}^*$ corresponds. By the definition of the transposed operator, we have

$$\left\langle J_{ef}^t f, u \right\rangle_e = \left\langle f, J_{ef} u \right\rangle_f \quad \text{for all } u \in \mathbb{E} \text{ and } f \in \mathbb{F}^*. \tag{1.17}$$

However, if we identify $\mathbb{E}$ with its image in $\mathbb{F}$, i.e., with $J_{ef}\mathbb{E}$, and identify $\mathbb{F}^*$ with its image in $\mathbb{E}^*$, i.e., with $J_{ef}^t\mathbb{F}^*$, then (1.17) can be rewritten in the following simpler form:

$$\langle f, u \rangle_e = \langle f, u \rangle_f \quad \text{for all } u \in \mathbb{E} \text{ and } f \in \mathbb{F}^*; \tag{1.18}$$

moreover, the following dense embeddings hold:

$$\mathbb{E} \overset{ds}{\subset} \mathbb{F}, \quad \mathbb{F}^* \overset{ds}{\subset} \mathbb{E}^*. \tag{1.19}$$

Thus, the following theorem on the equality of duality brackets is valid.

Theorem 1.4. *Let a reflexive Banach space $\mathbb{E}$ be continuously and densely embedded in a Banach space $\mathbb{F}$. Then the following equality holds:*

$$\langle f, u \rangle_e = \langle f, u \rangle_f \quad \text{for all } u \in \mathbb{E} \text{ and } f \in \mathbb{F}^*. \tag{1.20}$$

1.5   Bibliographical Notes

The contents of this lecture is taken from [3] (see also [22]).

Lecture 2

Weak and ∗-Weak Convergence

2.1 Criteria of Weak and ∗-Weak Convergence

Let $\mathbb{B}$ be a Banach space with respect to the norm $\|\cdot\|$ and $\mathbb{B}^*$ be the dual Banach space with respect to the norm $\|\cdot\|_*$, which has the following explicit form:

$$\|f^*\|_* := \sup_{\|u\|=1} |\langle f^*, u \rangle|,$$

where $\langle \cdot, \cdot \rangle$ is the duality bracket between $\mathbb{B}$ and $\mathbb{B}^*$. Moreover, let $\mathbb{B}^{**} = (\mathbb{B}^*)^*$ be the second dual Banach space with respect to the duality bracket

$$\langle u^{**}, f^* \rangle_*, \quad u^{**} \in \mathbb{B}^{**}, \quad f^* \in \mathbb{B}^*,$$

endowed with the norm

$$\|u^{**}\|_{**} := \sup_{\|f^*\|_*=1} |\langle u^{**}, f^* \rangle_*|.$$

Definition 2.1. We say that a sequence $\{u_n\} \subset \mathbb{B}$ *weakly converges* to an element $u \in \mathbb{B}$ if for any $f^* \in \mathbb{B}^*$, the following limit relation holds:

$$\langle f^*, u_n \rangle \rightharpoonup \langle f^*, u \rangle \quad \text{as } n \to +\infty;$$

notation $u_n \rightharpoonup u$.

Definition 2.2. We say that a sequence $\{f_n^*\} \subset \mathbb{B}^*$ *∗-weakly converges* to an element $f^* \in \mathbb{B}^*$ if for any $u \in \mathbb{B}$, the following limit relation holds:

$$\langle f_n^*, u \rangle \rightharpoonup \langle f^*, u \rangle \quad \text{as } n \to +\infty;$$

notation $f_n \overset{*}{\rightharpoonup} f$.

Lemma 2.1. *The Banach space $\mathbb{B}^*$ is ∗-weakly dense.*

Proof. Let $\{f_n^*\} \subset \mathbb{B}^*$ be a ∗-weak fundamental[1] sequence, i.e.,

$$\langle f_n^* - f_m^*, u \rangle \to +0 \quad \text{as } n, m \to +\infty$$

---

[1] Fundamental sequences are also called Cauchy sequences.

11

for any $u \in \mathbb{B}$. Therefore, the sequence $\{\langle f_n^*, u \rangle\}$ is fundamental in $\mathbb{R}$ ($\mathbb{C}$). Therefore, the numerical sequence $\{\langle f_n^*, u \rangle\}$ converges and, hence, is bounded:

$$\sup_{n \in \mathbb{N}} |\langle f_n^*, u \rangle| \leq c(u) < +\infty.$$

By the Banach–Steinhaus theorem (the uniform boundedness principle; see [70]) we have

$$\sup_{n \in \mathbb{N}} \|f_n^*\|_* < +\infty. \tag{2.1}$$

Thus,

$$\langle f_n^*, u \rangle \to f^*(u) \quad \text{as } n \to +\infty. \tag{2.2}$$

We prove that the functional $f^*(u)$ is linear. Indeed,

$$\langle f_n^*, \alpha_1 u_1 + \alpha_2 u_2 \rangle = \alpha_1 \langle f_n^*, u_1 \rangle + \alpha_2 \langle f_n^*, u_2 \rangle. \tag{2.3}$$

Taking into account (2.2) and passing to the limit, we obtain the following equality:

$$f^*(\alpha_1 u_1 + \alpha_2 u_2) = \alpha_1 f^*(u_1) + \alpha_2 f^*(u_2). \tag{2.4}$$

Now we prove that the linear functional f is bounded. We have

$$|\langle f^*, u \rangle| = \lim_{n \to +\infty} |\langle f_n^*, u \rangle| \leq \sup_{n \in \mathbb{N}} \|f_n^*\|_* \|u\| \leq K_1 \|u\| \tag{2.5}$$

for any $u \in \mathbb{B}$. Thus,

$$\|f^*\|_* = \sup_{\|u\|=1} |\langle f^*, u \rangle| \leq K_1 < +\infty. \tag{2.6}$$

Lemma 2.1 is proved. □

Lemma 2.2. *Any reflexive Banach space is weakly complete.*

Proof. Let $\{u_n\} \subset \mathbb{B}$ be a weakly fundamental sequence, i.e., for any $f^* \in \mathbb{B}^*$, the numerical sequence $\{\langle f^*, u_n \rangle\}$ is fundamental. Therefore,

$$\langle f^*, u_n - u_m \rangle \to +0 \quad \text{as } n, m \to +\infty.$$

Let $J : \mathbb{B} \to \mathbb{B}^{**}$ be the isometric embedding operator. Then

$$\langle J(u_n - u_m), f^* \rangle_* = \langle f^*, u_n - u_m \rangle \to +0 \quad \text{as } n, m \to +\infty.$$

Therefore, $\{Ju_n\} \subset \mathbb{B}^{**}$ is a *-weakly fundamental sequence. Due to Lemma 2.1,

$$Ju_n \xrightarrow{*} u^{**} \quad \text{*-weakly in } \mathbb{B}^{**} \text{ as } n \to +\infty.$$

By the reflexivity of $\mathbb{B}$, we have $J(\mathbb{B}) = \mathbb{B}^{**}$. Therefore, there exists a unique element $u \in \mathbb{B}$ such that $Ju = u^{**}$ and hence

$$\langle f^*, u_n - u \rangle = \langle J(u_n - u), f^* \rangle_* = \langle Ju_n - u^{**}, f^* \rangle_* \to +0$$

as $n \to +\infty$, i.e.,

$$u_n \xrightarrow{} u \quad \text{weakly in } \mathbb{B} \text{ as } n \to +\infty.$$

Lemma 2.2 is proved. □

Theorem 2.1. *Let $\{u_n\}$ be a bounded (with respect to the norm) sequence of elements of a reflexive and separable Banach space $\mathbb{B}$. Then $\{u_n\}$ contains a subsequence $\{u_{n_n}\}$ weakly converging in $\mathbb{B}$:*

$$u_{n_n} \rightharpoonup u \quad \text{weakly in } \mathbb{B} \text{ as } n \to +\infty.$$

Proof. 1. Since the space $\mathbb{B}$ is reflexive, we can identify the space $\mathbb{B}$ with the second dual space $\mathbb{B}^{**}$. Thus, we have $\mathbb{B}^{**} = \mathbb{J}(\mathbb{B})$; since $\mathbb{B}$ is separable, the space $\mathbb{B}^{**}$ is also separable. This space is dual to the Banach space $\mathbb{B}^*$; therefore, the space $\mathbb{B}^*$ is separable.

2. Let $\{f_n\} \subset \mathbb{B}^*$ be a countable, everywhere dense in $\mathbb{B}^*$ set. Since the sequence $\{u_n\} \subset \mathbb{B}$ is bounded in norm, the numerical sequence $\langle f_1, u_n \rangle$ is bounded; therefore, is contains a converging subsequence $\langle f_1, u_{n_1} \rangle$. The numerical sequence $\langle f_2, u_{n_1} \rangle$ is also bounded and hence it contains a converging subsequence $\langle f_2, u_{n_2} \rangle$.

Continuing this process, we obtain a subsequence $\{u_{n_{k+1}}\}$ such that the sequences $\{\langle f_j, u_{n_{k+1}} \rangle\}$ converge for $j = 1, \ldots, k+1$. Therefore, the diagonal subsequence $\{u_{n_n}\}$ of the initial sequence $\{u_n\} \subset \mathbb{B}$ weakly converges on the countable, everywhere dense in $\mathbb{B}^*$ set $\{f_n\}$.

3. Now we prove that the subsequence constructed $\{u_{n_n}\} \subset \mathbb{B}$ weakly converges to a certain element $u_\infty \in \mathbb{B}$.

Let $f \in \mathbb{B}^*$ be an arbitrary but fixed element. Then we have

$$|\langle f, u_{n_n} \rangle - \langle f, u_{m_m} \rangle| \leq |\langle f, u_{n_n} \rangle - \langle f_k, u_{n_n} \rangle|$$
$$+ |\langle f_k, u_{n_n} \rangle - \langle f_k, u_{m_m} \rangle| + |\langle f, u_{m_m} \rangle - \langle f_k, u_{m_m} \rangle|$$
$$\leq \|f - f_k\|_* \|u_{n_n}\| + |\langle f_k, u_{n_n} \rangle - \langle f_k, u_{m_m} \rangle|$$
$$+ \|f - f_k\|_* \|u_{m_m}\|.$$

The first and last terms in the right-hand side of the last inequality can be made arbitrarily small due to the facts that $\{f_k\}$ is dense in the space $\mathbb{B}^*$ with respect to the strong convergence and the subsequence $\{u_{n_n}\} \subset \{u_n\} \subset \mathbb{B}$ is bounded. Finally, as we have proved above, the second term tends to zero as $n, m \to +\infty$.

4. We see that the numerical sequence $\{\langle f, u_{n_n} \rangle\}$ is fundamental for any $f \in \mathbb{B}^*$ and hence converges. By Lemma 2.2, there exists an element $u \in \mathbb{B}$ such that

$$u_{n_n} \rightharpoonup u \quad \text{weakly in } \mathbb{B} \text{ as } n \to +\infty.$$

Theorem 2.1 is proved. $\qquad\square$

Now we formulate (without proof) a sufficient condition of the *-weak convergence.

Theorem 2.2. *Let $\mathbb{B}$ be a separable Banach space and $\{f_n\}$ be a bounded (with respect to norm) sequence of elements of the Banach space $\mathbb{B}^*$. Then $\{f_n\}$ contains a subsequence $\{f_{n_n}\}$, which *-weakly converges in $\mathbb{B}^*$:*

$$f_{n_n} \overset{*}{\rightharpoonup} f \quad \text{*-weakly in } \mathbb{B}^* \text{ as } n \to +\infty.$$

Theorem 2.3. *The following two assertions are valid:*

(i) *each weakly converging sequence $\{u_n\}$ in a Banach space $\mathbb{B}$ is bounded;*
moreover, if $u_n \rightharpoonup u_\infty$ as $n \to +\infty$, then $\|u_\infty\| \le \liminf\limits_{n \to +\infty} \|u_n\|$;

(ii) *each ∗-weakly converging sequence $\{f_n\}$ in a Banach space $\mathbb{B}^*$ is bounded;*
moreover, if $f_n \overset{*}{\rightharpoonup} f_\infty$ as $n \to +\infty$, then $\|f_\infty\|_* \le \liminf\limits_{n \to +\infty} \|f_n\|_*$.

Proof. 1. First, we prove (ii). Obviously, the sequence $\{f_n\} \subset \mathscr{L}(\mathbb{B}, \mathbb{K})$ (here $\mathbb{K} = \mathbb{R}$ or $\mathbb{K} = \mathbb{C}$ is the underlying field of the vector space $\mathbb{B}$) satisfies the conditions of the Banach—Steinhaus theorem (see, e.g., [70]). Indeed, if for any $u \in \mathbb{B}$, the sequence $\langle f_n, u \rangle$ converges, then

$$\sup_{n \in \mathbb{N}} |\langle f_n, u \rangle| \le c(u) < +\infty.$$

By the Banach–Steinhaus theorem,

$$\sup_{n \in \mathbb{N}} \|f_n\|_* < +\infty.$$

2. Thus, we have proved the collective boundedness of the norms $\|f_n\|$. However, we need the following estimate:

$$\|f_\infty\|_* \le \liminf_{n \to +\infty} \|f_n\|_*. \tag{2.7}$$

Recall that

$$\|f\|_* = \sup_{\substack{u \in \mathbb{B}, \\ \|u\| = 1}} |\langle f, u \rangle|; \tag{2.8}$$

therefore, for each $u \in \mathbb{B}$ with $\|u\| = 1$ we have

$$\|f_n\|_* \ge |\langle f_n, u \rangle|.$$

Passing to the lower limit in the left-hand side of this inequality and to the ordinary limit in the right-hand side (recall that $f_n \overset{*}{\rightharpoonup} f_\infty$), we obtain

$$\liminf_{n \to +\infty} \|f_n\|_* \ge |\langle f_\infty, u \rangle|.$$

Taking supremum over $u \in \mathbb{B}$, $\|u\| = 1$, in the last inequality and taking into account (2.8), we finally arrive at the estimate (2.7).

3. Now we prove (i). Let

$$u_n \rightharpoonup u_\infty \quad \text{weakly in } \mathbb{B} \text{ as } n \to +\infty, \tag{2.9}$$

i.e.,

$$\langle f, u_n \rangle \to \langle f, u_\infty \rangle \quad \text{as } n \to +\infty \tag{2.10}$$

for any $f \in \mathbb{B}^*$. By the definition of a linear isometric embedding operator,

$$J : \mathbb{B} \to \mathbb{B}^{**}, \quad \langle Ju, f \rangle_* := \langle f, u \rangle,$$

and from (2.10) we obtain the following limit property:

$$\langle Ju_n, f \rangle_* \to \langle Ju_\infty, f \rangle_* \quad \text{as } n \to +\infty \tag{2.11}$$

for any $f \in \mathbb{B}^*$, i.e.,

$$Ju_n \overset{*}{\to} Ju_\infty \quad \text{*-weakly in } \mathbb{B}^{**} \text{ as } n \to +\infty. \tag{2.12}$$

Therefore, due to the limit property (ii) proved earlier, we conclude that the sequence $\{Ju_n\}$ is bounded in $\mathbb{B}^{**}$ and

$$\liminf_{n \to +\infty} \|Ju_n\|_{**} \geq \|Ju_\infty\|_{**}. \tag{2.13}$$

Since

$$\|Jv\|_{**} = \|v\| \quad \text{for all } v \in \mathbb{B},$$

we see that the sequence $\{u_n\}$ is bounded in $\mathbb{B}$ and the following limit property is valid:

$$\liminf_{n \to +\infty} \|u_n\| \geq \|u_\infty\|. \tag{2.14}$$

Theorem 2.3 is proved. □

2.2 Uniform Convexity of Banach Spaces

Definition 2.3. A Banach space $\mathbb{B}$ is said to be *uniformly convex* if for any $\varepsilon > 0$, there exists $\delta(\varepsilon) > 0$ such that the inequalities $\|u\| \leq 1$, $\|v\| \leq 1$, and $\|u-v\| \geq \varepsilon > 0$ imply

$$\|u + v\| \leq 2(1 - \delta(\varepsilon)). \tag{2.15}$$

Now we formulate a sufficient condition of strong convergence, which is useful in applications to nonlinear boundary-value problems.

Theorem 2.4. *If $\mathbb{B}$ is a uniformly convex Banach space, then the conditions*

$$u_n \rightharpoonup u \quad \text{weakly in } \mathbb{B} \text{ as } n \to +\infty,$$

$$\|u_n\| \to \|u\| \quad \text{as } n \to +\infty,$$

imply that

$$u_n \to u \quad \text{strongly in } \mathbb{B} \text{ as } n \to +\infty.$$

Proof. 1. Without loss of generality, we assume that $\|u\| = 1$ and $\|u_n\| \neq 0$. Introduce the following notation:

$$v_n = \frac{u_n}{\|u_n\|}.$$

Clearly, $\|v_n\| = 1$ and $v_n \rightharpoonup u$ as $n \to +\infty$. We set $\varepsilon_n = \|v_n - u\|$.

2. We must prove that $\varepsilon_n \to 0$. Assume the contrary. Then there exists a subsequence $\{\varepsilon_{n_k}\}$ and $\varepsilon_0 > 0$ such that the inequality $\varepsilon_{n_k} \geq \varepsilon_0$ holds. Removing unacceptable terms if necessary, we can assume that for the initial sequence, the inequality $\varepsilon_n \geq \varepsilon_0 > 0$ is valid. Since

$$\|v_n - u\| \geq \varepsilon_n \geq \varepsilon_0 > 0,$$

due to (2.15) we have

$$\|v_n + u\| \leq 2(1 - \delta(\varepsilon_0)) =: C < 2. \tag{2.16}$$

3. By the definition of the norm of a linear functional, for any $f \in \mathbb{B}$ such that $\|f\|_* \leq 1$, the following inequality holds:

$$|\langle f, v_n + u \rangle| \leq \|v_n + u\|,$$

or, taking into account (2.16),

$$|\langle f, v_n + u \rangle| \leq 2(1 - \delta(\varepsilon_0)) = C < 2.$$

4. Since $v_n \rightharpoonup u$, we may pass to the limit as $n \to +\infty$:

$$|\langle f, 2u \rangle| \leq 2(1 - \delta(\varepsilon_0)) = C < 2,$$

or

$$|\langle f, u \rangle| \leq \frac{C}{2} < 1.$$

Taking the supremum over all $f \in \mathbb{B}^*$ such that $\|f\|_* \leq 1$, we obtain

$$\|u\| = \sup_{\substack{f \in \mathbb{B}^*, \\ \|f\|_{\mathbb{B}^*} = 1}} |\langle f, u \rangle| \leq \frac{C}{2} < 1;$$

this contradicts the equality $\|u\| = 1$. Therefore, the assumption is invalid and $\varepsilon_n \to 0$, i.e., $v_n \to u$. Thus, we obtain that

$$u_n = \|u_n\| v_n \to u \quad \text{strongly in } \mathbb{B} \text{ as } n \to +\infty$$

since $\|u_n\| \to \|u\| = 1$. Theorem 2.4 is proved. $\qquad\square$

2.3   Bibliographical Notes

The contents of this lecture is taken from [70].

Lecture 3

Bochner Integral

3.1 Basic Definitions

In this lecture, we introduce the notion of the Bochner integral, which is a generalization of the Lebesgue integral to the case of Banach-valued functions.

As in the case of the Lebesgue integral, let $([0, T], \mathscr{A}, \mu)$ be a measurable space, which consists of the segment $[0, T] \subset \mathbb{R}^1_+$, the σ-algebra $\mathscr{A}$ of its subsets, and the Lebesgue measure μ defined on $\mathscr{A}$. Similarly to the Lebesgue integral, we can introduce the Bochner integral not only for sets on $\mathbb{R}^1$; however, in what follows, we need only the Bochner integral over the "time" segment $[0, T]$. In this and the following lecture, we assume that the Banach space $\mathbb{B}$ is *separable*.

Definition 3.1. A *simple function* $h(t)$ on a segment $[0, T]$ with values in a Banach spaces $\mathbb{B}$ is a function of the form

$$h(t) = \sum_{i=1}^{N} \chi_i(t) b_i, \quad b_i \in \mathbb{B}, \tag{3.1}$$

where $\chi_i(t)$, $i = \overline{1, N}$, are the characteristic functions of certain sets $S_i \in \mathscr{A}$:

$$\chi_i(t) \stackrel{\text{def}}{=\joinrel=} \begin{cases} 1, & t \in S_i, \\ 0, & t \in [0, T] \backslash S_i; \end{cases}$$

moreover, $S_i \cap S_j = \varnothing$ for $i \neq j$.

Definition 3.2. The Bochner integral of a simple functions $h(t)$ over the segment $[0, T]$ is defined as follows:

$$\int_0^T h(t) \, d\mu \stackrel{\text{def}}{=\joinrel=} \sum_{i=1}^{N} \mu(S_i) b_i. \tag{3.2}$$

Definition 3.3. The Bochner integral of an arbitrary $\mathbb{B}$-valued function $f(t)$ is defined as follows:

$$\int_0^T f(t) \, d\mu \stackrel{\text{def}}{=\joinrel=} \lim_{n \to +\infty} \int_0^T h_n(t) d\mu, \tag{3.3}$$

17

where the limit in the Banach space $\mathbb{B}$ is meant in the strong sense under the following condition: there exists a sequence $\{h_n(t)\}$ of simple functions, which μ-almost everywhere on $[0, T]$ strongly converges in $\mathbb{B}$ to the function $f(t)$, such that

$$\lim_{n \to +\infty} \int_0^T \|f(t) - h_n(t)\| \, d\mu = 0. \tag{3.4}$$

Note that we have not yet proved that the scalar-valued integrand function $\|f(t) - h_n(t)\|$ of the Lebesgue integral (3.4) is μ-measurable on the segment $[0, T]$. To prove this, we must develop the theory of measurability of $\mathbb{B}$-valued functions. Next, we must prove that in the Banach space $\mathbb{B}$ there exists the limit

$$\lim_{n \to +\infty} \int_0^T h_n(t) d\mu,$$

which is meant in the strong sense. Moreover, we must prove the independence of the limit (3.3) of the choice of the sequence $\{h_n(t)\}$.

3.2   Strong and Weak Measurability

Definition 3.4. A $\mathbb{B}$-valued function $f(t)$ is said to be μ-*weakly measurable* on the segment $[0, T]$ if for any $f^* \in \mathbb{B}^*$, the function $\langle f^*, f(t) \rangle$ is μ-measurable on the segment $[0, T]$.

Definition 3.5. A $\mathbb{B}$-valued function $f(t)$ is said to be μ-*strongly measurable* on the segment $[0, T]$ if there exists a sequence $\{h_n(t)\}$ of simple functions on the segment $[0, T]$, which μ-almost everywhere on the segment $[0, T]$ strongly converges in $\mathbb{B}$ to $f(t)$, i.e.,

$$\|f(t) - h_n(t)\| \to +0 \quad \text{as } n \to +\infty \text{ for almost all } t \in [0, T].$$

Remark 3.1. Below we will use the fact that in the definition of a real-valued measurable function $f(t)$, the requirement of the measurability of the set $A_c := \{t \in [0, T] : f(t) < c\}$ can be replaced by the requirement of the measurability of the set $B_c := \{t \in [0, T] : f(t) \le c\}$ for any $c \in \mathbb{R}^1$. Indeed, if the set A is measurable, then the set

$$B_c = \bigcap_{n=1}^{+\infty} A_{c+1/n}$$

is also measurable as the intersection of a countable number of measurable sets. Conversely, if the set B_c is measurable, then the set

$$A_c = \bigcup_{n=1}^{+\infty} B_{c-1/n}$$

is measurable as the union of a countable number of measurable sets. Therefore, to prove the measurability of the function $f(t)$, it is necessary and sufficient to prove the measurability either of the sets A_c or of the sets B_c for any $c \in \mathbb{R}^1$.

Theorem 3.1 (Pettis theorem; see [70]). *If $\mathbb{B}$ is a separable space, then the μ-weak measurability of a $\mathbb{B}$-valued function $f(t)$ implies the μ-measurability of $\|f(t)\|$ on the segment $[0, T]$.*

Proof. 1. Consider the sets
$$A := \{t: \|f(t)\| \le c_1\}, \quad A_{f^*} := \{t: |\langle f^*, f(t)\rangle| \le c_1\}.$$
Note that, due to the μ-weak measurability of the function $f(t)$, the set A_{f^*} is μ-measurable on the segment $[0, T]$. Since
$$|\langle f^*, f(t)\rangle| \le \|f^*\|_* \|f(t)\|,$$
the following embedding holds:
$$A \subseteq \bigcap_{\|f^*\|_* \le 1} A_{f^*}. \tag{3.5}$$

2. Now we apply the following corollary of the Hahn–Banach theorem.

Corollary of the Hahn–Banach theorem

For each element $f \in \mathbb{B}$, there exists $f^* \in \mathbb{B}^*$ such that
$$\|f^*\|_* = 1, \quad \|f\| = \langle f^*, f\rangle.$$

Therefore, for each $t \in [0, T]$, there exists an element $f^*(t) \in \mathbb{B}^*$ such that the following equality holds:
$$\|f(t)\| = \langle f^*(t), f(t)\rangle, \quad \|f^*(t)\|_* = 1 \quad \Rightarrow \quad \|f^*(t)\|_* \le 1.$$
Hence the converse embedding also holds:
$$A \supseteq \bigcap_{\|f^*\|_* \le 1} A_{f^*}. \tag{3.6}$$
Indeed, we have
$$t \in \bigcap_{\|f^*\|_* \le 1} A_{f^*} \Rightarrow t \in \left\{t: \sup_{\|f^*\|_* \le 1} |\langle f^*, f(t)\rangle| \le c_1\right\} = \{t: \|f(t)\| \le c_1\} = A.$$
Therefore, from (3.5) and (3.6) we obtain
$$A = \bigcap_{\|f^*\|_* \le 1} A_{f^*}. \tag{3.7}$$

3. However, this is insufficient for the proof of the required result since we consider the intersection
$$\bigcap_{\|f^*\|_* \le 1} A_{f^*}$$
of uncountable family of set. To overcome this difficulty, we use the separability of the space $\mathbb{B}$.

Lemma 3.1 (see [70]). *Let $\mathbb{B}$ be a separable Banach space. There exists a sequence $\{f_n^*\} \subset \mathbb{B}^*$, $\|f_n^*\|_* \le 1$, such that for any $f^* \in \mathbb{B}^*$, $\|f^*\|_* \le 1$, there exists a subsequence $\{f_{n'}^*\} \subset \{f_n^*\}$ for which*
$$\lim_{n' \to +\infty} \langle f_{n'}^*, f\rangle = \langle f^*, f\rangle \quad \text{for all } f \in \mathbb{B}. \tag{3.8}$$

Using this lemma, we can prove the equality

$$A = \bigcap_{n=1}^{+\infty} A_{f_n^*}, \qquad (3.9)$$

where $\{f_n^*\}$ is the sequence from Lemma 3.1.

Indeed, consider the embedding

$$A \subset \bigcap_{n=1}^{+\infty} A_{f_n^*},$$

which is valid since if $t \in A$, then $\|f(t)\| \leq c_1$ and hence

$$|\langle f_n^*, f(t) \rangle| \leq \|f_n^*\|_* \|f(t)\| \leq c_1 \quad \Rightarrow \quad t \in A_{f_n^*} \quad \text{for all } n \in \mathbb{N}.$$

We prove the converse embedding. Let

$$t \in \bigcap_{n=1}^{+\infty} A_{f_n^*} \quad \Rightarrow \quad t \in \left\{ t : \sup_{n \in \mathbb{N}} |\langle f_n^*, f(t) \rangle| \leq c_1 \right\}.$$

Assume that $f^* \in \mathbb{B}^*$ is an arbitrary fixed functional such that $\|f^*\|_* \leq 1$. Then there exists a subsequence $\{f_{n'}^*\} \subset \{f_n^*\}$ such that

$$|\langle f^*, f \rangle| \leq |\langle f^* - f_{n'}^*, f \rangle| + |\langle f_{n'}^*, f \rangle| \leq \delta_{n'} + c_1,$$

where $\delta_{n'} \to +0$ as $n' \to +\infty$. Passing to the limit, we obtain, due to Eq. (3.7),

$$t \in \{t : |\langle f^*, f \rangle| \leq c_1 \quad \text{for any } \|f^*\|_* \leq 1\} = \bigcap_{\|f^*\|_* \leq 1} A_{f^*} = A.$$

However, countable intersections of μ-measurable sets are also μ-measurable sets; therefore, the set A is μ-measurable. $\qquad \square$

Lemma 3.2. *A strongly μ-measurable function on a segment $[0, T]$ is a weakly μ-measurable function of the same segment.*

Proof. 1. Let a $\mathbb{B}$-valued function $f(t)$ be strongly μ-measurable on the segment $[0, T]$. Then there exists a sequence of $\mathbb{B}$-valued simple functions $\{h_n(t)\}$, which strongly in $\mathbb{B}$ converges to the function $f(t)$ almost everywhere with respect to the Lebesgue measure μ on $[0, T]$, i.e.,

$$\|f(t) - h_n(t)\| \to +0 \quad \text{for almost all } t \in [0, T] \text{ as } n \to +\infty.$$

Therefore, this sequence converges to the function $f(t)$ weakly in $\mathbb{B}$ μ-almost everywhere on $[0, T]$, i.e., for any $f^* \in \mathbb{B}^*$, we have

$$\langle f^*, f(t) - h_n(t) \rangle \to 0 \quad \text{for almost all } t \in [0, T] \text{ as } n \to +\infty.$$

The equality

$$\langle f^*, h(t) \rangle = \sum_{i=1}^{N} \langle f^*, b_i \rangle \chi_{S_i}(t),$$

which holds for simple functions

$$h(t) = \sum_{i=1}^{N} b_i \chi_{S_i}(t),$$

implies the weak measurability of $h(t)$.

2. Consider the following sets:

$$A(f^*) := \{t : |\langle f^*, f(t) \rangle| < c_1\},$$

$$A_{k,m}(f^*) := \left\{t : |\langle f^*, h_k(t) \rangle| < c_1 - \frac{1}{m}\right\}.$$

Up to a set of zero Lebesgue measure on $[0, T]$, the following equality holds:

$$A(f^*) = \bigcup_{m=1}^{+\infty} \bigcup_{n=1}^{+\infty} \bigcap_{k=n}^{+\infty} A_{k,m}(f^*). \tag{3.10}$$

Indeed, let $t \in A(f^*)$; then there exists $0 < \varepsilon < c_1$ such that

$$|\langle f^*, f(t) \rangle| = \varepsilon. \tag{3.11}$$

We have

$$|\langle f^*, h_k(t) \rangle| \leq |\langle f^*, f(t) - h_k(t) \rangle| + |\langle f^*, f(t) \rangle| \leq \delta_k + \varepsilon < c_1 - \frac{1}{m_0} \tag{3.12}$$

for $k \geq n_0$ and sufficiently large $m_0, n_0 \in \mathbb{N}$ so that the inequality

$$\delta_k + \varepsilon + \frac{1}{m_0} < c_1, \quad k \geq n_0,$$

holds by the definition of $\varepsilon \in (0, c_1)$ and the fact that $\delta_k = |\langle f^*, f(t) - h_k(t) \rangle| \to +0$ as $k \to +\infty$. Therefore,

$$t \in \bigcap_{k=n_0}^{+\infty} A_{k,m_0}(f^*) \subset \bigcup_{m=1}^{+\infty} \bigcup_{n=1}^{+\infty} \bigcap_{k=n}^{+\infty} A_{k,m}(f^*).$$

Now let

$$t \in \bigcup_{m=1}^{+\infty} \bigcup_{n=1}^{+\infty} \bigcap_{k=n}^{+\infty} A_{k,m}(f^*).$$

Then there exist $m_0, n_0 \in \mathbb{N}$ such that

$$t \in \bigcap_{k=n_0}^{+\infty} A_{k,m_0} \subset \bigcap_{k=n_1}^{+\infty} A_{k,m_0} \quad \text{for any } n_1 \geq n_0. \tag{3.13}$$

We have

$$|\langle f^*, h_k(t) \rangle| < c_1 - \frac{1}{m_0} \quad \text{for all } k \geq n_1. \tag{3.14}$$

The inequalities

$$|\langle f^*, f(t) \rangle| \leq \delta_k + |\langle f^*, h_k(t) \rangle| < \delta_k + c_1 - \frac{1}{m_0} < c_1 \tag{3.15}$$

hold under the condition that $n_1 \in \mathbb{N}$ is so large that

$$\delta_k := |\langle f^*, f(t) - h_k(t) \rangle| < \frac{1}{m_0} \quad \text{for all } k \geq n_1.$$

Thus, $t \in A(f^*)$.

Therefore, the set $A(f^*)$ is μ-measurable due to the closedness of the family of measurable sets with respect to the operations of countable union and countable intersection. $\qquad\square$

The following question naturally arises: When does the weak measurability of a function $f(t) : [0, T] \to \mathbb{B}$ imply its strong measurability?

Theorem 3.2 (see [23, Theorem 2.1.3]). *A function $f(t) \colon [0, T] \to \mathbb{B}$ is strongly measurable if and only if it is weakly measurable and separable-valued almost everywhere on the segment.*

Proof. *Necessity.* Let $f(t) : [0, T] \to \mathbb{B}$ be a strongly measurable function. By Lemma 3.2, the function $f(t)$ is weakly measurable. By the definition of strong measurability (Definition 3.5), there exists a sequence of $\mathbb{B}$-valued simple functions $\{h_n(t)\}$ such that

$$h_n(t) \to f(t) \quad \text{strongly in } \mathbb{B} \text{ for all } t \in [0, T] \backslash A, \tag{3.16}$$

where $\mu(A) = 0$. (Recall that μ is the classical Lebesgue measure on $[0, T]$.)

Let $\{y_m\} \subset \mathbb{B}$ be the set of values of all simple functions from the sequence $\{h_n(t)\}$. Consider the set

$$Y := \overline{\text{span}\{y_m\}}. \tag{3.17}$$

Clearly, Y is a closed, linear, separable subspace of the Banach space $\mathbb{B}$. We prove that $f([0, T] \backslash A) \subset Y$. Indeed, if $Y = \mathbb{B}$, then the assertion is valid. Let Y be a proper, closed, linear subspace in $\mathbb{B}$. Assume that there exists $t_0 \in [0, T] \backslash A$ such that $f(t_0) \notin Y$, i.e., $f(t_0) \in \mathbb{B} \backslash Y$. Due to the closedness of Y, there exists $\varepsilon_0 > 0$ such that

$$O(f(t_0); \varepsilon_0) := \{x \in \mathbb{B} : \|x - f(t_0)\| < \varepsilon_0\} \subset \mathbb{B} \backslash Y;$$

moreover,

$$h_n(t_0) \to f(t_0) \quad \text{strongly in } \mathbb{B} \text{ as } n \to +\infty.$$

Therefore, on the one hand, there exists $N_0 \in \mathbb{N}$ such that

$$h_n(t_0) \in O(f(t_0); \varepsilon_0) \quad \text{for } n \geq N_0.$$

On the other hand, by the construction of Y, we have

$$\{h_n(t_0)\}_{n=1}^{+\infty} \subset Y.$$

This contradiction proved the assertion.

Thus, f is a separable-valued function for almost all $t \in [0, T]$.

Sufficiency. We prove the sufficiency for the case where $\mathbb{B}$ is separable. Let $f : [0, T] \to \mathbb{B}$ be weakly measurable. By Lemma 3.1, there exists a countable set $\{f_n^*\}$, which is dense in the unit ball

$$\{f^* \in \mathbb{B}^* : \|f^*\|_* \leq 1\} \tag{3.18}$$

in the $*$-weak topology of the space $\mathbb{B}^*$. Then

$$\|f(t)\| = \sup_{n \in \mathbb{N}} |\langle f_n^*, f(t) \rangle|. \tag{3.19}$$

Indeed, by the corollary of the Hahn–Banach theorem mentioned above, for any $t \in [0, T]$ there exists $f^*(t) \in \mathbb{B}^*$ such that

$$\|f(t)\| = |\langle f^*(t), f(t) \rangle|, \quad \|f^*(t)\|_* = 1.$$

Then there exists a subsequence $\{f_{n'}^*(t)\} \subset \{f_n^*\}$ such that

$$f_{n'}^*(t) \overset{*}{\rightharpoonup} f^*(t) \quad \text{*-weakly as } n' \to +\infty.$$

Therefore, we have

$$\|f(t)\| = \lim_{n' \to +\infty} |\langle f_{n'}^*(t), f(t) \rangle| = \sup_{n'} |\langle f_{n'}^*(t), f(t) \rangle| = \sup_{n \in \mathbb{N}} |\langle f_n^*, f(t) \rangle|,$$

where the last equality is valid since $\{f_{n'}^*(t)\} \subset \{f_n^*\}$ and

$$|\langle f_n^*, f(t) \rangle| \leq \|f_n^*\| \|f(t)\| \leq \|f(t)\|.$$

Note that the numerical functions $\varphi_n(t) := \langle f_n^*, f(t) \rangle$ are measurable on $[0, T]$ and hence the function

$$\varphi(t) = \sup_{n \in \mathbb{N}} \varphi_n(t)$$

is also measurable on $[0, T]$. Therefore, the function $\|f(t)\|$ is measurable on $[0, T]$. Consider the following set:

$$C_0 := \{t \in [0, T] : \|f(t)\| > 0\}. \tag{3.20}$$

Clearly, C_0 is a Lebesgue-measurable set. Then the function

$$h(t) := \|f(t) - y\|, \quad y \in \mathbb{B},$$

is measurable on the Lebesgue-measurable set $C_0 \cap [0, T]$.

Let $\{z_n\}$ be a countable, everywhere dense set in $f([0, T]) \subset \mathbb{B}$. For any $\varepsilon > 0$, we introduce the following set:

$$D_n := \{t \in C_0 : \|f(t) - z_n\| < \varepsilon\}, \tag{3.21}$$

which is Lebesgue-measurable and, moreover,

$$C_0 := \bigcup_{n=1}^{+\infty} D_n. \tag{3.22}$$

Let

$$E_n := D_n \backslash \bigcup_{i=1}^{n-1} D_i \quad \Rightarrow \quad E_{n_1} \cap E_{n_2} = \varnothing \quad \text{for } n_1 \neq n_2. \tag{3.23}$$

From (3.22) and (3.23) we obtain that

$$C_0 = \bigcup_{n=1}^{+\infty} E_n. \tag{3.24}$$

Introduce the function

$$f_\varepsilon(t) = \begin{cases} z_n, & t \in E_n, \ n \geq 1, \\ 0, & t \in [0, T] \backslash C_0, \end{cases} \tag{3.25}$$

which is defined on the whole segment $[0, T]$. The set of values of the function $f_\varepsilon(t)$ is no more than countable; therefore, due to (3.21), the following inequality holds:

$$\|f(t) - f_\varepsilon(t)\| < \varepsilon \quad \forall t \in [0, T]. \tag{3.26}$$

Setting $\varepsilon = 1/k$, we obtain

$$\|f(t) - f_{1/k}(t)\| < \frac{1}{k} \quad \forall t \in [0, T], \tag{3.27}$$

i.e., the function $f(t)$ is a uniform limit on $[0, T]$ of countable-valued functions, each of which can be considered as the pointwise limit of a sequence of simple functions. Indeed, for any $\varepsilon > 0$, there exists $k_0 \in \mathbb{N}$ such that

$$\|f(t) - f_{1/k_0}(t)\| < \frac{\varepsilon}{2}. \tag{3.28}$$

For this $k_0 \in \mathbb{N}$, there exists a sequence of simple functions $\{h_{k_0,n}(t)\}$ and a number $N_0 \in \mathbb{N}$ such that

$$\|f_{1/k_0}(t) - h_{k_0,n}(t)\| < \frac{\varepsilon}{2} \quad \forall n \geq N_0. \tag{3.29}$$

The inequalities (3.28) and (3.29) imply the following inequality:

$$\|f(t) - h_{k_0,n}(t)\| < \varepsilon \quad \forall n \geq N_0. \tag{3.30}$$

It remains to renumber the countable family of simple functions $\{h_{k_0,n}(t)\}$ and conclude that there exists a family of simple functions $\{\hat{h}_m(t)\}$, which strongly converges to the function $f(t)$ in $\mathbb{B}$ for almost all $t \in [0, T]$. Therefore, $f(t)$ is strongly measurable. $\qquad\square$

Remark 3.2. Now we can prove that the function $\|f(t) - h_n(t)\|$ (see Definition 3.3) is μ-measurable on the segment $[0, T]$. Indeed, due to Definition 3.5, the function $f(t)$ is strongly μ-measurable. Lemma 3.2 implies that it is weakly μ-measurable. Clearly, simple functions $h_n(t)$ are weakly μ-measurable. Further, the difference of two weakly measurable functions $f(t) - h_n(t)$ is obviously a weakly measurable function. Now it remains to apply Theorem 3.1.

Remark 3.3. The strong limit

$$s - \lim_{n \to +\infty} \int_0^T h_n(t) d\mu$$

exists since the following inequalities hold:

$$\left\| \int_0^T h_n(t) d\mu - \int_0^T h_k(t) d\mu \right\| = \left\| \int_0^T (h_n(t) - h_k(t)) \, d\mu \right\| \leq \int_0^T \|h_n(t) - h_k(t)\| \, d\mu,$$

where the last inequality follows from the explicit definition (3.2) of the Bochner integral of a simple function. Indeed, for a simple function

$$h(t) = \sum_{i=1}^N b_i \chi_{S_i}(t),$$

the following equalities hold:

$$\left\| \int_0^T h(t)\,d\mu \right\| \le \sum_{i=1}^N \|b_i\| \mu(S_i), \quad \|h(t)\| = \sum_{i=1}^N \|b_i\| \chi_{S_i}(t),$$

since $S_i \cap S_j = \varnothing$ for $i \ne j$. Therefore, we have

$$\left\| \int_0^T h(t)\,d\mu \right\| \le \int_0^T \|h(t)\|\,d\mu.$$

Let

$$h(t) = \sum_{j=1}^{N_1} b_j \chi_{B_j}, \quad g(t) = \sum_{k=1}^{N_2} a_k \chi_{A_k}$$

be two simple functions. Consider the joint partition into disjoint sets:

$$S_l := A_k \cap B_j, \quad l = l_{(k,j)} \in \overline{1,N}, \quad k = \overline{1,N_2}, \quad j = \overline{1,N_1},$$

$$\bullet \quad h(t) = \sum_{l=1}^N \overline{b}_l \chi_{S_l}, \quad g(t) = \sum_{l=1}^N \overline{a}_l \chi_{S_l},$$

where

$$\overline{b}_l = \begin{cases} b_l, & \text{if } S_l \in \bigcup_{j=1}^{N_1} B_k, \\ 0, & \text{if } S_l \notin \bigcup_{j=1}^{N_1} B_k; \end{cases} \qquad \overline{a}_l = \begin{cases} a_l, & \text{if } S_l \in \bigcup_{k=1}^{N_2} A_k, \\ 0, & \text{if } S_l \notin \bigcup_{k=1}^{N_2} A_k. \end{cases}$$

Then

$$\left\| \int_0^T [g(t) + h(t)]\,d\mu \right\| = \left\| \sum_{l=1}^N (\overline{a}_l + \overline{b}_l)\mu(S_l) \right\|$$

$$\le \sum_{l=1}^N \|\overline{a}_l + \overline{b}_l\| \mu(S_l) = \int_0^T \|g(t) + h(t)\|\,d\mu.$$

It remains to apply the relation

$$\int_0^T \|h_n(t) - h_k(t)\|\,d\mu \le \int_0^T \|h_n(t) - f(t)\|\,d\mu + \int_0^T \|f(t) - h_k(t)\|\,d\mu \to +0$$

as $n, k \to +\infty$, which holds due to (3.4).

Remark 3.4. We noted above that there is a subtle point in Definition 3.3; namely, we must prove that the Bochner integral of a function $f(t)$ is independent of the choice of an approximating sequence of simple functions $\{h_n(t)\} \subset \mathbb{B}$. Indeed, let two sequences of simple function approximate the same function $f(t)$. From elements of these sequences, one can construct another approximating sequence for the function $f(t)$.

3.3   Integrability in the Bochner Sense

Theorem 3.3 (S. Bochner). *A strongly μ-measurable function $f(t)$ is integrable in the Bochner sense on the segment $[0, T]$ if and only if the numerical function $\|f(t)\|$ is μ-integrable on this segment.*

Proof. 1. *Necessity.* The following inequality (the so-called triangle inequality) holds:

$$\|f(t)\| \le \|h_n(t)\| + \|f(t) - h_n(t)\|. \tag{3.31}$$

Note that $\|h_n(t)\|$ are μ-measurable on the segment $[0, T]$. Due to Remark 3.1, the functions $\|f(t) - h_n(t)\|$ are μ-measurable.

The inequality (3.31) implies the μ-integrability of the function $\|f(t)\|$. To show this, we must prove the integrability of the norm of an arbitrary simple function. Indeed, let

$$h(t) = \sum_{i=1}^{N} b_i \chi_{S_i}(t), \quad S_i \cap S_j = \varnothing \quad \text{for } i \ne j.$$

The following equality holds:

$$\left\| \sum_{i=1}^{N} b_i \chi_{S_i}(t) \right\| = \|b_j\|, \quad t \in S_j, \quad j \in \overline{1, n},$$

$$\left\| \sum_{i=1}^{N} b_i \chi_{S_i}(t) \right\| = 0, \quad t \notin \bigcup_{i=1}^{N} S_i.$$

These equalities imply

$$\int_0^T \|h(t)\|\, dt = \sum_{i=1}^{N} \|b_i\| \mu(S_i).$$

Therefore, both terms on the right-hand side of (3.31) are μ-integrable on the segment $[0, T]$ in the Lebesgue sense and hence the left-hand side of this inequality is also integrable.

2. *Sufficiency.* Let $\{h_n(t)\}$ be a sequence of simple functions, which μ-almost everywhere on the segment $[0, T]$ strongly in $\mathbb{B}$ converges to the function $f(t)$. Therefore, there exists a set $E \subset [0, T]$ such that $\mu(E) = 0$ and, in addition,

$$\|h_n(t)\| \to \|f(t)\| \quad \text{as } n \to +\infty \tag{3.32}$$

for all $t \in [0, T] \backslash E$. Consider the following sequence of simple functions:

$$w_n(t) \stackrel{\text{def}}{=} \begin{cases} h_n(t), & \|h_n(t)\| \le 2\|f(t)\|, \\ 0, & \|h_n(t)\| > 2\|f(t)\|. \end{cases}$$

Note that the following inequality holds:

$$\|w_n(t)\| \le 2\|f(t)\|.$$

Moreover, due to the limit property (3.32), we conclude that

$$w_n(t) \to f(t) \quad \text{strongly in } \mathbb{B} \text{ as } n \to +\infty \tag{3.33}$$

for all $t \in [0, T] \backslash E$ and hence almost everywhere. Moreover,

$$\|f(t) - w_n(t)\| \le \|f(t)\| + \|w_n(t)\| \le 3\|f(t)\|.$$

Owing to the μ-integrability of the function $\|f(t)\|$, we can apply the Lebesgue theorem on the passage to the limit in the Lebesgue integral and obtain

$$\int_0^T \|f(t) - w_n(t)\| \, d\mu \to 0 \quad \text{as } n \to +\infty.$$

Theorem 3.3 is proved. $\qquad \square$

Consider some more properties of the Bochner integral. First, the Bochner integral possesses the linearity property.

Lemma 3.3. *For any Bochner-integrable functions $f_1(t)$ and $f_2(t)$ and any constants $\alpha_1, \alpha_2 \in \mathbb{C}^1$, the following equality holds:*

$$\int_0^T [\alpha_1 f_1(t) + \alpha_2 f_2(t)] \, d\mu = \alpha_1 \int_0^T f_1(t) \, d\mu + \alpha_2 \int_0^T f_2(t) \, d\mu.$$

Proof. It suffices to take corresponding approximating sequences and pass to the limit since for simple functions this equality is obviously valid. $\qquad \square$

Lemma 3.4. *Let a function $f(t)$ be Bochner μ-integrable on a segment $[0, T]$. Then the following inequality holds:*

$$\left\| \int_0^T f(t) \, d\mu \right\| \le \int_0^T \|f(t)\| \, d\mu.$$

Proof. Indeed, we have the following chain of inequalities:

$$\left\| \int_0^T f(t) \, d\mu \right\| \le \left\| \int_0^T f(t) \, d\mu - \int_0^T h_n(t) \, d\mu \right\| + \left\| \int_0^T h_n(t) \, d\mu \right\|$$

$$\le \left\| \int_0^T f(t) \, d\mu - \int_0^T h_n(t) \, d\mu \right\| + \int_0^T \|h_n(t)\| \, d\mu$$

$$\le \left\| \int_0^T f(t) \, d\mu - \int_0^T h_n(t) \, d\mu \right\| + \int_0^T \|f(t) - h_n(t)\| \, d\mu + \int_0^T \|f(t)\| \, d\mu.$$

It remains to pass to the limit as $n \to +\infty$ and apply Definition 3.3. $\qquad \square$

Definition 3.6. We say that a $\mathbb{B}$-valued function $f(t)$ *belongs to the set* $C([0,T];\mathbb{B})$ if the following limit relation holds:

$$\lim_{|t_1-t_2|\to+0} \|f(t_1) - f(t_2)\| = 0 \quad \text{for all } t_1, t_2 \in [0,T]. \tag{3.34}$$

Definition 3.7. The *derivative* of a $\mathbb{B}$-valued function $f(t)$ at a point $t_0 \in (0,T)$ is the $\mathbb{B}$-valued function $f'(t_0) \in \mathbb{B}$ satisfying the relation

$$\lim_{t\to t_0} \left\| \frac{f(t) - f(t_0)}{t - t_0} - f'(t_0) \right\| = 0. \tag{3.35}$$

Notation. If a function $f(t)$ is differentiable everywhere on the segment $[0,T]$ and its derivative $f'(t)$ is a continuous function on this segment, then we will use the notation $f(t) \in C^{(1)}([0,T];\mathbb{B})$. (Note that the derivatives at the points $t_0 = 0$ and $t_0 = T$ are meant as one-sided derivatives.)

The following lemma is an important generalization of the Lebesgue differentiation theorem.

Lemma 3.5. *Let a function* $f(t)$ *be Bochner-integrable on a segment* $[0,T]$. *Then the function* $u(t)$ *defined by the formula*

$$u(t) := \int_{t_0}^{t} f(s)d\mu, \quad t_0 \le t, \quad t_0 \in [0,T), \tag{3.36}$$

is strongly differentiable for almost all $t \in [0,T]$. *Moreover, at differentiability points, the equality* $u'(t) = f(t)$ *holds.*

Proof. 1. Let $\{h_n(t)\}$ be a sequence of simple functions from Definition 3.3. Without loss of generality, we assume that the inequality $\|h_n(t)\| \le 2\|f(t)\|$ holds for almost all $t \in [0,T]$ for sufficiently large $n \ge N_0 \in \mathbb{N}$ (cf. Theorem 3.3). By Definition 3.3, $h_n(t) \to f(t)$ strongly in $\mathbb{B}$ as $n \to +\infty$ μ-almost everywhere on the segment $[0,T]$. The following relations hold:

$$\frac{1}{t-t_0} \int_{t_0}^{t} f(s)\, d\mu - f(t_0)$$

$$= \frac{1}{t-t_0} \int_{t_0}^{t} [f(s) - h_n(s)]\, d\mu + \frac{1}{t-t_0} \int_{t_0}^{t} h_n(s)\, d\mu - h_n(t_0) + h_n(t_0) - f(t_0).$$

Thus, we have

$$\left\| \frac{1}{t-t_0} \int_{t_0}^{t} f(s)\, d\mu - f(t_0) \right\| \le \frac{1}{|t-t_0|} \int_{t_0}^{t} \|f(s) - h_n(s)\|\, d\mu$$

$$+ \left\| \frac{1}{t-t_0} \int_{t_0}^{t} h_n(s)\, d\mu - h_n(t_0) \right\| + \|h_n(t_0) - f(t_0)\| =: I_1 + I_2 + I_3. \tag{3.37}$$

2. First, we note that due to the fact that $h_n(t)$ is a simple function, I_2 is equal to zero μ-almost everywhere on the segment $[0, T]$ if the length of the segment $|t - t_0|$ is sufficiently small.

Passing to the limit as $t \to t_0$ in the expression for I_1, we obtain

$$\lim_{t \to t_0} I_1 = \|h_n(t_0) - f(t_0)\|$$

μ-almost everywhere on the segment $[0, T]$ due to the Lebesgue differentiation theorem. Thus, passing to the limit as $t \to t_0$, we obtain from (3.37) the following inequality:

$$\lim_{t \to t_0} \left\| \frac{1}{t - t_0} \int_{t_0}^{t} f(s)\, d\mu - f(t_0) \right\| \leq 2 \|h_n(t_0) - f(t_0)\| \tag{3.38}$$

μ-almost everywhere on the segment $[0, T]$.

3. Now we pass to the limit as $n \to +\infty$ and obtain from the inequality (3.38) the following equality:

$$\lim_{t \to t_0} \left\| \frac{1}{t - t_0} \int_{t_0}^{t} f(s)\, d\mu - f(t_0) \right\| = 0$$

μ-almost everywhere on the segment $[0, T]$. $\qquad\square$

Lemma 3.6. *Let $f(t)$ be a Bochner-integrable function. Then for any $f^* \in \mathbb{B}^*$, the following equality holds:*

$$\left\langle f^*, \int_0^T f(t)\, d\mu \right\rangle = \int_0^T \langle f^*, f(t) \rangle\, d\mu. \tag{3.39}$$

Proof. 1. Let $\{h_n(t)\}$ be a sequence of simple functions for the function $f(t)$ from Definition 3.3. For each function from this sequence, Eq. (3.39) holds:

$$\left\langle f^*, \int_0^T h_n(t)\, d\mu \right\rangle = \int_0^T \langle f^*, h_n(t) \rangle\, d\mu.$$

The following chain of equalities holds:

$$\left\langle f^*, \int_0^T f(t)\, d\mu \right\rangle = \left\langle f^*, \int_0^T [f(t) - h_n(t)]\, d\mu \right\rangle + \left\langle f^*, \int_0^T h_n(t)\, d\mu \right\rangle$$

$$= \left\langle f^*, \int_0^T [f(t) - h_n(t)]\, d\mu \right\rangle + \int_0^T \langle f^*, h_n(t) \rangle\, d\mu$$

$$= \left\langle f^*, \int_0^T [f(t) - h_n(t)]\, d\mu \right\rangle + \int_0^T \langle f^*, h_n(t) - f(t) \rangle\, d\mu + \int_0^T \langle f^*, f(t) \rangle\, d\mu.$$

$$\tag{3.40}$$

2. The following estimates hold:

$$\left| \left\langle f^*, \int_0^T [f(t) - h_n(t)] \, d\mu \right\rangle \right| \leq \|f^*\|_* \left\| \int_0^T [f(t) - h_n(t)] \, d\mu \right\| \to 0,$$

$$\left| \int_0^T \langle f^*, h_n(t) - f(t) \rangle \, d\mu \right| \leq \int_0^T |\langle f^*, h_n(t) - f(t) \rangle| \, d\mu$$

$$\leq \|f^*\|_* \int_0^T \|h_n(t) - f(t)\| \, d\mu \to 0$$

as $n \to +\infty$. Taking into account these estimates, passing to the limit as $n \to +\infty$, from (3.40) we obtain the required assertion. $\qquad\square$

Theorem 3.4. *Let*

$$u(t) := \int_{t_0}^t f(s) \, d\mu, \quad t_0, t \in [0, T]. \tag{3.41}$$

If $f(t) \in C([0, T]; \mathbb{B})$*, then* $u(t) \in C^{(1)}([0, T]; \mathbb{B})$*.*

Proof. Theorem 3.2 implies the integrability of the function $f(t)$ and, therefore, the validity of the definition (3.41). Lemma 3.5 implies that the function $u(t)$ is strongly differentiable μ-almost everywhere on the segment $[0, T]$ and $u'(t) = f(t)$. Since $f(t) \in C([0, T]; \mathbb{B})$, we conclude (changing if necessary the function on a subset with zero Lebesgue measure μ) that $u'(t) \in C([0, T]; \mathbb{B})$. $\qquad\square$

3.4   Bibliographical Notes

The contents of this lecture is taken from [22–24, 70].

Lecture 4

Spaces of $\mathbb{B}$-Valued Functions and Distributions

4.1 Spaces $L^p(0, T; \mathbb{B})$

Now we introduce classes of Bochner-integrable functions.

Definition 4.1. The *functional class* $\mathscr{L}^p([0, T]; \mathbb{B})$, $p \in [1, +\infty)$, consists of strongly measurable on the segment $[0, T]$, $\mathbb{B}$-valued functions such that $\|f(t)\|^p$ is integrable on the segment $[0, T]$.

Remark 4.1. Similarly to the case of scalar spaces $L^p(a, b)$, we must identify equivalent functions, since in the opposite case, under the natural normalization (4.2) (see below), all functions that differ from the zero element of the space $\mathbb{B}$ on a set of zero Lebesgue μ-measure will have zero norm.

Denote by $\mathscr{J}_0(0, T; \mathbb{B})$ the subset of the set $\mathscr{L}^p(0, T; \mathbb{B})$ consisting of $\mathbb{B}$-valued functions that vanish almost everywhere on $[0, T]$. *Note that these functions are strongly measurable.*

Indeed, let $f(t) = 0$ for almost all $t \in [0, T]$. Consider the following sequence of simple functions $\{h_n(t)\}$, $h_n(t) = 0$ for all $t \in [0, T]$ and all $n \in \mathbb{N}$. Then

$$\|h_n(t) - f(t)\| \to +0 \quad \text{as } n \to +\infty \text{ for almost all } t \in [0, T].$$

Consider the factor space

$$\mathscr{L}^p(0, T; \mathbb{B}) / \mathscr{J}_0(0, T; \mathbb{B}). \tag{4.1}$$

This factor space is a vector space since we have identified all functions that differ only on a set of zero Lebesgue measure. Therefore, all function that vanish almost everywhere (with respect to the Lebesgue measure) have also been identified. As any factor space, the factor space introduces consists of disjunct equivalence classes. We will use this fact in the sequel.

Definition 4.2. For $p \in [1, +\infty)$, we denote by $L^p(0, T; \mathbb{B})$ the factor space $\mathscr{L}^p(0, T; \mathbb{B}) / \mathscr{J}_0(0, T; \mathbb{B})$.

The case $p = +\infty$ is introduced separately.

Definition 4.3. We denote by $\mathscr{L}^\infty(0, T; \mathbb{B})$ the set of μ-measurable functions that are bounded μ-almost everywhere with respect to the norm of the space $\mathbb{B}$.

Similarly to Definition 4.2, we introduce the following notation.

Definition 4.4. We denote the factor space $\mathscr{L}^\infty(0,T;\mathbb{B})/\mathscr{J}_0(0,T];\mathbb{B})$ by $L^\infty(0,T;\mathbb{B})$.

The sets $L^p(0,T;\mathbb{B})$ introduced above are vector spaces for $p \in [1,+\infty]$. The following question naturally arises: Are these spaces Banach spaces with respect to certain norms?

Theorem 4.1. *The space $L^p(0,T;\mathbb{B})$ is a Banach space for $p \in [1,+\infty)$ with respect to the norm*

$$\|u\|_p \stackrel{\text{def}}{=} \left(\int_0^T \|u(t)\|^p \, dt \right)^{1/p}, \tag{4.2}$$

whereas the space $L^\infty(0,T;\mathbb{B})$ is a Banach space with respect to the norm

$$\|u\|_\infty \stackrel{\text{def}}{=} \text{ess.sup}_{t\in[0,T]}\|u(t)\|. \tag{4.3}$$

Remark 4.2. Note that there exists an equivalent definition of the norm (4.3):

$$\|u\|_\infty = \inf_{v \sim u} \sup_{t\in[0,T]} \|v(t)\|. \tag{4.4}$$

Lemma 4.1. *Let a Banach space $\mathbb{B}_1$ be continuously embedded into a Banach space $\mathbb{B}_2$ and $1 \leq q \leq p$. Then the Banach space $L^p(0,T;\mathbb{B}_1)$ is continuously embedded into $L^q(0,T;\mathbb{B}_2)$.*

Proof. Note that the continuous embedding $\mathbb{B}_1 \subset \mathbb{B}_2$ implies the existence of a constant $c_1 > 0$ such that

$$\|u\|_{\mathbb{B}_2} \leq c_1\|u\|_{\mathbb{B}_1} \quad \text{for all } u \in \mathbb{B}_1. \tag{4.5}$$

Let $u(t) \in L^p(0,T;\mathbb{B}_1)$. Therefore, in particular, $u_1(t)$ is a $\mathbb{B}_1$-strongly measurable function. Therefore, there exists a sequence of simple $\mathbb{B}_1$-valued functions $u_n(t)$ such that

$$\|u(t) - u_n(t)\|_{\mathbb{B}_1} \to +0 \quad \text{for almost all } t \in [0,T].$$

Then, due to the inequality (4.5), we obtain

$$\|u(t) - u_n(t)\|_{\mathbb{B}_2} \to +0 \quad \text{for almost all } t \in [0,T],$$

i.e., $u(t)$ is a $\mathbb{B}_2$-strongly measurable function.

For any function $u(t) \in L^p(0,T;\mathbb{B}_1)$, the following chain of inequalities holds:

$$\left(\int_0^T \|u(t)\|_{\mathbb{B}_2}^q \, dt \right)^{1/q} \leq c_1 \left(\int_0^T \|u(t)\|_{\mathbb{B}_1}^q \, dt \right)^{1/q} \leq c_1 T^{(p-q)/(pq)} \left(\int_0^T \|u(t)\|_{\mathbb{B}_1}^p \, dt \right)^{1/p}. \tag{4.6}$$

Lemma 4.1 is proved. $\qquad\qquad\qquad\qquad\qquad\qquad\qquad\qquad\qquad\qquad\qquad\qquad\square$

Introduce the following important functional space $AC(0,T;\mathbb{B})$.

Definition 4.5. We say that a function $u(t)$ belongs to the functional space $AC(0,T;\mathbb{B})$ if there exist a function $v(t) \in L^1(0,T;\mathbb{B})$ and an element $u_0 \in \mathbb{B}$ such that the following representation holds:

$$u(t) = u_0 + \int_0^t v(s)\,ds, \quad t \in [0,T], \tag{4.7}$$

where the integral is meant in the Bochner sense.

Theorem 4.2. *Let* $u(t) \in AC(0,T;\mathbb{B})$ *and* $\varphi(t) \in C^{(1)}[0,T]$. *Then for any* $0 \le t_1 < t_2 \le T$, *the following formula of integrating by parts in the Bochner integral is valid:*

$$\int_{t_1}^{t_2} u'(s)\varphi(s)\,ds = u(t_2)\varphi(t_2) - u(t_1)\varphi(t_1) - \int_{t_1}^{t_2} u(s)\varphi'(s)\,ds. \tag{4.8}$$

Proof. 1. Note that the embedding $AC(0,T;\mathbb{B}) \subset C([0,T];\mathbb{B})$ is valid. Indeed, if $u(t) \in AC(0,T;\mathbb{B})$, then, due to (4.7) and Lemma 3.4, we obtain

$$\|u(t_2) - u(t_1)\| \le \int_{t_1}^{t_2} \|v(s)\|\,ds \to +0 \quad \text{as } |t_2 - t_1| \to +0,$$

since $\|v(s)\| \in L^1(0,T)$, and hence

$$\int_0^t \|v(s)\|\,ds \in AC[0,T].$$

Due to this fact, we have $u(0) = u_0$ in Eq. (4.7).

2. Now let $u(t) \in AC(0,T;\mathbb{B})$ and $\varphi(t) \in C^{(1)}[0,T]$. Due to Lemma 3.5, the function $u(t)$ is strongly differentiable almost everywhere on $t \in [0,T]$. Therefore, the following equality holds almost everywhere:

$$(u(s)\varphi(s))' = u'(s)\varphi(s) + u(s)\varphi'(s) \in L^1(0,T;\mathbb{B}). \tag{4.9}$$

Indeed, the following relations hold:

$$\left\| \frac{u(t)\varphi(t) - u(t_0)\varphi(t_0)}{t - t_0} - \varphi(t_0)u'(t_0) - \varphi'(t_0)u(t_0) \right\|$$
$$\le |\varphi(t) - \varphi(t_0)| \left\| \frac{u(t) - u(t_0)}{t - t_0} \right\| + |\varphi(t_0)| \left\| \frac{u(t) - u(t_0)}{t - t_0} - u'(t_0) \right\|$$
$$+ \|u(t_0)\| \left| \frac{\varphi(t) - \varphi(t_0)}{t - t_0} - \varphi'(t_0) \right|. \tag{4.10}$$

Note that

$$\left\| \frac{u(t) - u(t_0)}{t - t_0} \right\| \le \left\| \frac{u(t) - u(t_0)}{t - t_0} - u'(t_0) \right\| + \|u'(t_0)\|. \tag{4.11}$$

Therefore, the right-hand side of (4.10) tends to zero as $|t - t_0| \to +0$ for all $t, t_0 \in [0, T]$.

Consider the function

$$g(t) := \int\limits_0^t (u(s)\varphi(s))' \, ds. \tag{4.12}$$

Due to Lemma 3.5, for almost all $t \in [0, T]$, the following equality holds:

$$g'(t) = (u(t)\varphi(t))' \quad \Leftrightarrow \quad (g(t) - u(t)\varphi(t))' = 0$$
$$\Leftrightarrow \quad g(t) - u(t)\varphi(t) = b \in \mathbb{B}, \tag{4.13}$$

where $b \in \mathbb{B}$ is a certain fixed element.

Remark 4.3. Note that if $h'(t) = \theta \in \mathbb{B}$ for almost all $t \in [0, T]$ and $h(t) \in AC([0, T]; \mathbb{B})$, then moreover for any $f \in \mathbb{B}^*$ we have

$$\frac{d}{dt}\langle f^*, h(t) \rangle = 0 \quad \text{for almost all } t \in [0, T]$$

and $\langle f^*, h(t) \rangle \in AC[0, T]$. Hence we have the equality

$$\langle f^*, h(t) \rangle = \langle f^*, h(0) \rangle + \int\limits_0^t \frac{d}{ds}\langle f^*, h(s) \rangle \, ds.$$

Then

$$\langle f^*, h(t) \rangle = \langle f^*, h(0) \rangle \quad \text{for all } f^* \in \mathbb{B}^*.$$

Therefore, $h(t) = h(0)$ for all $t \in [0, T]$.

In particular, the following equality holds:

$$g(t) - u(t)\varphi(t) = g(0) - u(0)\varphi(0) = -u(0)\varphi(0).$$

Then (4.12) implies the following equality:

$$u(t)\varphi(t) = u(0)\varphi(0) + \int\limits_0^t (u(s)\varphi(s))' \, ds, \quad b = -u(0)\varphi(0). \tag{4.14}$$

It remains to apply Eq. (4.9) and obtain the following expression:

$$u(t)\varphi(t) = u(0)\varphi(0) + \int\limits_0^t u'(s)\varphi(s) \, ds + \int\limits_0^t u(s)\varphi'(s) \, ds. \tag{4.15}$$

Applying this equality for $t = t_1$ and for $t = t_2$ under the condition $0 \le t_1 < t_2 \le T$, we obtain the following equalities:

$$u(t_1)\varphi(t_1) = u(0)\varphi(0) + \int\limits_0^{t_1} u'(s)\varphi(s) \, ds + \int\limits_0^{t_1} u(s)\varphi'(s) \, ds, \tag{4.16}$$

$$u(t_2)\varphi(t_2) = u(0)\varphi(0) + \int\limits_0^{t_2} u'(s)\varphi(s) \, ds + \int\limits_0^{t_2} u(s)\varphi'(s) \, ds. \tag{4.17}$$

Subtracting (4.16) from (4.17), we arrive at the required equality (4.8). $\qquad\square$

For the spaces $L^p(0, T; \mathbb{B})$, the following generalization of the Hölder inequality holds.

Lemma 4.2. *If* $u(t) \in L^p(0, T; \mathbb{B})$ *and* $v(t) \in L^{p'}(0, T; \mathbb{B}^*)$, *where* $1 \le p < +\infty$ *and* $p' = p/(p-1)$, *then*

$$\langle v(t), u(t) \rangle \in L^1(0, T), \tag{4.18}$$

$$\int_0^T |\langle v(t), u(t) \rangle| \, dt \le \|v\|_{L^{p'}(0,T;\mathbb{B}^*)} \|u\|_{L^p(0,T;\mathbb{B})}. \tag{4.19}$$

Proof. Let $u(t) \in L^p(0, T; \mathbb{B})$ and $v(t) \in L^{p'}(0, T; \mathbb{B}^*)$. By Definitions 3.5, 4.2, and 4.4, there exist sequences of $\mathbb{B}$-valued simple functions $h_n(t)$ and $\mathbb{B}^*$-valued simple function $h_n^*(t)$ such that

$$\|u(t) - h_n(t)\| \to +0, \quad \|v(t) - h_n^*(t)\|_* \to +0 \tag{4.20}$$

as $n \to +\infty$ for almost all $t \in [0, T]$. Then the sequence of simple real-valued functions

$$\langle h_n^*(t), h_n(t) \rangle \to \langle v(t), u(t) \rangle \quad \text{as } n \to +\infty \tag{4.21}$$

for almost all $t \in [0, T]$. Indeed, we have

$$|\langle h_n^*(t), h_n(t) \rangle - \langle v(t), u(t) \rangle|$$
$$\le |\langle h_n^*(t) - v(t), h_n(t) \rangle| + |\langle v(t), h_n(t) - u(t) \rangle| := I_{1n}(t) + I_{2n}(t). \tag{4.22}$$

The limit properties (4.20) imply the relation

$$\sup_{n \in \mathbb{N}} \|h_n(t)\| \le M(T) \quad \text{for almost all } t \in [0, T], \tag{4.23}$$

and also the inequalities

$$I_{1n}(t) \le \|h_n^*(t) - v(t)\|_* \|h_n(t)\| \le M(T) \|h_n^*(t) - v(t)\|_*, \tag{4.24}$$

$$I_{2n}(t) \le \|v(t)\|_* \|h_n(t) - u(t)\|. \tag{4.25}$$

From (4.24), (4.25), (4.20), and (4.22), the limit property (4.21) follows.

Since any simple function is measurable and the limit of a sequence of simple functions, which almost everywhere converges to a certain function, is also measurable, then the function $\langle v(t), u(t) \rangle$ is Lebesgue-measurable on $[0, T]$. It remains to note that the following inequality holds:

$$|\langle v(t), u(t) \rangle| \le \|v(t)\|_* \|u(t)\| \quad \text{for almost all } t \in [0, T]. \tag{4.26}$$

Therefore, owing to the inclusions

$$\|v(t)\|_* \in L^{p'}(0, T), \quad \|u(t)\| \in L^p(0, T) \tag{4.27}$$

and the Hölder inequality, we have

$$\int_0^T |\langle v(t), u(t) \rangle| \, dt \le \left(\int_0^T \|v(t)\|_*^{p'} \, dt \right)^{1/p'} \left(\int_0^T \|u(t)\|^p \, dt \right)^{1/p}$$

$$= \|v\|_{L^{p'}(0,T;\mathbb{B}^*)} \|u\|_{L^p(0,T;\mathbb{B})}. \tag{4.28}$$

Lemma 4.2 is proved. $\qquad\qquad\qquad\qquad\qquad\qquad\qquad\qquad\qquad\qquad\qquad\qquad\qquad\square$

Remark 4.4. By induction, one can also prove the following more general result is valid: *if $u_k \in L^{p_k}(X)$, where $p_k \in (1; +\infty)$, $k = 1, \ldots, n$, and*

$$\frac{1}{p_1} + \frac{1}{p_2} + \cdots + \frac{1}{p_n} = \frac{1}{r}, \quad r \in [1; +\infty),$$

then

$$\|u_1 u_2 \cdots u_n\|_{L^r(X)} \le \|u_1\|_{L^{p_1}(X)} \|u_2\|_{L^{p_2}(X)} \cdots \|u_n\|_{L^{p_n}(X)}. \tag{4.29}$$

Now we present without a proof the following important result.

Theorem 4.3 (see [23, Theorem 2.2.9]). *If $p \in [1, +\infty)$, $p' = p/(p-1)$, and the space $\mathbb{B}$ is reflexive, then*

$$(L^p(0, T; \mathbb{B}))^* = L^{p'}(0, T; \mathbb{B}^*)$$

and the following explicit expression for the duality bracket holds:

$$\langle g, f \rangle_{L^p(0,T;\mathbb{B})} = \int_0^T \langle g(t), f(t) \rangle \, dt \tag{4.30}$$

for all $f(t) \in L^p(0, T; \mathbb{B})$ and $g(t) \in L^{p'}(0, T; \mathbb{B}^*)$, where $\langle \cdot, \cdot \rangle$ is the duality bracket between $\mathbb{B}$ and $\mathbb{B}^*$.

4.2   Weak Derivatives and $\mathbb{B}$-Valued Sobolev Spaces

Definition 4.6. Let $u(t) \in L^1(0, T; \mathbb{B})$. A function $v(t) \in L^1(0, T; \mathbb{B})$ is called the *weak derivative* of $u(t)$ (notation $v(t) = u'(t)$) if

$$\int_0^T \varphi'(t) u(t) \, dt = - \int_0^T \varphi(t) v(t) \, dt \tag{4.31}$$

for an arbitrary test function $\varphi(t) \in \mathscr{D}(0, T)$, where the integral is meant as the $\mathbb{B}$-valued Bochner integral.

Now we introduce the corresponding analog of the Sobolev space $W^{1,p}(0, T)$ for $\mathbb{B}$-valued functions.

Definition 4.7. The Sobolev space $W^{1,p}(0, T; \mathbb{B})$ consists of all functions $u(t) \in L^p(0, T; \mathbb{B})$ such that the weak derivative $u'(t)$ exists and belongs to $L^p(0, T; \mathbb{B})$.

The spaces $W^{1,p}(0, T; \mathbb{B})$ are complete with respect to the following norms:

$$\|u\|_{W^{1,p}(0,T;\mathbb{B})} = \left[\int_0^T \left(\|u(t)\|^p + \|u'(t)\|^p \right) dt \right]^{1/p}, \quad p \in [1 + \infty), \tag{4.32}$$

$$\|u\|_{W^{1,\infty}(0,T;\mathbb{B})} = \operatorname*{ess.sup}_{t \in [0,T]} \left(\|u(t)\| + \|u'(t)\| \right). \tag{4.33}$$

Let Ω be a bounded domain. Consider the space

$$W_{pp'}(0,T) := \left\{ u(t) \in L^p(0,T; W_0^{1,p}(\Omega)), \right.$$

$$\left. u'(t) \in L^{p'}(0,T; W^{-1,p'}(\Omega)) \right\}, \quad p > 1. \qquad (4.34)$$

This definition has the following sense. Consider and arbitrary function

$$u(t) \in L^1(0,T; W^{-1,p'}(\Omega))$$

possessing the generalized derivative

$$u'(t) \in L^1(0,T; W^{-1,p'}(\Omega))$$

in the sense of Definition 4.6 and assume, in addition, that

$$u(t) \in L^p(0,T; W_0^{1,p}(\Omega)), \quad u'(t) \in L^{p'}(0,T; W^{-1,p'}(\Omega)).$$

Functions satisfying these conditions form the space $W_{pp'}(0,T)$.

One can first consider a $W^{-1,p'}(\Omega)$-valued function with the $W^{-1,p'}(\Omega)$-valued derivative and then note that for almost all t, the values of the function $u(t)$ belong to $W_0^{1,p}(\Omega) \subset W^{-1,p'}(\Omega)$; moreover, $u(t), u'(t) \in L^1(0,T; W^{-1,p'}(\Omega))$.

The space $W_{pp'}(0,T)$ is a Banach space with respect to the norm

$$\|u\|_{pp'} := \|u\|_{L^p(0,T;W_0^{1,p}(\Omega))} + \|u'\|_{L^{p'}(0,T;W^{-1,p'}(\Omega))}. \qquad (4.35)$$

Lemma 4.3 (see [22]). *The vector space* $C^{(1)}([0,T]; W_0^{1,p}(\Omega))$ *is dense in the Banach space* $W_{pp'}(0,T)$.

In the sequel, we need a series of results collected in the following theorem.

Theorem 4.4. *Let* $u(t) \in W_{pp'}(0,T)$, $p \geq 2$.

 (i) *The function* $u(t)$ *belongs to* $C([0,T]; L^2(\Omega))$ *after redefining on a set of zero Lebesgue measure;*

 (ii) *the following estimate holds:*

$$\max_{t \in [0,T]} \|u(t)\|_{L^2(\Omega)} \leq c_2 \left(\|u\|_{L^p(0,T;W_0^{1,p}(\Omega))} + \|u'\|_{L^{p'}(0,T;W^{-1,p'}(\Omega))} \right);$$

$$(4.36)$$

 (iii) *the relation*

$$\frac{d}{dt} \|u(t)\|_{L^2(\Omega)}^2 = 2\langle u'(t), u(t) \rangle$$

holds for almost all $t \in [0,T]$, *where* $\langle \cdot, \cdot \rangle$ *is the duality bracket between* $W_0^{1,p}(\Omega)$ *and* $W^{-1,p'}(\Omega)$.

Proof. 1. *Proof of* (i) *and* (ii). First, we note that due to Lemma 4.3, the following dense embedding holds:

$$C^{(1)}([0,T]; W_0^{1,p}(\Omega)) \overset{ds}{\subset} W_{pp'}(0,T).$$

For $u(t), v(t) \in C^{(1)}([0, T]; W_0^{1,p}(\Omega))$, the equality

$$\frac{d}{d\tau}(u(\tau), v(\tau))_{L^2(\Omega)} = (u'(\tau), v(\tau))_{L^2(\Omega)} + (u(\tau), v'(\tau))_{L^2(\Omega)} \qquad (4.37)$$

holds for all $\tau \in [0, T]$. Integrating both sides by $\tau \in (s, t)$, we obtain the equality

$$(u(t), v(t))_{L^2(\Omega)} - (u(s), v(s))_{L^2(\Omega)}$$

$$= \int_s^t \left[(u'(\tau), v(\tau)) + (u(\tau), v'(\tau)) \right] d\tau$$

$$= \int_s^t \left[\langle u'(\tau), v(\tau) \rangle + \langle v'(\tau), u(\tau) \rangle \right] d\tau, \qquad (4.38)$$

where we have used the dense and continuous embeddings

$$W_0^{1,p}(\Omega) \overset{ds}{\subset} L^2(\Omega) \overset{ds}{\subset} W^{-1,p'}(\Omega) \qquad (4.39)$$

and Theorem 1.4.

Introduce the auxiliary function

$$\varphi(\tau) \in C^{(1)}[0, T], \quad \varphi(0) = 0, \quad \varphi(T) = 1. \qquad (4.40)$$

Clearly, there exists a constant $K_1 > 0$ such that the following inequality holds:

$$|\varphi(\tau)| + |\varphi'(\tau)| \leq K_1(T) < +\infty, \quad \tau \in [0, T]. \qquad (4.41)$$

Consider the auxiliary functions

$$v(\tau) = \varphi(\tau)u(\tau), \quad w(\tau) = u(\tau) - \varphi(\tau)u(\tau). \qquad (4.42)$$

Note that

$$v(0) = 0, \quad w(T) = 0 \qquad (4.43)$$

and, moreover,

$$v'(\tau) = \varphi'(\tau)u(\tau) + \varphi(\tau)u'(\tau), \qquad (4.44)$$

$$w'(\tau) = u'(\tau) - \varphi'(\tau)u(\tau) - \varphi(\tau)u'(\tau). \qquad (4.45)$$

First, we apply the formula (4.38) to functions $v(\tau)$ and $u(\tau)$ for $s = 0$. Taking into account (4.44) and (4.45), we obtain the following formula:

$$(u(t), v(t))_{L^2(\Omega)} = \int_0^t [2\varphi(\tau)\langle u'(\tau), u(\tau) \rangle + \varphi'(\tau)\langle u(\tau), u(\tau) \rangle] \, d\tau. \qquad (4.46)$$

Now we apply the formula (4.38) to the functions $w(\tau)$ and $u(\tau)$ for $t = T$ and, taking into account (4.43) and (4.45), obtain the following equality:

$$-(u(t), w(t))_{L^2(\Omega)} = \int_t^T \Big\{ \langle u'(\tau), u(\tau) - \varphi(\tau)u(\tau) \rangle$$

$$+ \langle u'(\tau) - \varphi'(\tau)u(\tau) - \varphi(\tau)u'(\tau), u(\tau) \rangle \Big\} \, d\tau, \qquad (4.47)$$

which can be rewritten in the form

$$(u(t), w(t))_{L^2(\Omega)} = \int_t^T \left\{ -2(1 - \varphi(\tau))\langle u'(\tau), u(\tau)\rangle + \varphi'(\tau)\langle u(\tau), u(\tau)\rangle \right\} d\tau. \quad (4.48)$$

Adding Eqs. (4.46) and (4.48), we arrive at the following equality:

$$\|u(t)\|_{L^2(\Omega)}^2$$
$$= \int_0^T \left[2\varphi(\tau)\langle u'(\tau), u(\tau)\rangle + \varphi'(\tau)\langle u(\tau), u(\tau)\rangle \right] d\tau - 2\int_t^T \langle u'(\tau), u(\tau)\rangle d\tau. \quad (4.49)$$

Equality (4.49) implies the following inequality:

$$\|u(t)\|_{L^2(\Omega)}^2 \leq \int_0^T \left[2[|\varphi(\tau)| + 1]\|u'(\tau)\|_{W^{-1,p'}(\Omega)} \|u(\tau)\|_{W_0^{1,p}(\Omega)} \right.$$
$$\left. + |\varphi'(\tau)|\|u(\tau)\|_{L^2(\Omega)}^2 \right] d\tau. \quad (4.50)$$

Note that for $p \geq 2$, due to Lemma 4.1, the continuous embedding

$$L^p(0, T; W_0^{1,p}(\Omega)) \subset L^2(0, T; L^2(\Omega)) \quad (4.51)$$

holds and hence there exists a constant $c_1 > 0$ such that the inequality

$$\int_0^T \|w(\tau)\|_{L^2(\Omega)}^2 d\tau \leq c_1^2 \left(\int_0^T \|w(\tau)\|_{W_0^{1,p}(\Omega)}^p d\tau \right)^{2/p} \quad (4.52)$$

is fulfilled for all $w(t) \in L^p(0, T; W_0^{1,p}(\Omega))$. Moreover, due to the Hölder inequality, the following inequality is valid:

$$\int_0^T \|u'(\tau)\|_{W^{-1,p'}(\Omega)} \|u(\tau)\|_{W_0^{1,p}(\Omega)} d\tau$$
$$\leq \|u'\|_{L^{p'}(0,T;W^{-1,p'}(\Omega))} \|u\|_{L^p(0,T;W_0^{1,p}(\Omega))}. \quad (4.53)$$

Therefore, from (4.52), (4.53), (4.50), and (4.40) we obtain the inequality

$$\|u(t)\|_{L^2(\Omega)}^2 \leq K_1 \|u'\|_{L^{p'}(0,T;W^{-1,p'}(\Omega))} \|u\|_{L^p(0,T;W_0^{1,p}(\Omega))}$$
$$+ K_2 c_1^2 \|u\|_{L^p(0,T;W_0^{1,p}(\Omega))}^2 \leq K_3 \|u\|_{pp'}^2 \quad (4.54)$$

for all $u(t) \in C^{(1)}([0, T]; W_0^{1,p}(\Omega))$. From (4.54) we obtain the inequality

$$\|u\|_{C([0,T];L^2(\Omega))} := \sup_{t \in [0,T]} \|u(t)\|_{L^2(\Omega)} \leq c_2 \|u\|_{pp'} \quad (4.55)$$

for all $u(t) \in C^{(1)}([0, T]; W_0^{1,p}(\Omega))$. As was mentioned above,

$$C^{(1)}([0, T]; W_0^{1,p}(\Omega)) \overset{ds}{\subset} W_{pp'}(0, T)$$

and hence for any $u(t) \in W_{pp'}(0,T)$, there exists a functional sequence $\{u_m(t)\} \subset C^{(1)}([0,T]; W_0^{1,p}(\Omega))$ such that

$$u_m(t) \to u(t) \quad \text{strongly in } W_{pp'}(0,T) \text{ as } m \to +\infty. \tag{4.56}$$

Note that due to (4.55) the following inequality holds:

$$\|u_n(t) - u_m(t)\|_{C([0,T];L^2(\Omega))} \leq c_2 \|u_n(t) - u_m(t)\|_{pp'}. \tag{4.57}$$

Since the space $C([0,T]; L^2(\Omega))$ is dense, we conclude that

$$u_m(t) \to v(t) \quad \text{strongly in } C([0,T];L^2(\Omega)) \text{ as } m \to +\infty. \tag{4.58}$$

Now we note that both spaces $W_{pp'}$ and $C([0,T];L^2(\Omega))$ are continuously embedded in $L^p(0,T;L^2(\Omega))$; therefore, we obtain from (4.56) and (4.58)

$$u_m(t) \to u(t) \quad \text{strongly in } L^p(0,T;L^2(\Omega)) \text{ as } m \to +\infty,$$
$$u_m(t) \to v(t) \quad \text{strongly in } L^p(0,T;L^2(\Omega)) \text{ as } m \to +\infty.$$

Then, owing to the uniqueness of the limit in the space $L^p(0,T;L^2(\Omega))$ we conclude that $v(t) = u(t)$ and, therefore,

$$u_m(t) \to u(t) \quad \text{strongly in } C([0,T];L^2(\Omega)) \text{ as } m \to +\infty. \tag{4.59}$$

Passing to the limit as $m \to +\infty$ in the inequality (4.57), we obtain the following inequality:

$$\|u_n(t) - u(t)\|_{C([0,T];L^2(\Omega))} \leq c_2 \|u_n(t) - u(t)\|_{pp'}. \tag{4.60}$$

Thus, the following chain of inequalities holds:

$$\|u(t)\|_{C([0,T];L^2(\Omega))} \leq \|u_n(t) - u(t)\|_{C([0,T];L^2(\Omega))} + \|u_n(t)\|_{C([0,T];L^2(\Omega))}$$
$$\leq c_2 \|u_n(t) - u(t)\|_{pp'} + c_2 \|u_n(t)\|_{pp'} \to c_2 \|u(t)\|_{pp'} \tag{4.61}$$

as $n \to +\infty$, where we have used the inequality (4.60) and the limit relation (4.56), which implies that

$$\|u_n(t)\|_{pp'} \to \|u(t)\|_{pp'} \quad \text{as } n \to +\infty.$$

Therefore, the continuous embedding $W_{pp'}(0,T) \subset C([0,T];L^2(\Omega))$ holds and the inequality (4.36) is valid for functions from $W_{pp'}(0,T)$.

    2. *Proof of* (iii). Let $u(t) \in W_{pp'}(0,T)$. Then there exists a sequence $\{u_m(t)\} \subset C^{(1)}([0,T]; W_0^{1,p}(\Omega))$ such that

$$u_m(t) \to u(t) \quad \text{strongly in } W_{pp'}(0,T) \text{ as } m \to +\infty. \tag{4.62}$$

Obviously, due to (4.62) and the continuous and dense embeddings (4.39), we have the equalities

$$\frac{d}{dt}\|u_m(t)\|_{L^2(\Omega)}^2 = 2(u_m'(t), u_m(t))_{L^2(\Omega)} = 2\langle u_m'(t), u_m(t)\rangle \tag{4.63}$$

for all $t \in [0,T]$. Integrating this equality, we obtain

$$\|u_m(t)\|_{L^2(\Omega)}^2 = \|u_m(0)\|_{L^2(\Omega)}^2 + 2\int_0^t \langle u_m'(\tau), u_m(\tau)\rangle \, d\tau. \tag{4.64}$$

Using the continuous embedding of $W_{pp'}(0, T)$ in $C([0, T]; L^2(\Omega))$ proved above, we conclude from (4.62) that as $m \to +\infty$

$$u_m(t) \to u(t) \quad \text{strongly in } L^2(\Omega) \text{ for all } t \in [0, T], \tag{4.65}$$

$$u_m(t) \to u(t) \quad \text{strongly in } L^p(0, T; W_0^{1,p}(\Omega)), \tag{4.66}$$

$$u_m'(t) \to u'(t) \quad \text{strongly in } L^{p'}(0, T; W^{-1, p'}(\Omega)). \tag{4.67}$$

From (4.65)–(4.67) and (4.64), passing to the limit as $m \to +\infty$, we obtain the following equality:

$$\|u(t)\|_{L^2(\Omega)}^2 = \|u(0)\|_{L^2(\Omega)}^2 + 2 \int_0^t \langle u'(\tau), u(\tau) \rangle \, d\tau \tag{4.68}$$

for all $t \in [0, T]$. Note that $u(t) \in W_{pp'}(0, T)$ and hence

$$\langle u'(\tau), u(\tau) \rangle \in L^1(0, T).$$

Therefore, by the Lebesgue differentiation theorem, from (4.68) we obtain the equality

$$\frac{d}{dt}\|u(t)\|_{L^2(\Omega)}^2 = 2\langle u'(t), u(t) \rangle \quad \text{for almost all } t \in [0, T]. \tag{4.69}$$

Theorem 4.4 is proved.   □

4.3   Space of $\mathbb{B}$-Valued Distributions $\mathscr{D}'(a, b; \mathbb{B})$

In this section, we consider Banach-valued functions defined on finite intervals (a, b) and Banach-valued distributions defined on the space of test functions $\mathscr{D}(a, b)$. Obviously, all results obtained in previous sections remain valid (with replacement of $(0, T)$ by (a, b)).

Let $\mathscr{D}(a, b)$ be the space of test functions with the standard topology τ of the strict inductive limit. Generalized functions (distributions) are defined as elements of the space of linear continuous mappings

$$\mathscr{D}'(a, b; \mathbb{R}^1) = \mathscr{L}\left(\mathscr{D}(a, b); \mathbb{R}^1\right).$$

The action of a generalized function $f \in \mathscr{L}\left(\mathscr{D}(a, b); \mathbb{R}^1\right)$ on a test function $\varphi(t) \in \mathscr{D}(a, b)$ is denoted as follows:

$$\langle\!\langle f, \varphi \rangle\!\rangle \in \mathbb{R}^1.$$

A generalization of the space $\mathscr{D}'(a, b; \mathbb{R}^1)$ is the space of linear and continuous mappings $\mathscr{L}(\mathscr{D}(a, b); \mathbb{B})$, where $\mathbb{B}$ is an arbitrary Banach space. We will use the notation

$$\mathscr{D}'(a, b; \mathbb{B}) = \mathscr{L}\left(\mathscr{D}(a, b); \mathbb{B}\right).$$

The action of a $\mathbb{B}$-valued distribution $f \in \mathscr{D}'(a, b; \mathbb{B})$ on a test function $\varphi(t) \in \mathscr{D}(a, b)$ is denoted as follows:

$$\langle\!\langle f, \varphi \rangle\!\rangle \in \mathbb{B}.$$

Let $f(t) \in L^p(a, b; \mathbb{B})$, where $p \geq 1$. Then this function generates a $\mathbb{B}$-valued distribution by the formula

$$\langle\!\langle f(t), \varphi(t) \rangle\!\rangle := \int_0^T f(t)\varphi(t)\, dt, \quad \varphi(t) \in \mathscr{D}(a, b),$$

where the integral is meant in the Bochner sense.

Definition 4.8. The *derivative* of a $\mathbb{B}$-valued generalized function $f(t) \in \mathscr{D}'(a, b; \mathbb{B})$ is defined as follows:

$$\left\langle\!\!\left\langle \frac{d}{dt} f(t), \varphi(t) \right\rangle\!\!\right\rangle = -\langle\!\langle f(t), \varphi'(t) \rangle\!\rangle \tag{4.70}$$

for any $\varphi(t) \in \mathscr{D}(a, b)$.

Theorem 4.5. *Let $t_0 \in (a, b) \subset \mathbb{R}^1$ be a finite interval and $f(t) \in C^{(1)}([a, t_0]; \mathbb{B}) \cap C^{(1)}([t_0, b]; \mathbb{B})$ (note that the strong derivatives at the points $t = a$, $t = t_0$, and $t = b$ are meant in the sense of one-sided limits). Then the following equality holds in the sense of $\mathbb{B}$-valued distributions $\mathscr{D}'(a, b; \mathbb{B})$:*

$$\frac{df(t)}{dt} = \{f'(t)\} + [f]_{t=t_0}\delta(t - t_0), \quad [f]_{t=t_0} = f(t_0 + 0) - f(t_0 - 0), \tag{4.71}$$

where d/dt is the derivative in the sense of generalized functions and the function $\{f'(t)\}$ coincides with the corresponding strong derivatives for $t \neq t_0$ and at the point $t = t_0$ it is defined arbitrarily.

Proof. Note that $f(t) \in L^p(a, b; \mathbb{B})$ and hence the following equalities hold:

$$\left\langle\!\!\left\langle \frac{df(t)}{dt}, \varphi(t) \right\rangle\!\!\right\rangle = -\langle\!\langle f(t), \varphi'(t) \rangle\!\rangle = -\int_a^b f(t)\varphi'(t)\, dt$$

$$= -\int_a^{t_0} f(t)\varphi'(t)\, dt - \int_{t_0}^b f(t)\varphi'(t)\, dt$$

$$= -f(t)\varphi(t)\Big|_{t=a}^{t=t_0} - f(t)\varphi(t)\Big|_{t=t_0}^{t=b} + \int_a^b \{f'(t)\}\varphi(t)\, dt$$

$$= \left\langle\!\!\left\langle [f]_{t=t_0}\delta(t - t_0) + \{f'(t)\}, \varphi(t) \right\rangle\!\!\right\rangle. \tag{4.72}$$

Theorem 4.5 is proved.   □

Theorem 4.6. *Let $t_0 \in (a, b)$ and $f(t) \in W_{pp'}(a, t_0) \cap W_{pp'}(t_0, b)$. Then the following equality holds in the sense of $W^{-1,p'}(\Omega)$-valued distributions:*

$$\frac{df(t)}{dt} = \{f'(t)\} + [f]_{t=t_0}\delta(t - t_0), \quad [f]_{t=t_0} = f(t_0 + 0) - f(t_0 - 0), \tag{4.73}$$

where d/dt is the derivative in the sense of generalized function from the space $\mathscr{D}'(a, b; W^{-1,p'}(\Omega))$ and the function $\{f'(t)\}$ coincides with the weak derivatives of the function $f(t)$ on the intervals (a, t_0) and (t_0, b).

Proof. Note that $W_{pp'}(a, t_0) \cap W_{pp'}(t_0, b)$ is a Banach space with respect to the norm

$$\|f\| := \|f\|_{W_{pp'}(a,t_0)} + \|f\|_{W_{pp'}(t_0,b)}. \tag{4.74}$$

Moreover, the following embeddings hold:

$$W_{pp'}(a, t_0) \cap W_{pp'}(t_0, b) \subset L^p(a, b; W_0^{1,p}(\Omega)), \tag{4.75}$$

$$W_{pp'}(a, t_0) \subset C([a, t_0]; L^2(\Omega)), \quad W_{pp'}(t_0, b) \subset C([t_0, b]; L^2(\Omega)), \tag{4.76}$$

where the last two embeddings are valid due to Theorem 4.4.

By Lemma 4.3, the dense embedding

$$C^{(1)}([a, t_0]; W_0^{1,p}(\Omega)) \cap C^{(1)}([t_0, b]; W_0^{1,p}(\Omega)) \overset{ds}{\subset} W_{pp'}(a, t_0) \cap W_{pp'}(t_0, b)$$

holds; therefore, for any $f(t) \in W_{pp'}(a, t_0) \cap W_{pp'}(t_0, b)$, there exists a sequence $\{f_m(t)\} \in C^{(1)}([a, t_0]; W_0^{1,p}(\Omega)) \cap C^{(1)}([t_0, b]; W_0^{1,p}(\Omega))$ such that

$$\|f_m - f(t)\| \to +0 \quad \text{as } m \to +\infty, \tag{4.77}$$

where the norm is defined by (4.74). By Theorem 4.5, the function $f_m(t)$ satisfies the relation

$$\frac{df_m(t)}{dt} = \{f'_m(t)\} + [f_m]_{t=t_0}\delta(t - t_0), \tag{4.78}$$

which is meant in the sense of distributions $\mathscr{D}'(a, b; W^{-1,p'}(\Omega))$. We prove that the equality

$$\frac{df(t)}{dt} = \{f'(t)\} + [f]_{t=t_0}\delta(t - t_0) \tag{4.79}$$

holds as $m \to +\infty$, where $\{f'(t)\}$ coincides with the corresponding weak derivatives on (a, t_0) and (t_0, b).

Indeed, the following equalities hold:

$$\left\langle\!\!\left\langle \frac{df_m(t)}{dt} - \frac{df(t)}{dt}, \varphi(t) \right\rangle\!\!\right\rangle = -\left\langle\!\!\left\langle f_m(t) - f(t), \varphi'(t) \right\rangle\!\!\right\rangle$$

$$= -\int_a^b (f_m(t) - f(t))\varphi'(t)\,dt$$

$$= -\int_a^{t_0} (f_m(t) - f(t))\varphi'(t)\,dt$$

$$- \int_{t_0}^b (f_m(t) - f(t))\varphi'(t)\,dt. \tag{4.80}$$

We have the estimates

$$\left\| \int_a^{t_0} (f_m(t) - f(t))\varphi'(t)\, dt \right\|_{W^{-1,p'}(\Omega)}$$

$$\leq \int_a^{t_0} \|f_m(t) - f(t)\|_{W^{-1,p'}(\Omega)} |\varphi'(t)|\, dt$$

$$\leq K_1 \int_a^{t_0} \|f_m(t) - f(t)\|_{W_0^{1,p}(\Omega)} |\varphi'(t)|\, dt$$

$$\leq K_1 \|f_m - f\|_{L^p(a,b;W_0^{1,p}(\Omega))} \|\varphi'\|_{L^{p'}(a,t_0)} \to +0 \qquad (4.81)$$

as $m \to +\infty$. Similarly, we can prove that

$$\int_{t_0}^b (f_m(t) - f(t))\varphi'(t)\, dt \to \theta \quad \text{strongly in } W^{-1,p'}(\Omega) \text{ as } m \to +\infty. \qquad (4.82)$$

Thus, the right-hand side of Eq. (4.80) converges strongly in $W^{-1,p'}(\Omega)$ as $m \to +\infty$. Therefore,

$$\left\langle\!\!\left\langle \frac{df_m(t)}{dt}, \varphi(t) \right\rangle\!\!\right\rangle \to \left\langle\!\!\left\langle \frac{df(t)}{dt}, \varphi(t) \right\rangle\!\!\right\rangle \qquad (4.83)$$

strongly in $W^{-1,p'}(\Omega)$ for any function $\varphi(t) \in \mathscr{D}(a,b)$ as $m \to +\infty$.

The following chain of equalities holds:

$$\langle\!\langle \{f_m'(t)\}, \varphi(t) \rangle\!\rangle = \int_a^b \{f_m'(t)\}\varphi(t)\, dt = \int_a^{t_0} f_m'(t)\varphi(t)\, dt + \int_{t_0}^b f_m'(t)\varphi(t)\, dt. \qquad (4.84)$$

From (4.77) and the definition of the norm (4.74), we obtain the following inequalities:

$$\left\| \int_a^{t_0} [f_m'(t) - f'(t)]\, \varphi(t)\, dt \right\|_{W^{-1,p'}(\Omega)}$$

$$\leq \|f_m' - f'\|_{L^{p'}(a,t_0;W^{-1,p'}(\Omega))} \|\varphi(t)\|_{L^p(a,t_0)} \to +0 \qquad (4.85)$$

as $m \to +\infty$. Similarly, we can prove that

$$\left\| \int_{t_0}^b [f_m'(t) - f'(t)]\, \varphi(t)\, dt \right\|_{W^{-1,p'}(\Omega)} \to +0 \quad \text{as } m \to +\infty. \qquad (4.86)$$

Therefore,

$$\langle\!\langle \{f_m'(t)\}, \varphi(t) \rangle\!\rangle \to \langle\!\langle \{f'(t)\}, \varphi(t) \rangle\!\rangle \quad \text{strongly in } W^{-1,p'}(\Omega) \text{ as } m \to +\infty. \qquad (4.87)$$

Finally, due to the continuous embeddings (4.76), we have the following limit properties:

$$f_m(t_0 + 0) \to f(t_0 + 0), \quad f_m(t_0 - 0) \to f(t_0 - 0) \tag{4.88}$$

as $m \to +\infty$ in the strong sense of the Banach space $W^{-1,p'}(\Omega)$.

Thus, the relations (4.83), (4.87), and (4.88) imply that we may pass to the limit as $m \to +\infty$ in Eq. (4.78) and obtain Eq. (4.79), which is meant in the sense of distributions from $\mathscr{D}'(a, b; W^{-1,p'}(\Omega))$. $\qquad\square$

The following assertion is an immediate consequence of Theorem 4.6.

Corollary 4.1. *Let* $\{t_j\}_{j=1}^n \subset (a, b)$ *be a finite set of points that partition the interval* (a, b) *into the intervals*

$$(a, t_1), \ (t_1, t_2), \quad \ldots, \quad (t_{n-1}, t_n), \ (t_n, b)$$

(we set $t_0 = a$ *and* $t_{n+1} = b$*) and let* $f(t) \in W_{pp'}(t_{j-1}, t_j)$ *on each of these intervals,* $j = \overline{1, n+1}$. *Then the following formula is valid in the sense of* $\mathscr{D}'(a, b; W^{-1,p'}(\Omega))$:

$$\frac{df(t)}{dt} = \{f'(t)\} + \sum_{j=1}^n [f]_{t=t_j} \delta(t - t_j),$$

where

$$[f]_{t=t_j} = f(t_j + 0) - f(t_j - 0), \quad j = \overline{1, n},$$

and the function $\{f'(t)\}$ *coincides with the weak derivative* $f'(t)$ *on the union* $(a, t_1) \cup (t_1, t_2) \cup \cdots \cup (t_{n-1}, t_n) \cup (t_n, b)$ *of these intervals.*

4.4 Lions–Aubin Theorem

Let X, Y, and Z be three Banach space and, moreover, let X and Z be reflexive. Assume that the continuous embeddings $X \subset Y \subset Z$ hold. Consider the space

$$W(0, T) := \{u(t) \in L^p(0, T; X), \ u'(t) \in L^r(0, T; Z)\}, \quad p, r > 1, \tag{4.89}$$

where the derivatives $u'(t)$ is meant in the weak sense.

Lemma 4.4. *The vector space* $W(0, T)$ *is a Banach space with respect to the norm*

$$\|u\|_W := \|u\|_{L^p(0,T;X)} + \|u'\|_{L^r(0,T;Z)}.$$

Remark 4.5. Note that the following continuous embeddings hold:

$$W(0, T) \subset L^p(0, T; X) \subset L^p(0, T; Y). \tag{4.90}$$

Theorem 4.7 (Lions–Aubin theorem). *Let* X, Y, *and* Z *be Banach spaces, where* X *and* Z *are reflexive, the embedding* $X \subset Y \subset Z$ *be continuous, and the embedding* $X \subset Y$ *be compact. Then the embedding* $W(0, T) \subset L^p(0, T; Y)$ *is compact.*

Proof. We present a proof of the Lions–Aubin theorem following [23, pp. 144–147]. Let a sequence $\{u_n\} \subset W(0, T)$ be bounded. We prove that there exists a subsequence $\{u_{n_k}\} \subset \{u_n\}$, which strongly converges in $L^p(0, T; Y)$.

Note that the reflexivity of X and Z implies the reflexivity of the Banach space $W(0, T)$ under the condition $p, r > 1$. Indeed, let

$$V := L^p(0, T; X) \otimes L^r(0, T; Z).$$

Then the vector space V is a Banach space with respect to the norm

$$\|w\|_V := \|u\|_{L^p(0,T;X)} + \|v\|_{L^r(0,T;Z)}, \quad w = (u, v) \in V.$$

We introduce the following mapping from $W(0, T)$ to V as follows:

$$P : W(0, T) \to V, \quad Pu(t) = w(t) = (u(t), u'(t)) \in V. \tag{4.91}$$

Clearly, the mapping P is linear and, moreover, since

$$\|Pu\|_V = \|u\|_W = \|u\|_{L^p(0,T;X)} + \|u'\|_{L^r(0,T;Z)}, \tag{4.92}$$

it is an isometry from $W(0, T)$ to a closed linear subspace in V. Note that

$$V^* = L^{p'}(0, T; X^*) \otimes L^{r'}(0, T; Z^*), \quad V^{**} = L^p(0, T; X^{**}) \otimes L^r(0, T; Z^{**})$$

since $p, r > 1$. Due to the reflexivity of X and Z, there exists a linear isometric operator J such that

$$J(V) = V^{**}. \tag{4.93}$$

We need the following auxiliary assertion.

Lemma 4.5 (see [6, Corollary 6.10.11]). *Let $\mathbb{B}$ be a reflexive Banach space and $\mathbb{D} \subset \mathbb{B}$ be a closed linear subspace. Then $\mathbb{D}$ is also a reflexive Banach space.*

Since $PW(0, T)$ is a closed linear subspace of the Banach space V, due to Lemma 4.5, the Banach space $PW(0, T)$ is reflexive. It remains to prove that the Banach space $W(0, T)$ is also reflexive. Indeed, for each element $f^* \in (W(0, T))^*$, there exists a unique element $L^* \in (PW(0, T))^*$ such that

$$\langle f^*, u \rangle_W = \langle L^*, Pu \rangle_V \quad \Leftrightarrow \quad f^* = L^* P. \tag{4.94}$$

The following equality holds:

$$\langle f^{**}, f^* \rangle_{W^*} = \langle L^{**}, L^* P \rangle_{V^*} = \langle JPu, L^* P \rangle_{V^*} \quad \Leftrightarrow \quad f^{**} = JPu.$$

Therefore, for each $f^{**} \in (W(0, T))^{**}$ there exists an element $u \in W(0, T)$ such that $f^{**} = \hat{J}u$, where $\hat{J} = JP$ is a linear isometry (since it is the composition of two linear isometries). Thus, the Banach space $W(0, T)$ is reflexive.

We conclude that there exists a subsequence (we denote it by the same symbol $\{u_n\}$) such that

$$u_n \rightharpoonup u \quad \text{weakly in } W(0, T) \text{ as } n \to +\infty. \tag{4.95}$$

In particular, we conclude that

$$u_n \rightharpoonup u \quad \text{weakly in } L^p(0,T;X) \text{ as } n \to +\infty, \tag{4.96}$$

$$u_n' \rightharpoonup u' \quad \text{weakly in } L^r(0,T;Z) \text{ as } n \to +\infty. \tag{4.97}$$

Indeed, let $\{u_n(t)\} \subset W(0,T)$ weakly converge in $W(0,T)$ to a function $u(t) \in W(0,T)$:

$$\langle f^*, u_n - u \rangle_W \to +0 \quad \text{for any } f^* \in (W(0,T))^* \text{ as } n \to +\infty. \tag{4.98}$$

For any $f^* \in (W(0,T))^*$, there exists unique $L^* \in V^*$ such that Eq. (4.94) is valid. This equality implies

$$\langle f^*, u_n - u \rangle_W = \langle L^*, P(u_n - u) \rangle_V. \tag{4.99}$$

Note that

$$L^* = (L_1^*, L_2^*) \in L^{p'}(0,T;X^*) \otimes L^{r'}(0,T;Z^*). \tag{4.100}$$

Moreover, the following equality holds:

$$\langle L^*, P(u_n - u) \rangle_V = \langle L_1^*, u_n - u \rangle_{L^p(0,T;X)} + \langle L_2^*, u_n' - u' \rangle_{L^r(0,T;Z)}. \tag{4.101}$$

Thus, taking into account (4.99)–(4.101), we obtain from (4.98) the following equalities as $n \to +\infty$:

$$\langle L_1^*, u_n - u \rangle_{L^p(0,T;X)} \to +0 \quad \text{for any } L_1^* \in L^{p'}(0,T;X^*), \tag{4.102}$$

$$\langle L_2^*, u_n' - u' \rangle_{L^r(0,T;Z)} \to +0 \quad \text{for any } L_2^* \in L^{r'}(0,T;Z^*). \tag{4.103}$$

Similarly to Theorem 4.4, we can prove the following general assertion.

Lemma 4.6. *The continuous embedding $W(0,T) \subset C([0,T];Z)$ holds.*

Let $v_n := u_n - u$ for $n \geq 1$. Since the sequence $\{u_n\}$ is bounded in $W(0,T) \subset C(0,T;Z)$, there exists a constant $c_2 > 0$ such that

$$\|v_n(t)\|_Z \leq \|v_n\|_{C(0,T;Z)} \leq c_1 \|v_n\|_W \leq c_2, \quad n \geq 1, \tag{4.104}$$

for all $t \in [0,T]$. We prove that

$$v_n(t) \to \theta \quad \text{strongly in } Z \text{ as } n \to +\infty \tag{4.105}$$

for all $t \in [0,T]$. We prove this assertion for $t = 0$; for arbitrary $t \in [0,T]$, the proof is the same.[1]

First, we prove the following equality in the sense of the Bochner Z-integral:

$$v_n(0) = v_n(t) - \int_0^t v_n'(\tau) \, d\tau. \tag{4.106}$$

Indeed, since X is continuously embedded into Z, the equality

$$\int_0^T v'(t)\varphi(t) \, dt = -\int_0^T v(t)\varphi'(t) \, dt \tag{4.107}$$

---

[1] Actually, one can prove the embedding $W(0,T) \subset AC(0,T;Z)$.

holds for any function $v(t) \in W(0,T)$ and any function $\varphi(t) \in \mathcal{D}(0,T)$. Then, due to [23, assertion 2.2.20], we have $v(t) \in AC(0,T;Z)$. It remains to apply Eq. (4.8) with $\varphi(t) = 1$.

Integrating both sides of this equality by $t \in [0,s]$, we obtain the following equality in the sense of the Banach space Z:

$$v_n(0) = \psi_n + \varphi_n, \quad \psi_n := \frac{1}{s} \int_0^s v_n(t)\, dt, \tag{4.108}$$

$$\varphi_n := -\frac{1}{s} \int_0^s \int_0^t v_n'(\tau)\, d\tau\, dt = -\frac{1}{s} \int_0^s (s-t)v_n'(t)\, dt. \tag{4.109}$$

Let $\varepsilon > 0$ be an arbitrary fixed number. For φ_n we have the estimate

$$\|\varphi_n\|_Z \leq \int_0^s \|v_n'(t)\|_Z\, dt. \tag{4.110}$$

Since $\{v_n'(t)\}$ is bounded in $L^r(0,T;Z) \subset L^1(0,T;Z)$, for sufficiently small $s \in (0,T]$, we obtain from (4.110) the inequality

$$\|\varphi_n\|_Z \leq \frac{\varepsilon}{2} \quad \text{for all } n \geq 1. \tag{4.111}$$

Note that, due to (4.96) and the definition of the function $v_n(t)$, the following limit property holds:

$$v_n \rightharpoonup \theta \quad \text{weakly in } L^p(0,T;X) \text{ as } n \to +\infty. \tag{4.112}$$

Therefore, for $s \in (0,T]$ fixed above, we obtain the limit property

$$\psi_n = \frac{1}{s} \int_0^s v_n(t)\, dt \rightharpoonup \theta \quad \text{weakly in } X \text{ as } n \to +\infty. \tag{4.113}$$

Indeed, the limit property (4.112) means that

$$\int_0^T \langle f^*(t), v_n(t) \rangle\, dt \to 0 \quad \text{as } n \to +\infty \tag{4.114}$$

for any function $f^*(t) \in L^{p'}(0,T;X^*)$. We choose this function as follows:

$$f^*(t) = \frac{1}{s}\theta(s-t)h^*, \quad h^* \in X^*, \quad s > 0. \tag{4.115}$$

Substituting the function (4.115) into Eq. (4.114) and taking into account Lemma 3.6, we obtain the following equality:

$$\int_0^T \langle f^*(t), v_n(t) \rangle\, dt = \frac{1}{s} \int_0^s \langle h^*, v_n(t) \rangle\, dt = \left\langle h^*, \frac{1}{s} \int_0^s v_n(t)\, dt \right\rangle \to +0 \tag{4.116}$$

as $n \to +\infty$ for any $h^* \in X^*$. Therefore,

$$\frac{1}{s} \int_0^s v_n(t)\, dt \rightharpoonup \theta \quad \text{weakly in } X \text{ as } n \to +\infty. \tag{4.117}$$

Since, by the condition of the theorem, X is embedded into Y completely continuously and Y is embedded in Z continuously, we conclude that X is embedded in Z completely continuously. Therefore, from (4.113) we see that

$$\psi_n \to \theta \quad \text{strongly in } Z \text{ as } n \to +\infty. \tag{4.118}$$

Therefore, for fixed $\varepsilon > 0$ for sufficiently large $n_0 \in \mathbb{N}$ we have

$$\|\psi_n\|_Z \leq \frac{\varepsilon}{2} \quad \text{for all } n \geq n_0. \tag{4.119}$$

Thus, from (4.108), (4.111), and (4.119) we obtain that

$$v_n(0) \to \theta \quad \text{strongly in } Z \text{ as } n \to +\infty. \tag{4.120}$$

Next, from (4.104), (4.105), and the Lebesgue theorem on passing to the limit in the Lebesgue integral we conclude that

$$\int_0^T \|v_n(t)\|_Z^p\, dt \to 0 \quad \text{as } n \to +\infty. \tag{4.121}$$

Thus,

$$v_n(t) \to \theta \quad \text{strongly in } L^p(0, T; Z) \text{ as } n \to +\infty. \tag{4.122}$$

Now we prove the following auxiliary assertion.

Lemma 4.7. *Let X, Y, and Z be Banach spaces such that the embeddings $X \subset Y \subset Z$ are continuous and the embedding $X \subset Y$ is compact. Then for each $\gamma > 0$, there exists $c(\gamma) > 0$ such that*

$$\|u\|_Y \leq \gamma \|u\|_X + c(\gamma) \|u\|_Z \quad \text{for all } u \in X. \tag{4.123}$$

Proof. Assume the contrary. Then there exists $\gamma > 0$ such that for any $n \in \mathbb{N}$, there exists $u_n \subset X$ such that

$$\|u_n\|_Y > \gamma \|u_n\|_X + n \|u_n\|_Z \quad \text{for all } n \geq 1. \tag{4.124}$$

Thus, there exists a sequence $\{u_n\} \subset X$ satisfying (4.124). We set

$$w_n := \frac{u_n}{\|u_n\|_X} \quad \text{for all } n \geq 1. \tag{4.125}$$

Then from (4.124) we obtain the following inequality:

$$\|w_n\|_Y > \gamma + n \|w_n\|_Z \quad \text{for all } n \geq 1. \tag{4.126}$$

Note that, on the one hand, $\|w_n\|_X = 1$ and the embedding $X \subset Y$ is continuous. Therefore, there exists a constant $M > 0$ such that

$$\|w_n\|_Y \leq M < +\infty \quad \text{for all } n \geq 1. \tag{4.127}$$

Then from (4.126) we obtain that

$$\|w_n\|_Z \to 0 \quad \text{as } n \to +\infty. \tag{4.128}$$

On the other hand, the compact embedding $X \subset Y$ holds and, due to the fact that $\|w_n\|_X = 1$, there exists a subsequence $\{w_{n_k}\} \subset \{w_n\}$ such that

$$w_{n_k} \to w \quad \text{strongly in } Y \text{ as } n_k \to +\infty. \tag{4.129}$$

Owing to the continuous embedding $Y \subset Z$, we conclude that

$$w_{n_k} \to w \quad \text{strongly in } Z \text{ as } n_k \to +\infty. \tag{4.130}$$

From (4.128) and (4.130) we obtain that $w = \theta$. The inequality (4.126) implies that

$$\|w_{n_k}\|_Y > \gamma > 0 \quad \text{for all } n \geq 1. \tag{4.131}$$

Passing to the limit as $n_k \to +\infty$, we obtain a contradiction:

$$0 = \|w\|_Y > \gamma > 0.$$

Lemma 4.7 is proved.   □

Now we return to the proof of the theorem. Let $w(t) \in L^p(0,T;X)$; then, due to Lemma 4.7, for any $\gamma > 0$ there exists $c(\gamma) > 0$ such that the following inequality holds:

$$\|w(t)\|_Y \leq \gamma \|w(t)\|_X + c(\gamma)\|w(t)\|_Z \quad \text{for almost all } t \in [0,T]. \tag{4.132}$$

By the triangle inequality for the norm, we obtain from (4.132) the following chain of inequalities:

$$\|w\|_{L^p(0,T;Y)}$$
$$= \left(\int_0^T \|w(t)\|_Y^p \, dt \right)^{1/p} \leq \left(\int_0^T [\gamma\|w(t)\|_X + c(\gamma)\|w(t)\|_Z]^p \, dt \right)^{1/p}$$
$$\leq \gamma\|w\|_{L^p(0,T;X)} + c(\gamma)\|w\|_{L^p(0,T;Z)}. \tag{4.133}$$

We apply this inequality to functions from the sequence $\{v_n(t)\} \subset L^p(0,T;X)$:

$$\|v_n\|_{L^p(0,T;Y)} \leq \gamma\|v_n\|_{L^p(0,T;X)} + c(\gamma)\|v_n\|_{L^p(0,T;Z)}$$
$$\leq \gamma c_3 + c(\gamma)\|v_n\|_{L^p(0,T;Z)}. \tag{4.134}$$

Due to the limit property (4.122), we obtain the following limit property from (4.134):

$$\limsup_{n \to +\infty} \|v_n\|_{L^p(0,T;Y)} \leq \gamma c_3. \tag{4.135}$$

Due to the arbitrariness of $\gamma > 0$, we conclude that

$$v_n \to \theta \quad \text{strongly in } L^p(0,T;Y) \text{ as } n \to +\infty. \tag{4.136}$$

Theorem 4.7 is proved.   □

4.5 Bibliographical Notes

The contents of this lecture is taken from [6, 19, 22–24, 46, 70].

PART 2

Nonlinear Operators: Continuity, Differentiability, and Compactness

Lecture 5

Nonlinear Operators

In this lecture, we introduce the important notions of differentiability in the Gâteaux and Fréchet sense. These two notions are fundamental in the study of variational and nonlinear boundary-valued problems. We prove an important theorem on the relationship between these two types of differentiability.

5.1 Gâteaux and Fréchet Derivatives of Nonlinear Operators

Let $\mathbb{B}_1$ and $\mathbb{B}_2$ be two Banach space with respect to the norms $\|\cdot\|_1$ and $\|\cdot\|_2$, respectively, and let $\langle \cdot, \cdot \rangle_1$ and $\langle \cdot, \cdot \rangle_2$ be the corresponding duality brackets.

For an operator (generally, nonlinear)

$$F : \mathbb{B}_1 \to \mathbb{B}_2,$$

we introduce the notion of the Gâteaux differentiability of the operator F.

Definition 5.1. An operator F is said to be differentiable in the Gâteaux sense (Gâteaux-differentiable) at a point $u \in \mathbb{B}_1$ if for any $h \in \mathbb{B}_1$, the following limit equality holds:

$$\lim_{\lambda \to 0} \left\| \frac{F(u + \lambda h) - F(u)}{\lambda} - F'_g(u)h \right\|_2 = 0, \tag{5.1}$$

where $F'_g(u) : \mathbb{B}_1 \to \mathbb{B}_2$ for fixed $u \in \mathbb{B}_1$. In this case, the operator $F'_g(u)$ (generally speaking, nonlinear by $u \in \mathbb{B}_1$) is called the Gâteaux derivative of the operator F.

Remark 5.1. For all $u, h \in \mathbb{B}_1$ and $\lambda \in \mathbb{R}^1$, we introduce the $\mathbb{B}_2$-valued function

$$\varphi(\lambda) := F(u + \lambda h).$$

It is easy to see that, due to Definition 5.1, the following equality holds:

$$F'_g(u)h = \frac{d\varphi(\lambda)}{d\lambda}\bigg|_{\lambda=0}.$$

Indeed, the following equality holds:

$$\frac{\varphi(\lambda) - \varphi(0)}{\lambda} = \frac{F(u + \lambda h) - F(u)}{\lambda}. \tag{5.2}$$

On the one hand, under the condition that the limit exists, the following limit equality holds:

$$\lim_{\lambda \to 0} \left\| \frac{\varphi(\lambda) - \varphi(0)}{\lambda} - \frac{d\varphi(\lambda)}{d\lambda} \Big|_{\lambda=0} \right\|_2 = 0. \tag{5.3}$$

On the other hand, under the condition that the limit exists, we have

$$\lim_{\lambda \to 0} \left\| \frac{F(u + \lambda h) - F(u)}{\lambda} - F_g'(u)h \right\|_2 = 0. \tag{5.4}$$

Therefore, due to (5.2), we obtain from (5.3) and (5.4) that under the condition that the Gâteaux derivative $F_g'(u)$ exist at the point $u \in \mathbb{B}_1$, the following equality holds:

$$\frac{d\varphi(\lambda)}{d\lambda} \Big|_{\lambda=0} = F_g'(u)h.$$

Now we consider several examples of Gâteaux derivatives.

Example 5.1. Consider a bounded linear operator

$$F : \mathbb{B}_1 \to \mathbb{B}_2.$$

Obviously, due to the linearity of this mapping, the following equality holds:

$$\frac{F(u + \lambda h) - F(u)}{\lambda} = Fh \quad \Rightarrow \quad F_g'(u) = F.$$

Thus, we conclude that a linear operator from $\mathbb{B}_1$ into $\mathbb{B}_2$ is an infinitely differentiable in the Gâteaux sense and the Gâteaux derivative coincides with the operator itself.

Example 5.2. Consider the mapping

$$F = (F_1, ..., F_n) : \mathbb{R}_m \to \mathbb{R}_n,$$

where $\mathbb{R}_m$ and $\mathbb{R}_n$ are Euclidean spaces of row matrices. Obviously, these spaces are Banach spaces with respect to the norms

$$\|u\|_1 := \left(|u_1|^2 + \cdots + |u_m|^2 \right)^{1/2}, \quad \|v\|_2 := \left(|v_1|^2 + \cdots + |v_n|^2 \right)^{1/2},$$

where $u = (u_1, \ldots, u_m) \in \mathbb{R}_m$ and $v = (v_1, \ldots, v_n) \in \mathbb{R}_n$. Calculate the Gâteaux derivative of the mapping F:

$$\left\| \frac{F(u + \lambda h) - F(u)}{\lambda} - F_g'(u)h \right\|_2 = \left| \sum_{j=1}^{n} \left(\frac{F_j(u + \lambda h) - F_j(u)}{\lambda} - \left(F_g'(u)h \right)_j \right)^2 \right|^{1/2}.$$

Note that the following equality holds:

$$\lim_{\lambda \to +0} \left| \frac{F_j(u + \lambda h) - F_j(u)}{\lambda} - \sum_{k=1}^{m} \frac{\partial F_j(u)}{\partial u_k} h_k \right| = 0.$$

Comparing the last two formulas, we conclude that

$$F_g'(u) = \begin{pmatrix} \dfrac{\partial F_1(u)}{\partial u_1} & \cdots & \dfrac{\partial F_n(u)}{\partial u_1} \\ \vdots & \ddots & \vdots \\ \dfrac{\partial F_1(u)}{\partial u_m} & \cdots & \dfrac{\partial F_n(u)}{\partial u_m} \end{pmatrix}.$$

Remark 5.2. In connection with Example 5.2 we can consider the Gâteaux derivative as a generalization of a the partial derivative of a function of several variables to the infinite-dimensional case.

Example 5.3. Consider the Hammerstein operator

$$G(u) := \int_0^1 k(x,y)g(u(y),y)\,dy, \quad y \in [0,1].$$

As the Banach spaces $\mathbb{B}_1$ and $\mathbb{B}_2$, we take $C[0,1]$ with the standard norm and assume that

$$k(x,y) \in C([0,1] \times [0,1]), \quad \frac{\partial g}{\partial u}(u,x) \in C(\mathbb{R}^1 \times [0,1]).$$

On the one hand, due to these assumptions, the limit equality

$$\lim_{\lambda \to 0} \frac{g(u+\lambda h, x) - g(u,x)}{\lambda} = \frac{\partial g}{\partial u}(u,x)h(x)$$

holds. On the other hand, we have

$$\frac{G(u+\lambda h) - G(u)}{\lambda} = \int_0^1 k(x,y)\frac{g(u(y)+\lambda h(y),y) - g(u(y),y)}{\lambda}\,dy.$$

Therefore, the Gâteaux derivative of the Hammerstein operator has the form

$$G'_g(u)h = \int_0^1 k(x,y)\frac{\partial g}{\partial u}(u(y),y)h(y)\,dy \quad \text{for all } h(x) \in C[0,1].$$

Now we introduce the notion of the Fréchet derivative.

Definition 5.2. An operator F is said to be differentiable in the Fréchet sense (Fréchet differentiable) at a point $u \in \mathbb{B}_1$ if in a certain neighborhood of this point, the following equality holds for any $h \in \mathbb{B}_1$:

$$F(u+h) = F(u) + F'_f(u)h + \omega(u,h), \tag{5.5}$$

where $\omega(u,h)$ satisfies the condition

$$\lim_{\|h\|_1 \to 0} \frac{\|\omega(u,h)\|_2}{\|h\|_1} = 0. \tag{5.6}$$

The operator

$$F'_f(u) : \mathbb{B}_1 \to \mathbb{B}_2$$

is called the Fréchet derivative of the operator F; it is bounded and linear for fixed $u \in \mathbb{B}_1$.

Example 5.4. Each bounded linear operator is infinitely differentiable in the Fréchet sense and the Fréchet derivative coincides with the operator itself (cf. Example 5.1). Prove this fact yourself.

Remark 5.3. We note that the existence of the Fréchet derivative of an operator $F : \mathbb{B}_1 \to \mathbb{B}_2$ implies the existence of the Gâteaux derivative and the equality $F'_g(u) = F'_f(u)$ for all $u \in \mathbb{B}_1$, where the Fréchet derivative exists.

Indeed, due to (5.5), we have

$$F(u + \lambda h) = F(u) + F'_f(u)\lambda h + \omega(u, \lambda h).$$

Due to the linearity of the operator $F'_f(u)$ for fixed $u \in \mathbb{B}_1$, this implies the equality

$$\frac{F(u + \lambda h) - F(u)}{\lambda} - F'_f(u)h = \frac{\omega(u, \lambda h)}{\lambda}. \tag{5.7}$$

We need to consider only the case where $\|h\|_1 \neq 0$. In this case, Eq. (5.7) implies the following relation:

$$\left\| \frac{F(u + \lambda h) - F(u)}{\lambda} - F'_f(u)h \right\|_2 = \|h\|_1 \frac{\|\omega(u, \lambda h)\|_2}{|\lambda| \|h\|_1}. \tag{5.8}$$

It remains to apply the property (5.6).

Remark 5.4. Let a Banach space $\mathbb{B}_3$ be *continuously* embedded into a Banach space $\mathbb{B}_1$, i.e., there exists an injective operator $J \in \mathscr{L}(\mathbb{B}_3; \mathbb{B}_1)$. Then the Fréchet derivative of the operator

$$F(J\cdot) : \mathbb{B}_3 \to \mathbb{B}_2$$

is the operator

$$F'_f(J\cdot)J : \mathbb{B}_3 \to \mathscr{L}(\mathbb{B}_3; \mathbb{B}_2),$$

where

$$F'_f(\cdot) : \mathbb{B}_1 \to \mathscr{L}(\mathbb{B}_1; \mathbb{B}_2)$$

is the Fréchet derivative of the operator

$$F(\cdot) : \mathbb{B}_1 \to \mathbb{B}_2.$$

Prove this assertion yourself.

Now we consider an important particular case. Let $\psi(u)$ be a functional

$$\psi(u) : \mathbb{H} \to \mathbb{C}^1,$$

where $\mathbb{H}$ is a Hilbert space with the inner product $(\cdot, \cdot)$. In this case, $\mathbb{B}_1 = \mathbb{H}$, $\mathbb{B}_2 = \mathbb{C}^1$, and $\mathbb{C}^1$ is a Banach space with respect to the absolute value $|\cdot|$. We introduce the notion of the *gradient of the functional* $\psi(u)$; note that sometimes the notion of the gradient on a Hilbert space is confused with the notion of the Fréchet derivative of this functional.

Definition 5.3. The *gradient* of a functional $\psi(u)$ is defined as follows:

$$\operatorname{grad} \psi(u) \stackrel{\text{def}}{=} J(\psi'_f(u)) : \mathbb{H} \to \mathbb{H},$$

where $J : \mathbb{H}^* \to \mathbb{H}$ is the Riesz operator.

Example 5.5. Consider the mapping defined by the formula

$$F : \mathbb{R}^2 \to \mathbb{R}^1, \quad F(x) = \begin{cases} \dfrac{x_1^3 x_2}{x_1^4 + x_2^2}, & x = (x_1, x_2) \neq \theta = (0,0); \\ 0, & x = (x_1, x_2) = \theta = (0,0). \end{cases}$$

Prove that this mapping is Gâteaux-differentiable at the point $(0,0)$.

Indeed, we have

$$\frac{F(\theta + \lambda h) - F(\theta)}{\lambda} = \frac{1}{\lambda} \frac{\lambda^4 h_1^3 h_2}{\lambda^4 h_1^4 + \lambda^2 h_2^2} = \lambda \frac{h_1^3 h_2}{\lambda^2 h_1^4 + h_2^2} \to 0 \quad \text{as } \lambda \to 0.$$

Thus, the Gâteaux derivative of this mapping at the point $(0,0)$ is the zero mapping: $F_g'(0) = \Theta$.

Assume that the Fréchet derivative of this mapping exists at the point $(0,0)$. Since the Fréchet derivative (if it exists) coincides with the Gâteaux derivative, we must accept that it is equal to the zero mapping Θ. Note that due to the definition of the Fréchet derivative (Definition 5.2) and the explicit form of the mapping F, we have

$$F(\theta + h) - F(\theta) = F_f'(\theta)h + \omega(\theta, h),$$

which implies the following equality:

$$F(h) = \omega(\theta, h), \quad \lim_{\|h\| \to 0} \frac{\|\omega(\theta, h)\|}{\|h\|} = 0 \quad \text{as } \|h\| \to +0.$$

Therefore, we obtain

$$\frac{\|F(h)\|}{\|h\|} \to 0.$$

However, if we consider the case where the endpoint of the vector $h \in \mathbb{R}^2$ tends to the point $(0,0)$ along the parabola $h_2 = h_1^2$, we obtain

$$\frac{\|F(h)\|}{\|h\|} = \frac{|h_1|^3 |h_2|}{h_1^4 + h_2^2} \frac{1}{\sqrt{h_1^2 + h_2^2}} = \frac{|h_1| h_1^4}{h_1^4 + h_1^4} \frac{1}{\sqrt{h_1^2 + h_1^4}} = \frac{1}{2} \frac{1}{\sqrt{1 + h_1^2}} \to \frac{1}{2} \neq 0$$

as $\|h\| \to 0$. The limit equality obtained means that the Fréchet derivative at the point $(0,0)$ does not exist.

Thus, *in general, the existence of the Gâteaux derivative at a point does not imply the existence of the Fréchet derivative at this point.*

The natural question arises: Under which additional conditions, the existence of the Gâteaux derivative at a certain point implies the existence of the derivative Fréchet?

Theorem 5.1. *Let an operator $F : \mathbb{B}_1 \to \mathbb{B}_2$ be differentiable in the Gâteaux sense in a certain neighborhood of a point $u \in \mathbb{B}_1$ and let the Gâteaux derivative $F_g'(\cdot)$ be continuous at the point $u \in \mathbb{B}_1$. Then the operator F is differentiable in the Fréchet sense at this point $u \in \mathbb{B}_1$ and*

$$F_g'(u) = F_f'(u).$$

Proof. Introduce the notation

$$\omega(u, h) := F(u + h) - F(u) - F'_g(u)h.$$

Let $f^* \in \mathbb{B}_2^*$; then we have

$$\langle f^*, \omega(u, h) \rangle_2 = \langle f^*, F(u + h) - F(u) \rangle_2 - \langle f^*, F'_g(u)h \rangle_2.$$

Introduce the functional

$$\psi(u) := \langle f^*, F(u) \rangle, \quad \psi(u) : \mathbb{B}_1 \to \mathbb{R}^1.$$

The difference relation

$$\frac{\psi(v + \lambda h) - \psi(v)}{\lambda} = \left\langle f^*, \frac{F(v + \lambda h) - F(v)}{\lambda} \right\rangle_2$$

implies that $\psi'_g(v)$ exists and

$$\langle \psi'_g(v), h \rangle_1 = \langle f^*, F'_g(v)h \rangle_2.$$

Consider the auxiliary real-valued function

$$\varphi(\lambda) := \psi(u + \lambda h), \quad \lambda \in \mathbb{R}^1 \text{ is sufficiently small.}$$

By the Lagrange formula, for any fixed $u, h \in \mathbb{B}_1$ and $f^* \in \mathbb{B}_2^*$, there exists a number $\lambda = \lambda(u, h, f^*) \in (0, 1)$ such that

$$\varphi(1) - \varphi(0) = \varphi'(\lambda) \quad \Rightarrow \quad \psi(u + h) - \psi(u) = \langle \psi'_g(u + \lambda h), h \rangle_2.$$

Otherwise,

$$\langle f^*, F(u + h) - F(u) \rangle = \langle f^*, F'_g(u + \lambda h)h \rangle_2.$$

Therefore,

$$\langle f^*, \omega(u, h) \rangle_2 = \langle f^*, F'_g(u + \lambda h)h - F'_g(u)h \rangle_2.$$

By the Hahn–Banach theorem, for fixed $u, h \in \mathbb{B}_1$. there exists $f^* \in \mathbb{B}_2^*$ with $\|f^*\|_{*2} = 1$ such that

$$\|\omega(u, h)\|_2 = \langle f^*, \omega(u, h) \rangle_2$$

(recall that $f^* \in \mathbb{B}_2^*$ is arbitrary). Therefore, we have the inequality

$$\|\omega(u, h)\|_2 \leq \left\| F'_g(u + \lambda h) - F'_g(u) \right\|_{1 \to 2} \|h\|_1;$$

here we used the inequality

$$|\langle f^*, u \rangle| \leq \|f^*\|_* \|u\| \quad \text{for all } u \in \mathbb{B} \text{ and } f^* \in \mathbb{B}^*.$$

Therefore, due to the continuity of $F'_g(\cdot)$ at the point $u \in \mathbb{B}_1$, we have the inequality

$$\lim_{\|h\|_1 \to 0} \frac{\|\omega(u, h)\|_2}{\|h\|_1} \leq \lim_{\|h\|_1 \to 0} \left\| F'_g(u + \lambda h) - F'_g(u) \right\|_{1 \to 2} = 0.$$

Theorem 5.1 is proved.   □

Now we establish the connection between the notions of the Fréchet differentiability and the continuity of a mapping.

Theorem 5.2. *Let* $F : \mathbb{B}_1 \to \mathbb{B}_2$ *be a mapping differentiable in the Fréchet sense at a point* $u \in \mathbb{B}_1$. *Then the mapping* F *is continuous at this point.*

Proof. The Fréchet differentiability at the point $u \in \mathbb{B}_1$ implies the inequality

$$\left\| F(u+h) - F(u) - F'_f(u)h \right\|_2 \le \|h\|_1$$

for all $h \in \mathbb{B}_1$ with sufficiently small norm. Then the following inequalities hold:

$$\|F(u+h) - F(u)\|_2 \le \left\| F(u+h) - F(u) - F'_f(u)h \right\|_2 + \left\| F'_f(u)h \right\|_2$$
$$\le \left(1 + \left\| F'_f(u) \right\|_{1\to 2} \right) \|h\|_1.$$

Theorem 5.2 is proved. $\qquad\square$

Example 5.6. We give an example of a mapping, which is Gâteaux-differentiable at a certain point but is not continuous at this point. Let

$$F : \mathbb{R}^2 \to \mathbb{R}^1, \quad F(x) = \begin{cases} \dfrac{x_1^4 x_2}{x_1^6 + x_2^3}, & (x_1, x_2) \neq (0,0); \\ 0, & (x_1, x_2) = (0,0). \end{cases}$$

The expression $(F(x + \lambda h) - F(x))/\lambda$ at the point $x = (0,0)$ has the form

$$\frac{\lambda^5 h_1^4 h_2}{\lambda \left(\lambda^6 h_1^6 + \lambda^3 h_2^3 \right)} = \lambda \frac{h_1^4 h_2}{\lambda^3 h_1^6 + h_2^3} \to 0 \quad \text{as } \lambda \to 0$$

and it is identically vanish for $h_2 = 0$. Therefore, the Gâteaux derivative of this mapping exists at the point $x = (0,0)$ and is equal to the zero mapping $F'_g(\theta) = \Theta$. We prove that the mapping F is not continuous at the origin.

Indeed, consider the parabola $x_2 = \lambda x_1^2$ in $\mathbb{R}^2$, where $\lambda > 0$. If the point (x_1, x_2) tends to $(0,0)$ along this parabola, then

$$F(x)\Big|_{x_2 = \lambda x_1^2} = \frac{\lambda}{1 + \lambda^3} > 0.$$

Thus, the limit as $x \to (0,0)$ along the curve $x_2 = \lambda x_1^2$ depends on the parameter $\lambda > 0$ and hence the mapping F is not continuous at the point $(0,0)$.

Theorem 5.3. *Consider operators* $F : \mathbb{B}_1 \to \mathbb{B}_2$ *and* $G : \mathbb{B}_2 \to \mathbb{B}_3$, *where the operator* F *is Fréchet-differentiable at a point* $u \in \mathbb{B}_1$ *and the operator* G *is Fréchet differentiable at the point* $F(u)$. *Then their composition*

$$K \overset{\text{def}}{=\joinrel=} G \circ F$$

is Fréchet different at the point $u \in \mathbb{B}_1$ *and the following equality holds:*

$$K'_f(u) = G'_f(F(u))F'_f(u). \tag{5.9}$$

Proof. We have

$$\left\|K(u+h) - K(u) - G'_f(F(u))F'_f(u)h\right\|_3$$

$$\leq \left\|G(F(u+h)) - G(F(u)) - G'_f(F(u))\left[F(u+h) - F(u)\right]\right\|_3$$

$$+ \left\|G'_f(F(u))\left[F(u+h) - F(u) - F'_f(u)h\right]\right\|_3$$

$$\leq \left\|\omega_1(F(u), F(u+h) - F(u))\right\|_3 + \left\|G'_f(F(u))\right\|_{2\to 3}\|\omega_2(u,h)\|_3.$$

Note that due to the Fréchet differentiability of the operator F, we have the estimate

$$\|F(u+h) - F(u)\|_2 \leq c\|h\|_1.$$

Therefore, the following limit equality holds:

$$\lim_{\|h\|_1 \to 0} \frac{\|\omega_1(F(u), F(u+h) - F(u))\|_3}{\|h\|_1}$$

$$= \lim_{\|h\|_1 \to 0} \frac{\|\omega_1(F(u), F(u+h) - F(u))\|_3}{\|F(u+h) - F(u)\|_2} \frac{\|F(u+h) - F(u)\|_2}{\|h\|_1} = 0.$$

Moreover, we have the limit equality

$$\lim_{\|h\|_1 \to 0} \frac{\|\omega_2(u,h)\|_3}{\|h\|_1} = 0,$$

which implies the validity of the theorem. $\square$

5.2   Fréchet Derivatives of Some Important Functionals

In this section, we calculate the Fréchet derivative of the following functionals:
(recall that $D_x = (\partial_{x_1}, \ldots, \partial_{x_N})$).

Lemma 5.1. *Let $p \in (1, +\infty)$. Then the Fréchet derivatives of the functionals*

$$\psi_1(u) : L^p(\Omega) \to \mathbb{R}^1, \qquad \psi_1(u) = \frac{1}{p}\int_\Omega |u|^p \, dx,$$

$$\psi_2(u) : W_0^{1,p}(\Omega) \to \mathbb{R}^1, \qquad \psi_2(u) = \frac{1}{p}\int_\Omega |D_x u|^p \, dx \tag{5.10}$$

are equal respectively to

$$\psi'_{1f}(u) = |u|^{p-2}u \in L^{p'}(\Omega),$$

$$\psi'_{2f}(u) = -\Delta_p u \stackrel{\text{def}}{=} -\operatorname{div}(|D_x u|^{p-2} D_x u) \in W^{-1,p'}(\Omega). \tag{5.11}$$

Proof. 1. First, we calculate the Fréchet derivative of the functional $\psi_1(u)$. We separately consider the cases $p \geq 2$ and $p \in (1, 2)$.

In the case $p \geq 2$, applying the Taylor formula with Lagrange's residual, we obtain

$$|u+h|^p = |u|^p + p|u|^{p-2}uh + \frac{p(p-1)}{2}|\xi|^{p-2}h^2, \quad u(x) < \xi(x) < u(x) + h(x); \tag{5.12}$$

moreover,

$$u = u(x) \in L^p(\Omega), \quad h = h(x) \in L^p(\Omega) \quad \Rightarrow \quad \xi = \xi(x) \in L^p(\Omega).$$

Dividing both sides of Eq. (5.12) by p and integrating by $x \in \Omega$, we arrive at the equality

$$\psi_1(u + h) = \psi_1(u) + \int_\Omega |u(x)|^{p-2} u(x) h(x) \, dx + \omega_1(u, h), \qquad (5.13)$$

where

$$\omega_1(u, h) = \frac{p-1}{2} \int_\Omega |\xi(x)|^{p-2} h^2(x) \, dx. \qquad (5.14)$$

Applying the Hölder inequality with the exponents

$$q_1 = \frac{p}{p-2}, \quad q_2 = \frac{p}{2}, \quad \frac{1}{q_1} + \frac{1}{q_2} = 1,$$

to the expression $\omega_1(u, h)$, we obtain

$$|\omega_1(u, h)| \leq \frac{p-1}{2} \left(\int_\Omega |\xi(x)|^p \, dx \right)^{(p-2)/p} \left(\int_\Omega |h(x)|^p \, dx \right)^{2/p} = c_1 \|h\|_p^2$$

$$\Rightarrow \quad \lim_{\|h\|_p \to +0} \frac{|\omega_1(u, h)|}{\|h\|_p} = 0.$$

Therefore,

$$\int_\Omega |u(x)|^{p-2} u(x) h(x) \, dx = \langle \psi'_{1f}(u), h \rangle \quad \Rightarrow \quad \psi'_{1f}(u) = |u|^{p-2} u.$$

Now we consider the case $p \in (1, 2)$. We apply the Lagrange's mean-value theorem:

$$|u + h|^p = |u|^p + p|u + \theta h|^{p-2} (u + \theta h) h$$
$$= |u|^p + p|u|^{p-2} u h + \omega(u, h), \quad \theta \in (0, 1), \qquad (5.15)$$
$$\omega(u, h) = \frac{p}{\theta} \left(|u + \theta h|^{p-2} (u + \theta h) - |u|^{p-2} u \right) \theta h.$$

Using the inequality

$$\left(|b|^{p-2} b - |a|^{p-2} a, b - a \right) \leq \gamma(p) |b - a|^p, \quad 1 < p \leq 2, \qquad (5.16)$$

we obtain the estimate

$$0 \leq \left(|u + \theta h|^{p-2} (u + \theta h) - |u|^{p-2} u \right) \theta h \leq \gamma(p) |\theta h|^p,$$

which implies

$$|\omega(u, h)| \leq p\gamma(p) |h|^p |\theta|^{p-1} \leq p\gamma(p) |h|^p. \qquad (5.17)$$

Now, dividing both sides of Eq. (5.15) by p and integrating the resulting equality by $x \in \Omega$, we arrive at the equality

$$\psi_1(u+h) = \psi_1(u) + \int_\Omega |u(x)|^{p-2} u(x) h(x) \, dx + \omega_1(u, h), \qquad (5.18)$$

where

$$\omega_1(u, h) = \int_\Omega \omega(u(x), h(x)) \, dx.$$

Due to (5.17), we conclude that the following inequality holds:

$$|\omega_1(u, h)| \le \gamma(p) \int_\Omega |h(x)|^p \, dx = \gamma(p) \|h\|_p^p \quad \Rightarrow \quad \lim_{\|h\|_p \to +0} \frac{|\omega_1(u, h)|}{\|h\|_p} = 0, \quad p > 1.$$

2. For calculating the Fréchet derivative of the functional $\psi_2(u)$, we must use the corresponding Taylor formulas for real-valued functions of several variables with Lagrange's residuals. Arguments used above can be applied here, but for $p \in (1, 2)$, we use the formula (5.16) and the following equality for the generalized p-Laplacian:

$$\int_\Omega \left(|D_x u|^{p-2} D_x u, D_x h(x) \right) \, dx = \langle -\Delta_p u, h \rangle \quad \text{for all } h(x) \in W_0^{1,p}(\Omega).$$

Lemma 5.1 is proved. $\qquad\qquad\qquad\qquad\qquad\qquad\qquad\qquad\qquad\qquad\qquad \square$

5.3   Bibliographical Notes

The contents of this lecture is taken from [14, 23, 24, 35, 67].

Lecture 6

Nemytsky Operator

6.1 Basic Definitions

In this lecture, we introduce an important class of operators, namely, Nemytsky operators. First, we must introduce so-called *Carathéodory functions*.

Let $(\Omega, \mathcal{M}, \mu)$ be a complete, measurable, σ-finite space.

Definition 6.1. A function

$$f(x, u) : \Omega \times \mathbb{R}^1 \to \mathbb{R}^1$$

is called a *Carathéodory function* if it is μ-measurable on Ω for all $u \in \mathbb{R}^1$ and continuous by $u \in \mathbb{R}^1$ for μ-almost all $x \in \Omega$.

Definition 6.2. Let a function $f(x, u)$ be a Carathéodory function. The operator N_f, which to each function $u(x)$ defined on Ω assigns the function $N_f(u)(x) \overset{\text{def}}{=\!=} f(x, u(x))$ is called the Nemytsky operator.

Theorem 6.1 (see [12]). *Nemytsky operators maps Lebesgue-measurable functions into measurable functions.*

The importance of Nemytsky operators in the study of nonlinear boundary-value problems is due to the following important theorem.

Theorem 6.2 (Krasnosel'sky theorem). *A Nemytsky operator N_f is a bounded and continuous operator acting from $L^p(\Omega)$ into $L^q(\Omega)$, where $p, q \in [1, +\infty)$, if and only if the corresponding Carathéodory function $f(x, u)$ satisfies the estimate*

$$|f(x, u)| \leq a(x) + b|u|^{p/q}, \quad b > 0, \quad a(x) \geq 0,$$

for all $u \in \mathbb{R}^1$ and μ-almost all $x \in \Omega$, where $a(x) \in L^q(\Omega)$.

The proof of this theorem is sufficiently complicated; we split it into several separate assertions.

6.2   Measure Convergence

Definition 6.3. We say that a functional sequence $\{f_n(x)\}$ *converges with respect to a measure* μ on Ω if for any $\varepsilon > 0$, the following limit property is fulfilled:

$$\lim_{n \to +\infty} \mu\left(\{x \in \Omega : |f_n(x) - f(x)| \geq \varepsilon\}\right) = 0. \tag{6.1}$$

Let D_n be the set of all points $x \in \Omega$ at which the inequality

$$|f_n(x) - f(x)| < \varepsilon \tag{6.2}$$

holds. Then the condition (6.1) is equivalent to the following relation:

$$\lim_{n \to +\infty} \mu\left(\Omega \backslash D_n\right) = 0 \quad \text{for any } \varepsilon > 0. \tag{6.3}$$

Theorem 6.3. *Let* $\mu(\Omega) < +\infty$. *Then the Nemytsky operator maps any sequence of Lebesgue-measurable functions* $\{u_n(x)\}$ *converging with respect to a measure* μ *into a sequence* $\{f(x, u_n(x))\}$ *of measurable functions, which also converges with respect to the measure* μ.

Proof. Let

$$u_n(x) \overset{\mu}{\to} u_0(x) \quad \text{as } n \to +\infty.$$

Then, in particular, the function $u_0(x)$ is Lebesgue-measurable. Indeed, from a measure-converging sequence we can extract a sequence converging almost everywhere and the limit of a sequence of measurable functions, which converges almost everywhere, is also measurable.

1. We fix $\varepsilon > 0$ and denote by $x \in \Omega_m$ the set of all points $x \in \Omega$ for which the following implication holds:

$$|u_0(x) - u| < \frac{1}{m} \quad \Rightarrow \quad |f(x, u_0(x)) - f(x, u)| < \varepsilon. \tag{6.4}$$

Note that the set Ω_m is a Lebesgue-measurable set. Indeed,

$$\Omega \backslash \Omega_m = \left\{x \in \Omega : |u_0(x) - u| < \frac{1}{m} \text{ and } |f(x, u_0(x)) - f(x, u)| \geq \varepsilon\right\}.$$

Consider the sets

$$\{x \in \Omega : |u_0(x) - u| < 1/m\}, \quad \{x \in \Omega : |f(x, u_0(x)) - f(x, u)| \geq \varepsilon\}.$$

The first of them is measurable due to the measurability of the function u_0. To prove the measurability of the second set, we note that, due to Theorem 6.1, the measurability of the function u_0 implies the measurability of the function $f(x, u_0(x))$ and the measurability of the function $f(x, u)$ for an arbitrary constant function u immediately follows from Definition 6.1. Therefore, the intersection $\Omega \backslash \Omega_m$ of these sets is also measurable. Finally, the complement Ω_m of the measurable set $\Omega \setminus \Omega_m$ is measurable.

The following chain of embeddings holds:

$$\Omega_1 \subset \Omega_2 \subset \cdots \subset \Omega_m \subset \Omega_{m+1} \subset \cdots .$$

Indeed, for $x \in \Omega_m$ we have

$$|u_0(x) - u| < \frac{1}{m} \quad \Rightarrow \quad |f(x, u_0(x)) - f(x, u)| < \varepsilon.$$

In particular, if

$$|u_0(x) - u| < \frac{1}{m+1},$$

then, obviously, we have

$$|u_0(x) - u| < \frac{1}{m+1} < \frac{1}{m} \quad \Rightarrow \quad |f(x, u_0(x)) - f(x, u)| < \varepsilon,$$

i.e., $x \in \Omega_{m+1}$.

2. Since the function $f(x, u)$ is continuous by $u \in \mathbb{R}^1$ for almost all $x \in \Omega$, we have

$$\mu \left(\bigcup_{m=1}^{+\infty} \Omega_m \right) = \mu(\Omega), \quad \Omega_m \subset \Omega_{m+1}. \tag{6.5}$$

Indeed, by the definition of the Carathéodory function $f(x, u)$, there exists a set $E \subset \Omega$ such that $\mu(E) = 0$ and for any $x \in \Omega \backslash E$, there exists $m \in \mathbb{N}$ such that

$$|u_0(x) - u| < \frac{1}{m} \quad \Rightarrow \quad |f(x, u_0(x)) - f(x, u)| < \varepsilon,$$

i.e., $x \in \Omega_m$. Thus, the following embeddings hold:

$$\Omega = \bigcup_{m=1}^{+\infty} \Omega_m \cup E, \quad \bigcup_{m=1}^{+\infty} \Omega_m \subset \Omega. \tag{6.6}$$

It remains to apply the monotonicity of the Lebesgue measure and obtain from (6.6) the following two inequalities:

$$\mu(\Omega) \leq \mu \left(\bigcup_{m=1}^{+\infty} \Omega_m \right) + \mu(E), \quad \mu \left(\bigcup_{m=1}^{+\infty} \Omega_m \right) \leq \mu(\Omega).$$

This implies Eq. (6.5).

Therefore, the following limit property holds:

$$\lim_{m \to +\infty} \mu(\Omega_m) = \mu(\Omega). \tag{6.7}$$

Indeed, consider the sets

$$\Omega'_n = \Omega_n \backslash \Omega_{n-1} \quad \text{for } n \geq 2, \quad \Omega'_1 = \Omega_1.$$

Obviously, the sets Ω'_{n_1} and Ω'_{n_2} do not intersect for $n_1 \neq n_2$; moreover,

$$\Omega_m = \bigcup_{n=1}^{m} \Omega'_n.$$

The following equality holds:

$$\mu\left(\bigcup_{n=1}^{+\infty} \Omega'_n\right) = \sum_{n=1}^{+\infty} \mu(\Omega'_n) = \mu(\Omega).$$

Therefore, we have

$$\mu(\Omega_m) = \sum_{n=1}^{m} \mu(\Omega'_n) \to \sum_{n=1}^{+\infty} \mu(\Omega'_n) = \mu(\Omega)$$

as $m \to +\infty$.

Let $\varepsilon > 0$ and $\eta > 0$ be taken arbitrarily. We choose a sufficiently large number $m_0 \in \mathbb{N}$ so thus

$$\mu(\Omega) < \mu(\Omega_{m_0}) + \frac{\eta}{2}; \tag{6.8}$$

this is possible due to the limit property (6.7).

3. Let F_n be the set of all points $x \in \Omega$ such that

$$|u_0(x) - u_n(x)| < \frac{1}{m_0}. \tag{6.9}$$

The set F_n is Lebesgue-measurable since the functions $u_0(x)$ and $u_n(x)$ are Lebesgue-measurable. Since the sequence of functions $\{u_n(x)\}$ converges with respect to the measure μ on Ω, we have the limit property

$$\lim_{n \to +\infty} \mu(\Omega \backslash F_n) = 0. \tag{6.10}$$

Thus, we conclude that

$$\lim_{n \to +\infty} \mu(F_n) = \mu(\Omega). \tag{6.11}$$

Indeed, we have

$$\Omega = F_n \cup (\Omega \backslash F_n);$$

therefore,

$$0 \leq \mu(\Omega) - \mu(F_n) \leq \mu(\Omega \backslash F_n) \to +0 \quad \text{as } n \to +\infty.$$

Thus, for fixed $\eta > 0$, there exists a natural number $N \in \mathbb{N}$ such that for all $n \geq N$, the following inequality is valid:

$$\mu(\Omega) < \mu(F_n) + \frac{\eta}{2}. \tag{6.12}$$

4. Now we consider the sequence $\{f(x, u_n(x))\}$. Let D_n be the set of all points $x \in \Omega$ for which the inequality

$$|f(x, u_0(x)) - f(x, u_n(x))| < \varepsilon \tag{6.13}$$

holds. The set D_n is Lebesgue-measurable since the composite functions $f(x, u_0(x))$ and $f(x, u_n(x))$ are Lebesgue-measurable due to Theorem 6.1.

The following embedding holds:

$$\Omega_{m_0} \cap F_n \subset D_n. \tag{6.14}$$

Indeed, due to (6.4) and (6.9), the inequality (6.13) is valid for $x \in \Omega_{m_0}$ and $x \in F_n$.

The following consequence of the finite additivity of the measure μ is obvious:

$$\mu\left(\Omega_{m_0} \cap F_n\right) = \mu\left(\Omega_{m_0}\right) + \mu\left(F_n\right) - \mu\left(\Omega_{m_0} \cup F_n\right). \tag{6.15}$$

Moreover, the obvious inequality holds:

$$\mu\left(\Omega_{m_0} \cup F_n\right) \leq \mu\left(\Omega\right). \tag{6.16}$$

Due to the embedding (6.14), the relation (6.15), and the inequalities (6.16), (6.8), and (6.12), we arrive at the following chain of inequalities:

$$
\begin{aligned}
\mu(D_n) &\geq \mu(\Omega_{m_0} \cap F_n) = \mu\left(\Omega_{m_0}\right) + \mu\left(F_n\right) - \mu\left(\Omega_{m_0} \cup F_n\right) \\
&> \mu(\Omega) - \frac{\eta}{2} + \mu(\Omega) - \frac{\eta}{2} - \mu(\Omega) = \mu(\Omega) - \eta \\
&\Rightarrow \quad \mu(\Omega) - \mu(D_n) < \eta.
\end{aligned}
\tag{6.17}
$$

Note that

$$\mu(\Omega) = \mu(D_n) + \mu(\Omega\backslash D_n) \quad \Rightarrow \quad \mu(\Omega\backslash D_n) = \mu(\Omega) - \mu(D_n).$$

Thus, for any $\varepsilon > 0$ and any $\eta > 0$, there exists a natural number N such that for any $n \geq N$, the inequality $\mu(\Omega\backslash D_n) < \eta$ holds, i.e., for any $\varepsilon > 0$ we have

$$\lim_{n \to +\infty} \mu(\Omega\backslash D_n) = 0.$$

Therefore, the sequence $\{f(x, u_n(x))\}$ converges to the function $f(x, u_0(x))$ on Ω with respect to the measure μ. $\qquad\square$

6.3 Continuity of Nemytsky Operators

Any Nemytsky operator possesses the following interesting property.

Theorem 6.4. *Let a Nemytsky operator map each function from $L^p(\Omega)$ into a function from $L^q(\Omega)$ where $p, q \geq 1$. Then it is continuous with respect to the strong topologies of these Lebesgue spaces.*

We prove this theorem only for a finite domain $\Omega \in \mathbb{R}^N$ (i.e., $\mu(\Omega) < +\infty$).

Proof. Stage I. A particular case. First, let $f(x, 0) = 0$. We prove that in this particular case, the Nemytsky operator is continuous at the zero of the space $L^p(\Omega)$.

1. Assume that the Nemytsky operator is not continuous at the zero element of the space $L^p(\Omega)$. Then there exists a sequence $\{u_n(x)\} \subset L^p(\Omega)$ such that

$$u_n \to \theta \quad \text{strongly in } L^p(\Omega) \text{ as } n \to +\infty, \tag{6.18}$$

but

$$\int_\Omega |f(x, u_n(x))|^q \, dx \geq \alpha > 0, \tag{6.19}$$

where α is a number independent of $n \in \mathbb{N}$. If necessary, we may pass to a subsequence of the sequence $\{u_n(x)\}$ so that the following property be valid:

$$\sum_{n=1}^{+\infty} \int_\Omega |u_n(x)|^p \, dx < +\infty. \tag{6.20}$$

Indeed, due to (6.18), we have

$$\int_\Omega |u_n(x)|^p \, dx \to +0 \quad \text{as } n \to +\infty. \tag{6.21}$$

Passing to a subsequence $\{u_{n'}\} \subset \{u_n\}$, we can assume that the inequality

$$\int_\Omega |u_{n'}(x)|^p \, dx \leq \frac{K}{n'^2} \tag{6.22}$$

holds, where the constant $K > 0$ is independent of $n' \in \mathbb{N}$.

2. Now we construct sequences of numbers $\{\varepsilon_k\}$, functions $\{u_{n_k}(x)\} \subset \{u_n(x)\}$, and sets $\{\Omega_k\} \subset \Omega$ such that the following properties hold:

(i) $\varepsilon_{k+1} < \varepsilon_k/2$;
(ii) $\mu(\Omega_k) \leq \varepsilon_k$;
(iii) $\int_{\Omega_k} |f(x, u_{n_k})|^q \, dx > 2\alpha/3$;
(iv) for any measurable set $D \subset \Omega$, the condition $\mu(D) \leq 2\varepsilon_{k+1}$ implies that

$$\int_D |f(x, u_{n_k}(x))|^q \, dx < \frac{\alpha}{3}. \tag{6.23}$$

We construct these sequences by induction. Let

$$\varepsilon_1 = \mu(\Omega), \quad u_{n_1}(x) := u_1(x), \quad \Omega_1 := \Omega.$$

The property (ii) is trivial. The property (iii) holds due to the inequality (6.19).

If ε_k, $u_{n_k}(x)$, and Ω_k have already been constructed, then we choose as ε_{k+1} a number such that the condition (iv) be fulfilled (such $\varepsilon_{k+1} > 0$ can be chosen due to the property of absolute continuity of the Lebesgue integral). Then the inequality (i) holds.

Indeed, by the choice of ε_{k+1}, for any measurable set $D \subset \Omega$ such that $\mu(D) \leq 2\varepsilon_{k+1}$, the inequality (6.23) holds. Moreover, by the inductive hypothesis, the inequality (iii) holds. Therefore, by the property (iv) we have

$$\mu(\Omega_k) > 2\varepsilon_{k+1},$$

and by the inequality (ii) we have $\mu(\Omega_k) \leq \varepsilon_k$. Therefore, the inequality (i) is valid.

We emphasize the strong convergence of the functional sequence $\{f_n(x)\} \subset L^p(\Omega)$:

$$f_n(x) \to f(x) \quad \text{strongly in } L^p(\Omega) \text{ as } n \to +\infty. \tag{6.24}$$

This fact and the Tchebychev inequality imply the measure convergence of this sequence. Indeed, due the Tchebychev inequality, the following inequality holds for any $c > 0$:

$$\mu\left(\{x \in \Omega : |f_n(x) - f(x)|^p > c\}\right) \leq \frac{1}{c} \int_\Omega |f_n(x) - f(x)|^p \, dx; \qquad (6.25)$$

its right-hand side tends to zero as $n \to +\infty$.

Since $\{u_{n_k}(x)\} \subset \{u_n(x)\}$, this sequence converges to zero strongly in $L^p(\Omega)$ and hence converges by measure. Then, due to Theorem 6.3, the sequence $\{f(x, u_{n_k}(x))\} \subset L^q(\Omega)$ also converges to zero by measure. Therefore, for $\varepsilon_{k+1} > 0$, there exist a natural number n_{k+1} and a Lebesgue-measurable set $F_{k+1} \subset \Omega$ such that for all $x \in F_{k+1}$, the inequality

$$\left|f(x, u_{n_{k+1}}(x))\right| \leq \left(\frac{\alpha}{3\mu(\Omega)}\right)^{1/q} \qquad (6.26)$$

holds and

$$\mu(\Omega) - \mu(F_{k+1}) < \varepsilon_{k+1}. \qquad (6.27)$$

Let

$$\Omega_{k+1} := \Omega \backslash F_{k+1}. \qquad (6.28)$$

Then the property (ii) is fulfilled:

$$\mu(\Omega_{k+1}) < \varepsilon_{k+1}. \qquad (6.29)$$

Indeed, by (6.27) we have

$$\Omega = (\Omega \backslash F_{k+1}) \cup F_{k+1} \quad \Rightarrow \quad \mu(\Omega) = \mu(\Omega \backslash F_{k+1}) + \mu(F_{k+1})$$
$$\Rightarrow \quad \mu(\Omega_{k+1}) = \mu(\Omega) - \mu(F_{k+1}) < \varepsilon_{k+1}. \qquad (6.30)$$

Now we prove that the property (iii) hold. Indeed, due to (6.19) and (6.26), the following chain of inequalities is valid:

$$\int_{\Omega_{k+1}} \left|f(x, u_{n_{k+1}}(x))\right|^q \, dx$$

$$= \int_\Omega \left|f(x, u_{n_{k+1}}(x))\right|^q \, dx - \int_{F_{k+1}} \left|f(x, u_{n_{k+1}}(x))\right|^q \, dx$$

$$> \alpha - \frac{\alpha}{3\mu(\Omega)}\mu(\Omega) = \frac{2}{3}\alpha. \qquad (6.31)$$

Thus, the sequences of numbers $\{\varepsilon_k\}$, functions $\{u_{n_k}(x)\}$, and set $\{\Omega_k\}$ satisfying the properties (i)–(iv) have been constructed.

3. Introduce the Lebesgue-measurable sets

$$D_k := \Omega_k \backslash \bigcup_{i=k+1}^{+\infty} \Omega_i. \qquad (6.32)$$

Prove that these sets are disjoint. Indeed, let $k_1 \neq k_2$. For definiteness, assume that $k_2 > k_1$. Consider the sets

$$D_{k_1} = \Omega_{k_1} \setminus (\Omega_{k_1} \cup \cdots \cup \Omega_{k_2} \cup \ldots),$$

$$D_{k_2} = \Omega_{k_2} \setminus \bigcup_{i=k_2+1}^{+\infty} \Omega_i \subset \Omega_{k_2}. \tag{6.33}$$

From (6.33) it is obvious that D_{k_1} does not contain points of the set Ω_{k_2} and hence of its subset D_{k_2}. Therefore, $D_{k_1} \cap D_{k_2} = \varnothing$ for $k_1 \neq k_2$.

The properties (i) and (ii) imply the following inequality:

$$\mu \left(\bigcup_{i=k+1}^{+\infty} \Omega_i \right) < 2\varepsilon_{k+1}. \tag{6.34}$$

Indeed, due to the property (i), we have

$$\varepsilon_{k+1+l} < \frac{1}{2^l} \varepsilon_{k+1} \quad \Rightarrow \quad \sum_{i=k+1}^{+\infty} \varepsilon_i < \varepsilon_{k+1} \sum_{l=0}^{+\infty} \frac{1}{2^l} = \varepsilon_{k+1} \frac{1}{1-1/2} = 2\varepsilon_{k+1}. \tag{6.35}$$

From the property (ii) and the final inequality in (6.35) we obtain the inequality

$$\mu \left(\bigcup_{i=k+1}^{+\infty} \Omega_i \right) \leq \sum_{i=k+1}^{+\infty} \varepsilon_i < 2\varepsilon_{k+1}. \tag{6.36}$$

Introduce the following function:

$$\psi(x) = \begin{cases} u_{n_k}(x), & x \in D_k, \ k \in \mathbb{N}, \\ 0, & x \notin \bigcup_{i=1}^{+\infty} D_i. \end{cases} \tag{6.37}$$

Note that (6.32) implies

$$\Omega_k \setminus D_k = \bigcup_{i=k+1}^{+\infty} \Omega_i;$$

then from (6.34) we obtain

$$\mu \left(\Omega_k \setminus D_k \right) < 2\varepsilon_{k+1}. \tag{6.38}$$

Therefore, due to the property (iv) we arrive at the inequality

$$\int_{\Omega_k \setminus D_k} |f(x, u_{n_k}(x))|^q \, dx < \frac{\alpha}{3}. \tag{6.39}$$

Finally, from the property (iii) we obtain the following inequality:

$$\int_{\Omega_k} |f(x, u_{n_k}(x))|^q \, dx > \frac{2}{3}\alpha. \tag{6.40}$$

Due to the inequalities (6.39) and (6.40), we obtain the following chain of inequalities:

$$\int_{D_k} |f(x, \psi(x))|^q \, dx = \int_{D_k} |f(x, u_{n_k}(x))|^q \, dx$$

$$= \int_{\Omega_k} |f(x, u_{n_k}(x))|^q \, dx - \int_{\Omega_k \setminus D_k} |f(x, u_{n_k}(x))|^q \, dx$$

$$> \frac{2}{3}\alpha - \frac{\alpha}{3} = \frac{\alpha}{3}. \tag{6.41}$$

Due to the properties (6.20), we have

$$\int_{\Omega} |\psi(x)|^p \, dx = \sum_{k=1}^{+\infty} \int_{D_k} |u_{n_k}(x)|^p \, dx \leq \sum_{n=1}^{+\infty} \int_{\Omega} |u_n(x)|^p \, dx < +\infty, \tag{6.42}$$

i.e., $\psi(x) \in L^p(\Omega)$. On the one hand, by the condition of the theorem we have $f(x, \psi(x)) \in L^q(\Omega)$. On the other hand, the following lower estimate is valid:

$$+\infty > \int_{\Omega} |f(x, \psi(x))|^q \, dx \geq \sum_{k=1}^{+\infty} \int_{D_k} |f(x, \psi(x))|^q \, dx \geq \sum_{k=1}^{+\infty} \frac{\alpha}{3} = +\infty. \tag{6.43}$$

This contradicts the continuity of the Nemytsky operator at zero in the particular case considered.

At the following stage, we consider the general case.

Stage II. General case for $\mu(\Omega) < +\infty$.

We prove the continuity of the Nemytsky operator at a point $u_0(x) \in L^p(\Omega)$. Introduce the function

$$g(x, u(x)) := f(x, u_0(x) + u(x)) - f(x, u_0(x)), \tag{6.44}$$

which, obviously, satisfies the condition $g(x, 0) = 0$; the corresponding Nemytsky operator N_g maps $L^p(\Omega)$ into $L^q(\Omega)$. As was proved above, the operator N_g is continuous at the zero element $\theta(x) \in L^p(\Omega)$. Therefore, the Nemytsky operator generated by the function $f(x, u)$ is continuous at $u_0(x) \in L^p(\Omega)$.

Indeed, we must prove that

$$\left\| f(x, u_n(x)) - f(x, u_0(x)) \right\|_{L^q(\Omega)} \to +0 \quad \text{as } n \to +\infty$$

if

$$\left\| u_n(x) - u_0(x) \right\|_{L^p(\Omega)} \to +0 \quad \text{as } n \to +\infty.$$

The substitution $v_n(x) := u_n(x) - u_0(x)$ reduces the problem to the proof of the following assertion:

$$\left\| g(x, v_n(x)) \right\|_{L^q(\Omega)} = \left\| f(x, v_n(x) + u_0(x)) - f(x, u_0(x)) \right\|_{L^q(\Omega)} \to +0 \tag{6.45}$$

as $n \to +\infty$ if

$$\left\| v_n(x) \right\|_{L^p(\Omega)} \to +0 \quad \text{as } n \to +\infty.$$

However, this follows from the facts proved in the stage I. Thus, Theorem 6.4 is proved under the condition $\mu(\Omega) < +\infty$. $\qquad \square$

6.4   Boundedness of Nemytsky Operators

As is known, in the nonlinear case the continuity of an operator, generally speaking, does not imply its boundedness. Therefore, in is necessary to prove the boundedness of the Nemytsky operator.

Definition 6.4. An operator $A : \mathbb{B}_1 \to \mathbb{B}_2$ is said to be *bounded* if it maps a bounded set of the Banach space $\mathbb{B}_1$ into a bounded set of the Banach space $\mathbb{B}_2$.

Theorem 6.5. *Let a Nemytsky operator transform each function from $L^p(\Omega)$ into a function from $L^q(\Omega)$, where $p, q \geq 1$. Then the Nemytsky operator is bounded.*

Proof. Without loss of generality, we can assume that

$$f(x, 0) = 0. \tag{6.46}$$

In the opposite case, we use the representation (6.44) and take into account the fact that $f(x, u_0(x))$ is a fixed function from $L^q(\Omega)$. By Theorem 6.4, the Nemytsky operator is continuous at the zero element $\theta(x) \in L^p(\Omega)$. This means that there exists $R > 0$ such that

$$\int_\Omega |f(x, \varphi(x))|^q \, dx \leq 1 \tag{6.47}$$

if

$$\int_\Omega |\varphi(x)|^p \, dx \leq R^p. \tag{6.48}$$

Now assume that $u(x) \in L^p(\Omega)$ and for some $N \in \mathbb{N} \cup \{0\}$, the following inequality holds:

$$NR^p \leq \|u\|_p^p \leq (N+1)R^p. \tag{6.49}$$

We partition the set Ω into measurable parts $\Omega_1, \ldots, \Omega_{N+1}$ such that

$$\Omega = \bigcup_{n=1}^{N+1} \Omega_n, \quad \int_{\Omega_n} |u(x)|^p \, dx \leq R^p, \quad n = \overline{1, N+1}. \tag{6.50}$$

Indeed, since $u(x) \in L^p(\Omega)$, the function of a set

$$\nu(A) := \int_A |u(x)|^p \, dx, \quad A \subset \Omega$$

is a countable additive measure, which is absolutely continuous with respect to the initial Lebesgue measure. By (6.49), there exists $M \in (0, R^p)$ such that the equality

$$\nu(\Omega) \equiv \int_\Omega |u(x)|^p \, dx = (N+1)M$$

holds. Let $\Omega_1 \subset \Omega$ be a measurable subset such that

$$\nu(\Omega_1) = M \Rightarrow \nu(\Omega \backslash \Omega_1) = NM.$$

The existence of such a set follows from the absolute continuity of the measure ν. Further, by induction, we construct a family of Lebesgue-measurable, pairwise disjoint sets $\{\Omega_k\}_{k=1}^{N+1}$ such that

$$\nu(\Omega_k) = M, \quad \Omega = \bigcup_{k=1}^{N+1} \Omega_k.$$

Now we consider instead of the function $u(x)$ the functional family

$$u_k(x) := \chi_{\Omega_k}(x)u(x) \in L^p(\Omega),$$

where χ_A is the characteristic function of a measurable set $A \subset \Omega$:

$$\chi_A(x) := \begin{cases} 1, & x \in A, \\ 0, & x \notin A. \end{cases} \tag{6.51}$$

Then by (6.50) we have $\|u_k\|_p^p \leq R^p$ and hence, taking into account (6.46), (6.48), and (6.47), we arrive at the relations

$$\int_{\Omega} |f(x, u(x))|^q \, dx = \sum_{k=1}^{N+1} \int_{\Omega_k} |f(x, u(x))|^q \, dx$$

$$= \sum_{k=1}^{N+1} \int_{\Omega} |f(x, u_k(x))|^q \, dx \leq N + 1. \tag{6.52}$$

Then from (6.49) and (6.52) we obtain the inequality

$$\|f(x, u(x))\|_q = \left(\int_{\Omega} |f(x, u(x))|^q \, dx \right)^{1/q} \leq \left[\left(\frac{\|u\|_p}{R} \right)^p + 1 \right]^{1/q},$$

i.e., the Nemytsky operator is bounded. $\qquad\square$

6.5 Necessary and Sufficient Conditions of the Continuity of Nemytsky Operators

Theorem 6.6. *A Nemytsky operator acts from $L^p(\Omega)$ into $L^q(\Omega)$, where $p, q \geq 1$, if and only if there exist a nonnegative function $a(x) \in L^q(\Omega)$ and a positive constant b such that the following inequality holds:*

$$|f(x, u)| \leq a(x) + b|u|^{p/q}. \tag{6.53}$$

Proof. Consider the case where $f(x, 0) = 0$ for almost all $x \in \Omega$.

1. Sufficiency. The inequality (6.53) implies the inequality

$$\|f(x, u)\|_q \leq \|a(x)\|_q + b\|u\|_p^{p/q}. \tag{6.54}$$

Then the Nemytsky operator acts from $L^p(\Omega)$ into $L^q(\Omega)$.

2. *Necessity.* Let a Nemytsky operator act from $L^p(\Omega)$ into $L^q(\Omega)$. Then by Theorem 6.5, there exists a positive number b such that

$$\int_\Omega |f(x, u(x))|^q \, dx \leq b^q \tag{6.55}$$

if

$$\int_\Omega |u(x)|^p \, dx \leq 1. \tag{6.56}$$

Consider the function

$$\varphi(x, u) = \begin{cases} |f(x, u)| - b|u|^{p/q}, & |f(x, u)| \geq b|u|^{p/q}, \\ 0, & |f(x, u)| \leq b|u|^{p/q}; \end{cases} \tag{6.57}$$

it is measurable if the function $u(x)$ is measurable. Note that if $\varphi(x, u) \neq 0$, then

$$|\varphi(x, u)|^q \leq |f(x, u)|^q - b^q|u|^p. \tag{6.58}$$

Indeed, if $\varphi(x, u) \neq 0$, then we have

$$|\varphi(x, u)| + b|u|^{p/q} = |f(x, u)|$$

$$\Leftrightarrow \quad |f(x, u)|^q = \left(|\varphi(x, u)| + b|u|^{p/q}\right)^q \geq |\varphi(x, u)|^q + b^q|u|^p.$$

Let $u(x) \in L^p(\Omega)$ be an arbitrary fixed function such that

$$\|u(x)\|_{L^p(\Omega)} \neq 0. \tag{6.59}$$

Consider the set

$$\widetilde{\Omega} := \{x \in \Omega : \varphi(x, u(x)) > 0\}; \tag{6.60}$$

it is Lebesgue-measurable since the function $\varphi(x, u(x))$ is Lebesgue-measurable due to the measurability of the function $u(x) \in L^p(\Omega)$.

Note that for all $x \in \widetilde{\Omega}$, we have

$$\int_{\widetilde{\Omega}} |u(x)|^p \, dx \neq 0. \tag{6.61}$$

Indeed, let

$$\int_{\widetilde{\Omega}} |u(x)|^p \, dx = 0 \quad \Rightarrow \quad u(x) = 0 \quad \text{for almost all } x \in \widetilde{\Omega}. \tag{6.62}$$

Then for $x \in \widetilde{\Omega}$ we have

$$0 < |\varphi(x, u(x))| = |\varphi(x, 0)| = |f(x, 0)| - b|0|^{p/q} = 0 \tag{6.63}$$

since we consider the case $f(x, 0) = 0$.

Let $N \in \mathbb{N} \cup \{0\}$ and $\varepsilon \in [0, 1)$ be such that

$$\int_{\widetilde{\Omega}} |u(x)|^p \, dx = N + \varepsilon. \tag{6.64}$$

Then the set $\widetilde{\Omega}$ can be partitioned into pairwise disjoint, measurable sets $\Omega_1, \ldots, \Omega_{N+1}$ so that

$$\int_{\Omega_n} |u(x)|^p \, dx < 1 \quad \text{for } n \in \overline{1, N+1} \tag{6.65}$$

(see the arguments in the previous section). Then by (6.55) and (6.56), we obtain the following chain of inequalities:

$$\int_{\widetilde{\Omega}} |f(x, u(x))|^q \, dx \leq \sum_{n=1}^{N+1} \int_{\Omega_n} |f(x, u(x))|^q \, dx$$

$$= \sum_{n=1}^{N+1} \int_{\Omega_n} |f(x, u_n(x))|^q \, dx \leq b^q (N+1), \tag{6.66}$$

where $u_n(x) = u(x)\chi_{\Omega_n}(x)$, $\chi_{\Omega_n}(x)$ is the characteristic function of the set Ω_n (see (6.51)).

Indeed, we have

$$\int_{\Omega} |u_n(x)|^p \, dx = \int_{\Omega_n} |u(x)|^p \, dx < 1.$$

This relation and the relations (6.55), (6.56), and $f(x, 0) = 0$ imply the inequality

$$\int_{\Omega_n} |f(x, u(x))|^q \, dx = \int_{\Omega} |f(x, u_n(x))|^q \, dx \leq b^q.$$

From the inequalities (6.58), (6.64), and (6.66) we obtain the following chain of inequalities:

$$\int_{\Omega} |\varphi(x, u(x))|^q \, dx = \int_{\widetilde{\Omega}} |\varphi(x, u(x))|^q \, dx$$

$$\leq \int_{\widetilde{\Omega}} |f(x, u(x))|^q \, dx - b^q \int_{\widetilde{\Omega}} |u(x)|^p \, dx$$

$$\leq (N+1)b^q - b^q(N+\varepsilon) \leq b^q. \tag{6.67}$$

Now let $\{D_k\}_{k=1}^{+\infty}$ be a sequence of nested measurable sets ($\mu(D_k) < +\infty$) such that

$$\Omega = \bigcup_{k=1}^{+\infty} D_k.$$

Let

$$\alpha_k(x) := \max_{|u| \leq k} \varphi(x, u). \tag{6.68}$$

In [67], the existence of a sequence of measurable functions $\{u_k(x)\}$ such that

$$\varphi(x, u_k(x)) = \max_{|u| \leq k} \varphi(x, u), \quad x \in D_k, \tag{6.69}$$

and

$$u_k(x) = 0 \quad \text{for } x \notin D_k, \ |u_k(x)| \le k, \tag{6.70}$$

was proved. Indeed, for almost all $x \in \Omega$, the function $\varphi(x, u)$ is continuous by $u \in \mathbb{R}^1$. Then for almost all $x \in D_k$, there exists a number $u_k(x) \in \mathbb{R}^1$, which provides the maximum of (6.69). Thus, the function $u_k(x) : D_k \to \mathbb{R}^1$, $|u_k(x)| \le k$, is defined. We extend this function by zero outside D_k. In [67], the existence of a measurable function $u_k(x)$ was proved.

Obviously, $u_k(x) \in L^p(\Omega)$ (recall that the measures of all sets D_k are finite). Therefore, we can apply all arguments between the formulas (6.59) and (6.67) to all functions $u_k(x)$. We set

$$a(x) := \sup_{u \in \mathbb{R}^1} \varphi(x, u) = \lim_{k \to +\infty} \varphi(x, u_k(x)). \tag{6.71}$$

In particular, $a(x)$ is a measurable function since it is the pointwise limit of a sequence of measurable functions. The Fatou lemma[1] and the inequalities (6.67) imply the following inequalities:

$$\int_\Omega |a(x)|^q \, dx \le \sup_{k \in \mathbb{N}} \int_\Omega |\varphi(x, u_k(x))|^q \, dx \le b^q \quad \Rightarrow \quad a(x) \in L^q(\Omega). \tag{6.72}$$

By the definition of $a(x)$, we have

$$a(x) = \sup_{u \in \mathbb{R}^1} \varphi(x, u) \ge \sup_{u \in \mathbb{R}^1} \left\{ |f(x, u)| - b|u|^{p/q} \right\}$$

$$\ge |f(x, u)| - b|u|^{p/q} \quad \text{for all } u \in \mathbb{R}^1. \tag{6.73}$$

Thus, we arrive at the estimate (6.53). Theorem 6.6 is proved. $\qquad\square$

6.6 Potential of Nemytsky Operator

Now we consider an important result, which will be useful in the study of variational problems.

Let

$$f(x, u) : \Omega \times \mathbb{R}^1 \to \mathbb{R}^1$$

be a Carathéodory function. Introduce the so-called potential function

$$F(x, z) := \int_0^z f(x, \xi) \, d\xi \tag{6.74}$$

and the functional

$$\psi(u) := \int_\Omega F(x, u(x)) \, dx. \tag{6.75}$$

---

[1] Note that in the Fatou lemma, the condition of the finiteness of the limit almost everywhere is absent. This fact is essential since the *a priori* finiteness of the limit (6.71) almost everywhere is not obvious.

We assume that
$$|f(x,u)| \le a(x) + b|u|^{p/p'}, \quad p' = \frac{p}{p-1}, \quad p \in (1,+\infty),$$
where $a(x) \in L^{p'}(\Omega)$ and $a(x) \ge 0$ almost everywhere, $b > 0$. Then for the potential function $F(x,u)$ defined by the formula (6.74), due to the arithmetical Hölder inequality,
$$a_1 \cdot a_2 \le \frac{a_1^p}{p} + \frac{a_2^{p'}}{p'}, \quad \frac{1}{p} + \frac{1}{p'} = 1, \quad a_1, a_2 \ge 0,$$
the following inequality holds:
$$|F(x,u)| \le \left| \int_0^{u(x)} f(x,\xi)\, d\xi \right| \le a(x)|u| + \frac{b}{p}|u|^p$$
$$\le \frac{|a(x)|^{p'}}{p'} + \frac{|u|^p}{p} + \frac{b}{p}|u|^p = a_1(x) + c_1|u|^p, \tag{6.76}$$
where $a_1(x) \in L^1(\Omega)$ and $c_1 > 0$.

We prove the Fréchet differentiability of the functional
$$\psi(u) : L^p(\Omega) \to \mathbb{R}^1.$$
Consider the following expression:
$$\omega(u,v) := \psi(u+v) - \psi(u) - \langle N_f(u), v \rangle, \quad u, v \in L^p(\Omega),$$
$$|\omega(u,v)| \le \left| \int_\Omega [F(x, u(x) + v(x)) - F(x, u(x))]\, dx - \int_\Omega N_f(u)(x)v(x)\, dx \right|.$$
We have the following chain of equalities:
$$F(x, u(x) + v(x)) - F(x, u(x))$$
$$= \int_0^1 \frac{d}{dt} F(x, u(x) + tv(x))\, dt = \int_0^1 f(x, u(x) + tv(x))v(x)\, dt.$$
Therefore, we obtain the estimate
$$|\omega(u,v)| \le \int_0^1 dt \int_\Omega dx\, |N_f(u+tv)(x) - N_f(u)(x)|\, |v(x)|$$
$$\le \int_0^1 dt\, \|N_f(u+tv) - N_f(u)\|_{p'}\, \|v\|_p.$$
Therefore, due to the continuity of the Nemytsky operator N_f, we have the limit inequality
$$\lim_{\|v\|_p \to 0} \frac{|\omega(u,v)|}{\|v\|_p} \le \lim_{\|v\|_p \to 0} \int_0^1 dt\, \|N_f(u+tv) - N_f(u)\|_{p'} = 0.$$
Thus, we have proved the following lemma.

Lemma 6.1. *Under the conditions formulated above, the functional $\psi(u)$ defined by the formula (6.75) is Fréchet-differentiable and the following equality holds:*
$$\psi_f'(u) = N_f(u) \quad \text{for all } u \in L^p(\Omega),\ p \in (1,+\infty). \tag{6.77}$$

6.7   Bibliographical Notes

The contents of this lecture is taken from [35, 56, 67].

Lecture 7

Fréchet Derivatives of Implicit Functionals in Control Theory

In control problems, the task of calculating the gradients (Fréchet derivative in the general case of a Banach space) of certain functionals (usually quadratic) with respect to an implicit variable appears. In this lecture, we give a series of examples from the control theory and obtain explicit expressions for the gradients of such functionals under the assumption that they exist.

7.1 Elliptic Boundary-Value Control Problems

Example 7.1 (control by the right-hand side). Let $\Omega \subset \mathbb{R}^N$ be a bounded domain with sufficiently smooth boundary $\partial\Omega$. Consider the following problem:

$$Ay(x; u) = f(x) + u(x), \quad x \in \Omega, \tag{7.1}$$

$$y(x; u) = 0, \quad x \in \partial\Omega; \tag{7.2}$$

this is a Dirichlet problem for classical solutions $y(x; u) \in C^{(2)}(\Omega) \cap C(\overline{\Omega})$, where $u(x)$ is a control.[1] The solution $y(x; u)$ depends on u. The operator A has the form

$$Aw(x) := -\sum_{i,j=1,1}^{N,N} \frac{\partial}{\partial x_i}\left(a_{ij}(x)\frac{\partial w(x)}{\partial x_j}\right), \quad a_{ij}(x) = a_{ji}(x) \in C^{(1)}(\Omega), \tag{7.3}$$

where

$$m|\lambda|^2 \le \sum_{i,j=1,1}^{N,N} a_{ij}(x)\lambda_i\lambda_j \le M|\lambda|^2, \quad 0 < m \le M < +\infty, \tag{7.4}$$

for all $x \in \Omega$.

Now we introduce the notion of a weak solution of the Dirichlet problem (7.1), (7.2).

Definition 7.1. A function $y(x; u) \in H_0^1(\Omega)$ satisfying the equality

$$\langle Ay(x; u) - f(x) - u(x), \varphi(x)\rangle = 0 \quad \text{for all } \varphi(x) \in H_0^1(\Omega), \tag{7.5}$$

is called a *weak solution* of the Dirichlet problem (7.1), (7.2), where $f(x) \in H^{-1}(\Omega)$ is a given function, $u(x) \in L^2(\Omega)$ is a control, and $\langle \cdot, \cdot \rangle$ are the duality brackets between $H_0^1(\Omega)$ and $H^{-1}(\Omega)$.

---

[1] In the weak statement, we assume that $u(x) \in L^2(\Omega)$.

Here, the action of the operator A is meant in the following weak sense:

$$\langle A\psi(x), \varphi(x)\rangle := \sum_{i,j=1,1}^{N,N} \int_\Omega a_{ij}(x)\frac{\partial\psi(x)}{\partial x_j}\frac{\partial\varphi(x)}{\partial x_i}\,dx, \quad a_{ij}(x) \in L^\infty(\Omega), \tag{7.6}$$

for any $\varphi(x), \psi(x) \in H_0^1(\Omega)$.

Remark 7.1. Note that (7.6) implies the following equality:

$$\langle A\psi(x), \varphi(x)\rangle = \langle A\varphi(x), \psi(x)\rangle \tag{7.7}$$

for any $\varphi(x), \psi(x) \in H_0^1(\Omega)$.

Remark 7.2. Note that the following dense and continuous embeddings hold:

$$H_0^1(\Omega) \overset{ds}{\subset} L^2(\Omega) \overset{ds}{\subset} H^{-1}(\Omega). \tag{7.8}$$

Now we consider the following functional, which implicitly depends on the control $u(x) \in L^2(\Omega)$:

$$J(u) := \int_\Omega (y(x; u) - h(x))^2\,dx, \quad h(x) \in L^2(\Omega), \tag{7.9}$$

where $h(x)$ is a given function. Note that a weak solution $y(x; u)$ can be considered as an operator of the control $u \in L^2(\Omega)$:

$$y(x; u) : L^2(\Omega) \to H_0^1(\Omega). \tag{7.10}$$

Therefore, if the Fréchet derivative $y_f'(x; u)$ exists,[2] we have

$$y_f'(x; \cdot) : L^2(\Omega) \to \mathscr{L}(L^2(\Omega); H_0^1(\Omega)). \tag{7.11}$$

The Fréchet derivative of the functional $J(u)$ has the form

$$(J_f'(u), v - u) = 2\int_\Omega (y(x; u) - h(x))y_f'(x; u)(v(x) - u(x))\,dx \tag{7.12}$$

for any $u(x), v(x) \in L^2(\Omega)$. Due to the uniform ellipticity of the operator A (see the inequalities (7.4)), we conclude the existence of the inverse operator $A^{-1} : H^{-1}(\Omega) \to H_0^1(\Omega)$. Therefore, the equality

$$Ay(x; u) = f(x) + u(x), \tag{7.13}$$

which is meant in the sense of the space of functionals $H^{-1}(\Omega)$, we obtain the following equality:

$$
\begin{aligned}
y(x; u) &= A^{-1}(f + u)\\
&\Rightarrow \quad y_f'(x; u) \cdot \psi = A^{-1}\psi \quad \text{for any } \psi \in L^2(\Omega)\\
&\Rightarrow \quad y_f'(x, u) \cdot (v - u) = A^{-1}(v - u) = y(x; v) - y(x; u),
\end{aligned}
\tag{7.14}
$$

---

[2] Since the real Banach space $L^2(\Omega)$ is taken as the space of controls, the notions of the Fréchet derivative and the gradient coincide.

since

$$y(x;u) = A^{-1}(f+u), \quad y(x;v) = A^{-1}(f+v)$$

$$\Rightarrow \quad y(x;v) - y(x;u) = A^{-1}(v-u). \tag{7.15}$$

Thus, from (7.12) and (7.14) we obtain the following form of the Fréchet derivative:

$$(J'_f(u), v - u) = 2 \int_\Omega (y(x;u) - h(x))(y(x;v) - y(x;u)) \, dx. \tag{7.16}$$

Now we consider the so-called *adjoint problem* for the Dirichlet problem (7.1), (7.2):

$$A^* z(x;u) = y(x;u) - h(x), \quad x \in \Omega, \tag{7.17}$$

$$z(x;u) = 0, \quad\quad\quad\quad\quad\quad x \in \partial\Omega, \tag{7.18}$$

where

$$A^* w(x) = A w(x) = -\sum_{i,j=1,1}^{N,N} \frac{\partial}{\partial x_j}\left(a_{ij}(x)\frac{\partial w(x)}{\partial x_i}\right), \tag{7.19}$$

which is meant in the following weak sense.

Definition 7.2. A function $z(x;u) \in H_0^1(\Omega)$ satisfying the equality

$$\langle A^* z(x;u) + h(x) - y(x;u), \varphi(x)\rangle = 0 \quad \text{for all } \varphi(x) \in H_0^1(\Omega), \tag{7.20}$$

is called a *weak solution* of the Dirichlet problem (7.17), (7.18), where $h(x) \in L^2(\Omega)$ is a given function, $y(x;u) \in H_0^1(\Omega)$ is a weak solution of the Dirichlet problem (7.1), (7.2), and is the $\langle \cdot, \cdot \rangle$ duality bracket between $H_0^1(\Omega)$ and $H^{-1}(\Omega)$.

Taking into account Definitions 7.1 and 7.2, from Eqs. (7.16), (7.7), and (7.8), we obtain the following equality:

$$\left(J'_f(u), v - u\right) = 2 \int_\Omega A^* z(x;u)(y(x;v) - y(x;u)) \, dx$$

$$= 2 \int_\Omega Az(x;u)(y(x;v) - y(x;u)) \, dx = 2\langle Az(x;u), y(x;v) - y(x;u)\rangle$$

$$= 2\langle A(y(x;v) - y(x;u)), z(x;u)\rangle = 2 \int_\Omega A(y(x;v) - y(x;u))z(x;u) \, dx$$

$$= 2 \int_\Omega z(x;u)(v(x) - u(x)) \, dx. \tag{7.21}$$

Therefore, the following assertion is proved.

Theorem 7.1. *The Fréchet derivative of the functional $J(u)$ defined by Eq. (7.9), is equal to*

$$J'_f(u) = 2z(x;u), \tag{7.22}$$

where $z(x;u) \in H_0^1(\Omega)$ is a weak solution of the Dirichlet problem (7.17), (7.18).

Example 7.2 (boundary observations). Consider the Dirichlet problem (7.17), (7.18). Assume that the coefficients $a_{ij}(x)$ are sufficiently smooth so that the weak solution of the Dirichlet problem in the sense of Definition 7.1 lies in the class $y(x; u) \in H_0^1(\Omega) \cap H^2(\Omega)$. Then

$$\frac{\partial y(x; u)}{\partial \nu_x}\bigg|_{\partial\Omega} := \sum_{i,j=1,1}^{N,N} a_{ij}(x) \frac{\partial y(x; u)}{\partial x_j} \cos(n_x, e_i)\bigg|_{\partial\Omega} \in L^2(\partial\Omega).$$

Consider the quadratic functional

$$J(u) := \int_{\partial\Omega} \left(M(x) \frac{\partial y(x; u)}{\partial \nu_x} - h(x) \right)^2 dS_x, \tag{7.23}$$

where $h(x) \in L^2(\partial\Omega)$ and $M(x) \in L^\infty(\partial\Omega)$ are given functions. The Fréchet derivative of this functional has the following form:

$$\left(J_f'(u), v - u \right)$$

$$= 2 \int_{\partial\Omega} \left(M \frac{\partial y(x; u)}{\partial \nu_x} - h(x) \right) M \frac{\partial}{\partial \nu_x} \left(y_f'(u) \cdot (v(x) - u(x)) \right) dS_x. \tag{7.24}$$

Due to (7.14), we arrive at the following equality:

$$\left(J_f'(u), v - u \right)$$

$$= 2 \int_{\partial\Omega} \left(M \frac{\partial y(x; u)}{\partial \nu_x} - h(x) \right) \left(M \frac{\partial y(x; v)}{\partial \nu_x} - M \frac{\partial y(x; u)}{\partial \nu_x} \right) dS_x. \tag{7.25}$$

Now we consider the Dirichlet problem for the adjoint state. In the classical statement, this Dirichlet problem has the following form:

$$A^* z(x; u) = 0, \qquad\qquad\qquad x \in \Omega, \tag{7.26}$$

$$z(x; u) = -M \left(M \frac{\partial y(x; u)}{\partial \nu_x} - h(x) \right), \qquad x \in \partial\Omega, \tag{7.27}$$

where the operator A^* is defined by Eq. (7.19). We introduce the notion of a weak solution of the Dirichlet problem (7.26), (7.27).

Definition 7.3. A function $z(x; u) \in H^1(\Omega) \cap H^2(\Omega)$ is called a *weak solution* of the Dirichlet problem (7.26), (7.27) if the following equality holds:

$$\langle A^* z(x; u), \varphi(x) \rangle = 0 \quad \text{for all } \varphi(x) \in H_0^1(\Omega) \tag{7.28}$$

and

$$z(x; u) = -M(x) \left(M(x) \frac{\partial y(x; u)}{\partial \nu_x} - h(x) \right) \quad \text{for } x \in \partial\Omega. \tag{7.29}$$

Lemma 7.1. *The following equality is valid:*

$$\int_{\partial\Omega} \left(M \frac{\partial y(x; u)}{\partial \nu_x} - h(x) \right) \left(M \frac{\partial y(x; v)}{\partial \nu_x} - M \frac{\partial y(x; u)}{\partial \nu_x} \right) dS_x$$

$$= \int_\Omega z(x; u)(v(x) - u(x)) \, dx. \tag{7.30}$$

Proof. In Eq. (7.28), we set $\varphi(x) = y(x; v) - y(x; u)$ and integrate by parts; we obtain the following equality:

$$0 = \langle A^* z(x, u), y(x; v) - y(x; u) \rangle = -\int_{\partial\Omega} (y(x; v) - y(x; u)) \frac{\partial z(x; u)}{\partial \nu_x} dS_x$$

$$+ \int_{\partial\Omega} z(x; u) \left(\frac{\partial y(x; v)}{\partial \nu_x} - \frac{\partial y(x; u)}{\partial \nu_x} \right) dS_x + \int_{\Omega} A(y(x; v) - y(x; u)) z(x; u) \, dx.$$

(7.31)

Taking into account the Dirichlet problems (7.17), (7.18) and (7.26), (7.27), we obtain Eq. (7.30). $\qquad\square$

Thus, the following theorem is proved.

Theorem 7.2. *If the Fréchet derivative of the functional $J(u)$ defined by Eq. (7.23) exists, then it is equal to*

$$J_f'(u) = 2z(x; u), \tag{7.32}$$

where the function $z(x; u) \in H_0^1(\Omega) \cap H^2(\Omega)$ is a weak solution of the Dirichlet problem (7.26), (7.27).

7.2 Parabolic Mixed Control Problems

Example 7.3 (final observation for the Dirichlet problem). Let $\Omega \subset \mathbb{R}^N$ be a bounded domain with sufficiently smooth boundary $\partial\Omega$. Introduce the following notation:

$$D_T := \Omega \otimes (0, T), \quad S_T := \partial\Omega \otimes [0, T],$$
$$B = \Omega \otimes \{t = 0\}, \quad B_T := \Omega \otimes \{t = T\}.$$

Consider the following classical mixed problem:

$$\frac{\partial y(x, t; u)}{\partial t} + A(t)y(x, t; u) = f(x, t) + u(x, t), \quad (x, t) \in D_T \cup B_T, \tag{7.33}$$

$$y(x, t; u) = 0, \qquad (x, t) \in S_T, \tag{7.34}$$

$$y(x, 0; u) = y_0(x), \qquad x \in \Omega, \tag{7.35}$$

$$\tag{7.36}$$

where

$$A(t)w(x, t) := -\sum_{i,j=1,1}^{N,N} \frac{\partial}{\partial x_i} \left(a_{ij}(x, t) \frac{\partial w(x, t)}{\partial x_j} \right), \tag{7.37}$$

$$a_{ij}(x, t) = a_{ji}(x, t), \quad m|\lambda|^2 \le \sum_{i,j=1,1}^{N,N} a_{ij}(x, t)\lambda_i\lambda_j \le M|\lambda|^2 \tag{7.38}$$

for all $(x, t) \in D_T$ and the constants satisfy the condition $0 < m \le M < +\infty$.

Definition 7.4. A function of the class $y(x, t; u) \in L^\infty(0, T; H_0^1(\Omega))$, $y'(x, t, u) \in L^2(0, T; L^2(\Omega))$ is called a *weak solution* of the problem (7.33)–(7.35) if for any $\varphi(x, t) \in L^2(0, T; H_0^1(\Omega))$, the equality

$$\int_0^T \int_\Omega \left[y'(x, t; u) - f(x, t) - u(x, t) \right] \varphi(x, t) \, dx \, dt$$

$$+ \int_0^T \langle A(t) y(x, t; u), \varphi(x, t) \rangle \, dt = 0 \tag{7.39}$$

holds and

$$y(x, 0; u) = y_0(x) \in L^2(\Omega), \tag{7.40}$$

where $f(x, t) \in L^2(D_T)$ and $u(x, t) \in L^2(D_T)$ is a control.

Remark 7.3. By Lemma 4.4, the Banach space

$$W(0, T) := \left\{ w(x, t) \in L^\infty(0, T; H_0^1(\Omega)), \; w'(x, t) \in L^2(0, T; L^2(\Omega)) \right\} \tag{7.41}$$

with respect to the norm

$$\|w\|_{W(0,T)} := \operatorname{ess.sup}_{t \in (0,T)} \|w(t)\|_{H_0^1(\Omega)} + \|w'\|_{L^2(D_T)}$$

is continuously embedded into the space $C([0, T]; L^2(\Omega))$ and hence $y(x, T; u) \in L^2(\Omega)$.

Taking into account Remark 7.3, consider the following quadratic functional:

$$J(u) := \int_\Omega [y(x, T; u) - h(x)]^2 \, dx, \quad h(x) \in L^2(\Omega), \tag{7.42}$$

where the function $h(x)$ is given. Introduce the functions

$$\psi(x) := y(x, T; u) - h(x), \quad \varphi(x) := y(x, T; v) - y(x, T; u). \tag{7.43}$$

Then the following equalities hold:

$$J(v) - J(u) = \int_\Omega (\psi(x) + \varphi(x))^2 \, dx - \int_\Omega \psi^2(x) \, dx$$

$$= 2 \int_\Omega \psi(x) \varphi(x) \, dx + \int_\Omega \varphi^2(x) \, dx$$

$$= 2 \int_\Omega [y(x, T; u) - h(x)] \, (y(x, T; v) - y(x, T; u)) \, dx$$

$$+ \int_\Omega (y(x, T; v) - y(x, T; u))^2 \, dx$$

$$= \left(J'_f(u), v - u \right) + \omega(u, v - u). \tag{7.44}$$

If the Fréchet derivative exists, then it has the following form:

$$\left(J_f'(u), v - u\right) = 2 \int_\Omega [y(x, T; u) - h(x)] \, (y(x, T; v) - y(x, T; u)) \, dx. \qquad (7.45)$$

Now we consider the adjoint initial-boundary problem; in the classical statement, it has the form

$$-\frac{\partial z(x, t; u)}{\partial t} + A^*(t)z(x, t; u) = 0, \quad (x, t) \in D_T \cup B_T, \qquad (7.46)$$

$$z(x, t; u) = 0, \qquad\qquad\qquad (x, t) \in S_T, \qquad (7.47)$$

$$z(x, T; u) = y(x, T; u) - h(x), \qquad\quad x \in \Omega, \qquad (7.48)$$

where

$$A^*(t)w(x, t) := -\sum_{i,j=1,1}^{N,N} \frac{\partial}{\partial x_j} \left(a_{ij}(x, t)\frac{\partial w(x, t)}{\partial x_i}\right). \qquad (7.49)$$

Remark 7.4. Note that $A^*(t) = A(t)$.

Definition 7.5. A function of the class $z(x, t; u) \in L^\infty(0, T; H_0^1(\Omega))$, $z'(x, t; u) \in L^2(0, T; L^2(\Omega))$ is called a *weak solution* of the problem (7.46)–(7.48) if for any $\varphi(x, t) \in L^2(0, T; H_0^1(\Omega))$, the following equalities hold:

$$-\int_0^T \int_\Omega z'(x, t; u)\varphi(x, t) \, dx \, dt + \int_0^T \langle A^*(t)z(x, t; u), \varphi(x, t)\rangle \, dt = 0, \qquad (7.50)$$

$$z(x, T; u) = y(x, T; u) - h(x) \in L^2(\Omega). \qquad (7.51)$$

Remark 7.5. It follows from Remark 7.3 that $z(x, t; u) \in W(0, T) \subset C([0, T]; L^2(\Omega))$ and hence the value of the function $z = z(x, t; u)$ at $t = T$ is defined and moreover $z(x, T; u) \in L^2(\Omega)$.

Lemma 7.2. *The following equality holds:*

$$\int_\Omega (y(x, T; u) - h(x))(y(x, T; v) - y(x, T; u)) \, dx$$

$$= \int_0^T \int_\Omega z(x, t; u)(v(x, t) - u(x, t)) \, dx \, dt, \qquad (7.52)$$

where $z(x, t; u)$ is a weak solution of the mixed problem (7.46)–(7.48) in the sense of Definition 7.5.

Proof. We use the following formulas of integrating by parts:

$$\int_0^T \int_\Omega \frac{\partial z(x,t;u)}{\partial t} \left(y(x,t;v) - y(x,t;u) \right) dx \, dt$$

$$= \int_\Omega z(x,T;u) \left(y(x,T;v) - y(x,T;u) \right) dx$$

$$- \int_0^T \int_\Omega z(x,t;u) \left(\frac{\partial y(x,t;v)}{\partial t} - \frac{\partial y(x,t;u)}{\partial t} \right) dx \, dt \qquad (7.53)$$

and

$$\int_0^T \langle A^*(t) z(x,t;u), y(x,t;v) - y(x,t;u) \rangle \, dt$$

$$= \int_0^T \langle A \left[y(x,t;v) - y(x,t;u) \right], z(x,t;u) \rangle \, dt, \qquad (7.54)$$

where we have used the following equalities:

$$y(x,0;v) - y(x,0;u) = 0, \quad y(x,t;v) - y(x,t;u) \in L^\infty(0,T;H_0^1(\Omega)). \qquad (7.55)$$

Now we set

$$\varphi(x,t) = y(x,t;v) - y(x,t;u)$$

in Eq. (7.50). Then, due to Eqs. (7.53) and (7.54), we obtain the following equality:

$$\int_\Omega z(x,T;u) \left(y(x,T;v) - y(x,T;u) \right) dx$$

$$= \int_0^T \int_\Omega z(x,t;u) \left(\frac{\partial y(x,t;v)}{\partial t} - \frac{\partial y(x,t;u)}{\partial t} \right) dx \, dt$$

$$+ \int_0^T \langle A \left[y(x,t;v) - y(x,t;u) \right], z(x,t;u) \rangle \, dt \qquad (7.56)$$

Setting $\varphi(x,t) = z(x,t;u)$ in (7.39), we obtain

$$\int_0^T \int_\Omega \left(\frac{\partial y(x,t;u)}{\partial t} - f(x,t) - u(x,t) \right) z(x,t;u) \, dx \, dt$$

$$+ \int_0^T \langle A(t) y(x,t;u), z(x,t;u) \rangle \, dt = 0 \qquad (7.57)$$

and

$$\int\limits_0^T \int\limits_\Omega \left(\frac{\partial y(x,t;v)}{\partial t} - f(x,t) - v(x,t) \right) z(x,t;u)\, dx\, dt$$

$$+ \int\limits_0^T \langle A(t)y(x,t;v), z(x,t;u) \rangle\, dt = 0. \tag{7.58}$$

Subtracting Eq. (7.57) from Eq. (7.58), we arrive at the equality

$$\int\limits_0^T \int\limits_\Omega \left(\frac{\partial y(x,t;v)}{\partial t} - \frac{\partial y(x,t;u)}{\partial t} \right) z(x,t;u)\, dx\, dt$$

$$+ \int\limits_0^T \langle A(t)\left[y(x,t;v) - y(x,t;u)\right], z(x,t;u) \rangle\, dt$$

$$= \int\limits_0^T \int\limits_\Omega z(x,t;u)\left[v(x,t) - u(x,t)\right]\, dx\, dt. \tag{7.59}$$

From Eqs. (7.56) and (7.59) we obtain the required equality (7.52). □

Thus, we have proved the following assertion.

Theorem 7.3. *If the Fréchet derivative of the functional $J(u)$ defined by Eq. (7.42) exists, then it is equal to*

$$J'_f(u) = 2z(x,t;u), \tag{7.60}$$

where the function $z(x,t;u) \in W(0,T)$ is a weak solution of the Dirichlet problem (7.46)–(7.48).

Example 7.4 (nonquadratic functional). Consider the Dirichlet problem (7.33)–(7.35) whose solutions are meant in the weak sense of Definition 7.4. We examine nonquadratic functional

$$J(v) = \int\limits_0^T \int\limits_\Omega |y(x,t;v) - h(x,t)|^p\, dx\, dt, \quad p > 2, \tag{7.61}$$

where $h(x,t) \in L^p(D_T)$ is a given function.

The Fréchet derivative of the functional (7.61) on the Banach space $L^p(D_T)$ can be calculated as the Fréchet derivative of the composition of Fréchet-differentiable mappings as follows:

$$\langle J'_f(u), w \rangle = p \int\limits_0^T \int\limits_\Omega |y(x,t;u) - h(x,t)|^{p-2}$$

$$\times \left[y(x,t;u) - h(x,t) \right] y_f'(x,t;u)w \, dx \, dt, \qquad (7.62)$$

where $\langle \cdot, \cdot \rangle$ is the duality bracket between the Banach spaces $L^p(D_T)$ and $L^{p'}(D_T)$. (Note that in this case we use the term "Fréchet derivative" since the notion of gradient is not defined for $p \neq 2$.) Now we set

$$\psi(x,t;w) := y_f'(x,t;u)w \qquad (7.63)$$

and note that due to the linearity of Eq. (7.33) the function $\psi(x,t;w)$ satisfies the classical problem

$$\frac{\partial \psi(x,t;w)}{\partial t} + A(t)\psi(x,t;w) = w(x,t), \quad (x,t) \in D_T \cup B_T, \qquad (7.64)$$

$$\psi(x,t;w) = 0, \qquad (x,t) \in S_T, \qquad (7.65)$$

$$\psi(x,0;w) = 0, \qquad x \in \Omega, \qquad (7.66)$$

which is meant in the weak sense similarly to Definition 7.4. Then from (7.62) and (7.63) we obtain the following equality:

$$\langle J_f'(u), v - u \rangle = p \int_0^T \int_\Omega |y(x,t;u) - h(x,t)|^{p-2}$$

$$\times \left[y(x,t;u) - h(x,t) \right] \psi(x,t;v-u) \, dx \, dt. \qquad (7.67)$$

Now we consider the adjoint problem whose classical statement is as follows:

$$-\frac{\partial z(x,t;u)}{\partial t} + A^*(t)z(x,t;u) = |y(x,t;u) - h(x,t)|^{p-2}$$

$$\times \left[y(x,t;u) - h(x,t) \right], \quad (x,t) \in D_T \cup B_T, \quad (7.68)$$

$$z(x,t;u) = 0, \quad (x,t) \in S_T, \qquad (7.69)$$

$$z(x,T;u) = 0, \quad x \in \Omega. \qquad (7.70)$$

Lemma 7.3. *The following equality holds:*

$$\int_0^T \int_\Omega |y(x,t;u) - h(x,t)|^{p-2} \left[y(x,t;u) - h(x,t) \right] \psi(x,t;v-u) \, dx \, dt$$

$$= \int_0^T \int_\Omega z(x,t;u) \left[v(x,t) - u(x,t) \right] \, dx \, dt. \qquad (7.71)$$

Proof. Due to Eq. (7.68), we obtain the following equalities:

$$\int_0^T \int_\Omega |y(x,t;u) - h(x,t)|^{p-2} \left[y(x,t;u) - h(x,t) \right] \psi(x,t;v-u) \, dx \, dt$$

$$= \int_0^T \int_\Omega \left[-\frac{\partial z(x,t;u)}{\partial t} + A^*(t)z(x,t;u) \right] \psi(x,t;v-u) \, dx \, dt$$

$$= \int_0^T \int_\Omega \left[\frac{\partial \psi(x,t; v-u)}{\partial t} + A(t)\psi(x,t; v-u) \right] z(x,t; u)\, dx\, dt$$

$$= \int_0^T \int_\Omega z(x,t; u)\, [v(x,t) - u(x,t)]\, dx\, dt; \tag{7.72}$$

we used the boundary and initial conditions (7.65), (7.66), (7.69), and (7.70). $\quad\square$

Thus, the following assertion is proved.

Theorem 7.4. *If the Fréchet derivative of the functional $J(u)$ defined by Eq. (7.61) exists, then it is equal to*

$$J'_f(u) = pz(x,t; u), \tag{7.73}$$

where the function $z(x,t; u) \in W(0,T)$ is a weak solution of the Dirichlet problem (7.68)–(7.70).

7.3 Coefficient Control Problem for the Burgers Equation

Example 7.5 (final observation problem). Consider the following problem for the Burgers equation on a segment. Let

$$D_T = (0,1) \otimes (0,T), \quad S_T = (\{x=0\} \otimes [0,T]) \cup (\{x=1\} \otimes [0,T]),$$
$$B = (0,1) \otimes \{t=0\}, \quad B_T = (0,1) \otimes \{t=T\}.$$

The classical statement of the problem is as follows:

$$\frac{\partial y(x,t; u)}{\partial t} - \frac{\partial^2 y(x,t; u)}{\partial x^2} = \frac{1}{2} \frac{\partial y^2(x,t; u)}{\partial x}$$
$$+ u(x,t)y(x,t; u), \quad (x,t) \in D_T \cup B_T, \tag{7.74}$$

$$y(0,t; u) = y_0(t), \quad y(1,t; u) = y_1(t), \quad t \in [0,T], \tag{7.75}$$
$$y(x,0; u) = y_2(x), \quad x \in (0,1). \tag{7.76}$$

We will consider only classical solutions of this problem. Let

$$y_0(t),\ y_1(t) \in C^{(1)}[0,T], \quad y_2(x) \in C^{(2)}[0,1]$$

be given functions. We assume that a solution $y(x,t; u)$ of this problem belong to the class $C^{2,1}_{x,t}(\overline{D}_T)$. Consider the functional

$$J(v) = \int_0^1 (y(x,T; v) - h(x))^2\, dx, \quad h(x) \in C[0,1], \tag{7.77}$$

where $h(x)$ is a given function and $u(x,t) \in C([0,1] \otimes [0,T])$ is a control. From the chain rule for calculating the Fréchet derivative of the composition of Fréchet-differentiable operators, we obtain the following equality for the gradient:

$$\left(J_f'(u), v - u\right) = 2 \int_0^1 \left(y(x, T; u) - h(x)\right) y_f'(x, T; u)(v(x, T) - u(x, T)) \, dx. \quad (7.78)$$

Introduce the notation

$$\psi(x, t; v - u) := y_f'(x, t; u)(v(x, t) - u(x, t)). \quad (7.79)$$

Then Eq. (7.78) takes the following form:

$$\left(J_f'(u), v - u\right) = 2 \int_0^1 \left(y(x, T; u) - h(x)\right) \psi(x, T; v - u) \, dx. \quad (7.80)$$

Differentiating by Fréchet both sides of Eq. (7.74), we obtain the equation for the function $\psi(x, t; v - u)$:

$$\frac{\partial \psi(x, t; v - u)}{\partial t} - \frac{\partial^2 \psi(x, t; v - u)}{\partial x^2}$$
$$= \frac{\partial (y(x, t; u)\psi(x, t; v - u))}{\partial x} + u(x, t)\psi(x, t; v - u)$$
$$+ y(x, t; u)[v(x, t) - u(x, t)], \quad (x, t) \in D_T \cup B_T. \quad (7.81)$$

Moreover, obviously, $\psi(x, t; v - u)$ satisfies the homogeneous initial and boundary conditions:

$$\psi(0, t; v - u) = \psi(1, t; v - u) = 0, \quad t \in [0, T], \quad (7.82)$$
$$\psi(x, 0; v - u) = 0, \quad\quad\quad\quad\quad x \in (0, 1). \quad (7.83)$$

We assume that a solution of the problem (7.81)–(7.83) belongs to the class $C_{x,t}^{2,1}(\overline{D}_T)$. Consider the adjoint problem for the mixed problem (7.81)–(7.83):

$$-\frac{\partial z(x, t; u)}{\partial t} - \frac{\partial^2 z(x, t; u)}{\partial x^2}$$
$$= -y(x, t; u)\frac{\partial z(x, t; u)}{\partial x} + u(x, t)z(x, t; u), \quad (x, t) \in D_T \cup B_T, \quad (7.84)$$

$$z(0, t; u) = z(1, t; 0) = 0, \quad\quad t \in [0, T], \quad (7.85)$$
$$z(x, T; u) = y(x, T; u) - h(x), \quad x \in (0, 1). \quad (7.86)$$

We assume that a solution of this problem belongs to the class $C_{x,t}^{2,1}(\overline{D}_T)$. Taking into account the boundary condition (7.86), we obtain from (7.78) the following equality:

$$\left(J_f'(u), v - u\right) = 2 \int_0^1 z(x, T; u)\psi(x, T; v - u) \, dx. \quad (7.87)$$

Now we multiply Eq. (7.84) by the function $\psi(x, t; v - u)$ using the scalar product of the real Hilbert space $L^2(D_T)$ and integrate by parts taking into account the conditions (7.82), (7.83), (7.85), and (7.86). As a result, we obtain the following equality:

$$-\int_0^1 z(x,T;u)\psi(x,T;v-u)\,dx + \int_0^T\int_0^1 \left[\frac{\partial\psi(x,t;v-u)}{\partial t} - \frac{\partial^2\psi(x,t;v-u)}{\partial x^2}\right.$$
$$\left. - \frac{\partial}{\partial x}\left(y(x,t;u)\psi(x,t;v-u)\right) - u(x,t)\psi(x,t;v-u)\right] z(x,t;u)\,dx\,dt = 0. \quad (7.88)$$

Indeed, the following equalities hold:

$$-\int_0^T\int_0^1 \frac{\partial z(x,t;u)}{\partial t}\psi(x,t;v-u)\,dx\,dt$$

$$= -\int_0^1 z(x,T;u)\psi(x,T;v-u)\,dx + \int_0^T\int_0^1 z(x,t;u)\frac{\partial\psi(x,t;v-u)}{\partial t}\,dx\,dt,$$

$$-\int_0^T\int_0^1 \frac{\partial^2 z(x,t;u)}{\partial x^2}\psi(x,t;v-u)\,dx\,dt$$

$$= \int_0^T\int_0^1 z(x,t,u)\frac{\partial^2\psi(x,t;v-u)}{\partial x^2}\,dx\,dt,$$

$$-\int_0^T\int_0^1 y(x,t;u)\frac{\partial z(x,t;u)}{\partial x}\psi(x,t;v-u)\,dx\,dt$$

$$= \int_0^T\int_0^1 z(x,t;u)\frac{\partial}{\partial x}\left(y(x,t;u)\psi(x,t;v-u)\right)\,dx\,dt.$$

Taking into account Eq. (7.81), from (7.88) we obtain

$$\int_0^1 z(x,T;u)\psi(x,T;v-u)\,dx = \int_0^T\int_0^1 z(x,t;u)y(x,t;u)[v(x,t)-u(x,t)]\,dx\,dt.$$

Thus, the following theorem is valid.

Theorem 7.5. *If the Fréchet derivative of the functional $J(u)$ defined by Eq. (7.77) exists, then it is equal to*

$$J'_f(u) = 2z(x,t;u)y(x,t;u), \qquad (7.89)$$

where the function $y(x,t;u)$ is a smooth solution of the problem (7.74)–(7.76) and the function $z(x,t;u)$ is a smooth solution of the problem (7.84)–(7.86).

Problem 7.1. Prove that if a control $u = u(x)$ is independent of time and is considered as an element of the Banach space $L^2(0,1)$, then the Fréchet derivative of the functional (7.77) by $u(x) \in L^2(0,1)$, if it exists, has the following form:

$$J_f'(u)(x) = 2 \int_0^T z(x,t;u) y(x,t;u) \, dt,$$

where the function $y(x,t;u)$ is a smooth solution of the problem (7.74)–(7.76) and the function $z(x,t;u)$ is a smooth solution of the problem (7.84)–(7.86).

7.4　Coefficient Control Problem for the Korteweg–de Vries equation

Example 7.6 (Final observation problem). Consider the mixed boundary-value problem for the Korteweg–de Vries equation on the segment $x \in [0,1]$:

$$\frac{\partial y(x,t;u)}{\partial t} + u(x,t)\frac{\partial y(x,t;u)}{\partial x}$$
$$+ \frac{\partial^3 y(x,t;u)}{\partial x^3} + \frac{1}{2}\frac{\partial y^2(x,t;u)}{\partial x} = 0, \quad (x,t) \in D_T \cup B_T, \tag{7.90}$$

$$y(0,t;u) = y_0(t), \; y(1,t;u) = y_1(t), \; y_x(1,t;u) = y_2(t), \quad t \in [0,T], \tag{7.91}$$
$$y(x,0;u) = y_3(x), \quad x \in (0,1). \tag{7.92}$$

We assume that $y_0(t), y_1(t), y_2(t) \in C^{(1)}[0,T]$, $y_3(x) \in C^{(3)}[0,1]$, and $u(x,t) \in C^{(1)}([0,1] \otimes [0,T])$ is a control. We assume that there exists a solution of the problem (7.90)–(7.92) lying in the class $y(x,t;u) \in C_{x,t}^{3,1}(\overline{D}_T)$. Consider the functional

$$J(u) = \int_0^1 (y(x,T;u) - h(x))^2 \, dx, \tag{7.93}$$

where $h(x) \in C[0,1]$ is a given function. The Fréchet derivative of this functional has the form (7.78), where the function $\psi(x,t;v-u) \in C_{x,t}^{3,1}(\overline{D}_T)$ is a solution of the following problem:

$$\frac{\partial \psi(x,t;v-u)}{\partial t} + u(x,t)\frac{\partial \psi(x,t;v-u)}{\partial x}$$
$$+ \frac{\partial(y(x,t;u)\psi(x,t;v-u))}{\partial x} + \frac{\partial^3 \psi(x,t;v-u)}{\partial x^3}$$
$$+ \frac{\partial y(x,t;u)}{\partial x}[v(x,t) - u(x,t)] = 0, \quad (x,t) \in D_T \cup B_T, \tag{7.94}$$

$$\psi(0,t;v-u) = \psi(1,t;v-u) = \psi_x(1,t;v-u) = 0, \quad t \in [0,T], \tag{7.95}$$
$$\psi(x,0;v-u) = 0, \quad x \in (0,1). \tag{7.96}$$

Indeed, the following equalities hold:

$$\left(u\frac{\partial y(x,t;u)}{\partial x} \right)'_f \cdot (v-u) = y'_f(x,t;u)(v-u) + u\frac{\partial \psi(x,t;v-u)}{\partial x},$$

$$\left(\frac{\partial}{\partial x}\left(y^2(x,t;u)\right) \right)'_f \cdot (v-u) = \frac{\partial}{\partial x}\left(2y(x,t;u)\psi(x,t;v-u)\right).$$

Also, we consider the adjoint problem:

$$\frac{\partial z(x,t;u)}{\partial t} + \frac{\partial(u(x,t)z(x,t;u))}{\partial x}$$

$$+ \frac{\partial^3 z(x,t;u)}{\partial x^3} + y(x,t;u)\frac{\partial z(x,t;u)}{\partial x} = 0, \quad (x,t) \in D_T \cup B_T, \qquad (7.97)$$

$$z(0,t;u) = z(1,t;u) = z_x(0,t;u) = 0, \quad t \in [0,T], \qquad (7.98)$$

$$z(x,T;u) = -(y(x,T;u) - h(x)), \qquad x \in (0,1). \qquad (7.99)$$

We assume that a solution $z(x,t;u)$ of this problem belongs to the class $C_{x,t}^{3,1}(\overline{D}_T)$. Multiply both sides of Eq. (7.97) by the function $\psi(x,t;v-u)$ and integrate by parts. Using the conditions (7.95), (7.96), (7.98), and (7.99), we obtain the following equality:

$$\int_0^1 z(x,T;u)\psi(x,T;v-u)\,dx$$

$$= \int_0^T \int_0^1 \left[\frac{\partial\psi(x,t;v-u)}{\partial t} + u(x,t)\frac{\partial\psi(x,t;v-u)}{\partial x} \right.$$

$$\left. + \frac{\partial(y(x,t;u)\psi(x,t;v-u))}{\partial x} + \frac{\partial^3\psi(x,t;v-u)}{\partial x^3} \right] z(x,t;u)\,dx\,dt$$

$$= -\int_0^T \int_0^1 \frac{\partial y(x,t;u)}{\partial x}z(x,t;u)[v(x,t) - u(x,t)]\,dx\,dt. \qquad (7.100)$$

Therefore, from (7.78), (7.87), and (7.99), we obtain the following assertion.

Theorem 7.6. *If the Fréchet derivative of the functional $J(u)$ defined by Eq. (7.77) exists, then it is equal to*

$$J'_f(u) = 2z(x,t;u)\frac{\partial y(x,t;u)}{\partial x}, \qquad (7.101)$$

where the function $y(x,t;u)$ is a smooth solution of the problem (7.90)–(7.92) and the function $z(x,t;u)$ is a smooth solution of the problem (7.97)–(7.99).

7.5 Bibliographical Notes

The contents of this lecture is taken from [47].

Lecture 8

Compact, Completely Continuous, and Totally Continuous Operators

8.1 Compact Operators

The importance of considering so-called compact operators is due to the fact that this notion is widely used in topological methods of functional analysis for generalizing the notion of the degree of a finite-dimensional mapping. Consider an operator

$$F : \mathbb{B}_1 \to \mathbb{B}_2,$$

where $\mathbb{B}_1$ and $\mathbb{B}_2$ are two Banach spaces with the norms $\|\cdot\|_1$ and $\|\cdot\|_2$, respectively, and the corresponding duality brackets $\langle \cdot, \cdot \rangle_1$ and $\langle \cdot, \cdot \rangle_2$.

Definition 8.1. An operator F is said to be *compact* if for any bounded set $B \subset \mathbb{B}_1$, the closure of the set $F(B) \subset \mathbb{B}_2$ is compact in $\mathbb{B}_2$.

Remark 8.1. Recall that, on one side, a set $B \subset \mathbb{B}$ of a Banach space $\mathbb{B}$ is said to be *compact* if any of its coverings by open sets

$$B \subset \bigcup_{\alpha \in A} U_\alpha, \quad U_\alpha \in \mathbb{B},$$

contains a finite subcovering

$$B \subset \bigcup_{k=1}^{M} U_{\alpha_k}, \quad \alpha_k \in A \text{ for all } k = \overline{1, M}, \, M \in \mathbb{N}.$$

On the other hand, the following definition of a compact set is "more natural" from the standpoint of Banach spaces (actually, metric spaces): a set $B \subset \mathbb{B}$ of a Banach space $\mathbb{B}$ is said to be *compact* if from any sequence $\{u_n\} \subset B$, one can extract a subsequence $\{u_{n_m}\} \subset \{u_n\}$, which strongly converges in $\mathbb{B}$.

Definition 8.1 is equivalent (in the case of Banach spaces) to the following definition.

Definition 8.2. An operator F is said to be *compact* if for any bounded sequence $\{u_n\} \subset \mathbb{B}_1$, one can extract a strongly converging subsequence $\{F(u_{n_m})\} \subset \mathbb{B}_2$ from the corresponding sequence $\{F(u_n)\} \subset \mathbb{B}_2$:

$$F(u_{n_m}) \to v \quad \text{strongly in } \mathbb{B}_2 \text{ as } n_m \to +\infty.$$

As a rule, the following narrower concept is used in applications.

Definition 8.3. An operator F is said to be *completely continuous* or *compact* if it is continuous and compact.

The notion of a totally continuous operator is very important in the study of nonlinear boundary-value problems.

Definition 8.4. An operator F is said to be *totally continuous* if the condition

$$u_n \rightharpoonup u \quad \text{weakly in } \mathbb{B}_1$$

implies

$$F(u_n) \to F(u) \quad \text{strongly in } \mathbb{B}_2.$$

Definition 8.5. Probably, the term "total continuity" is not common. Thus, operators called *totally continuous* in the present monograph, are said to be *completely continuous* in [23]. As the authors know, the common usage of the term *complete continuity* meets Definition 8.3. So, we adhere to the terminology specified in Definitions 8.1–8.4.

Naturally, the question on the relationship between the notions of complete continuity and total continuity arises. The following theorem partially answers this question.

Theorem 8.1. *Any completely continuous operator* $L \in \mathscr{L}(\mathbb{B}_1, \mathbb{B}_2)$ *is totally continuous.*

Proof. 1. Let $u_n \rightharpoonup u$ weakly in $\mathbb{B}_1$; then this sequence is bounded in $\mathbb{B}_1$. Due to the compactness of L, we can extract from the sequence $\{u_n\}$ a subsequence $\{u_{n_k}\}$ such that

$$Lu_{n_k} \to v \quad \text{strongly in } \mathbb{B}_2 \text{ as } n_k \to +\infty.$$

Consider the transposed operator for L:

$$L^t : \mathbb{B}_2^* \to \mathbb{B}_1^*.$$

Since $L \in \mathscr{L}(\mathbb{B}_1, \mathbb{B}_2)$, i.e., it is linear and continuous, we conclude that $L^t \in \mathscr{L}(\mathbb{B}_2^*, \mathbb{B}_1^*)$; moreover, by the definition of the transposed operator (see Lecture 1), the following equality holds:

$$\langle L^t f^*, u \rangle_1 \overset{\text{def}}{=\!=} \langle f^*, Lu \rangle_2 \quad \text{for all } f^* \in \mathbb{B}_2^*, \, u \in \mathbb{B}_1.$$

We prove that

$$Lu_n \rightharpoonup Lu \quad \text{weakly in } \mathbb{B}_2.$$

Indeed, we have

$$\langle f^*, Lu_n - Lu \rangle_2 = \langle L^t f^*, u_n - u \rangle_1 \to 0 \quad \text{as } n \to +\infty,$$

since

$$u_n \rightharpoonup u \quad \text{weakly in } \mathbb{B}_1.$$

Thus, we conclude that

$$Lu_n \rightharpoonup Lu \quad \text{weakly in } \mathbb{B}_2. \tag{8.1}$$

Now we prove that actually

$$Lu_n \to Lu \quad \text{strongly in } \mathbb{B}_2.$$

Since the operator L is completely continuous and the sequence $\{u_n\}$ is bounded, there exists $v \in \mathbb{B}_2$ such that

$$Lu_{n_k} \to v \quad \text{strongly in } \mathbb{B}_2;$$

therefore,

$$Lu_{n_k} \rightharpoonup v \quad \text{weakly in } \mathbb{B}_2.$$

Thus, due to (8.1), we arrive at the equality $v = Lu$.

2. Now we assume that there exists a subsequence $\{u_{n_k}\} \subset \{u_n\}$ such that the inequality

$$\|Lu_{n_k} - Lu\|_2 \geq c > 0$$

holds for all $n_k \in \mathbb{N}$. On the other hand, this subsequence possesses a subsequence

$$\left\{u_{n_{k_l}}\right\} \subset \{u_{n_k}\}$$

such that

$$\left\|Lu_{n_{k_l}} - Lu\right\|_2 \to 0 \quad \text{as } l \to +\infty.$$

We obtain the chain of inequalities

$$0 < c \leq \|Lu_{n_k} - Lu\|_2 \leq \left\|Lu_{n_k} - Lu_{n_{k_l}}\right\|_2 + \left\|Lu_{n_{k_l}} - Lu\right\|_2.$$

We choose $l \in \mathbb{N}$ so large that the following inequality holds:

$$\left\|Lu_{n_{k_l}} - Lu\right\|_2 \leq \frac{c}{2}.$$

On the other hand, for each $l \in \mathbb{N}$ there exists $n_k \in \mathbb{N}$ such that

$$n_k = n_{k_l} \quad \Rightarrow \quad u_{n_k} = u_{n_{k_l}} \quad \Rightarrow \quad Lu_{n_{k_l}} = Lu_{n_k}$$

and hence

$$\left\|Lu_{n_k} - Lu_{n_{k_l}}\right\|_2 = 0;$$

we arrive at the contradiction $0 < c \leq c/2$, which proved the theorem. $\qquad \square$

Note that the converse assertion is in general invalid.

Example 8.1. As is known, the space

$$l_1 \overset{\text{def}}{=\!=\!=} \left\{ \{x_k\}_{k=1}^{+\infty} : \sum_{k=1}^{+\infty} |x_k| < +\infty, \ x_k \in \mathbb{C}^1 \right\}$$

possesses the Schur property, i.e., the condition

$$u_n = \{x_{nk}\}_{k=1}^{+\infty} \rightharpoonup u = \{x_k\}_{k=1}^{+\infty} \quad \text{weakly in } l_1 \text{ as } n \to +\infty$$

$$\Leftrightarrow \quad \langle f, u_n \rangle = \sum_{k=1}^{+\infty} f_k x_{nk} \to \langle f, u \rangle = \sum_{k=1}^{+\infty} f_k x_k \quad \text{as } n \to +\infty$$

for each fixed

$$f \in l_\infty \overset{\text{def}}{=\!=\!=} \left\{ \{f_k\}_{k=1}^{+\infty} : \sup_{k=\overline{1,+\infty}} |f_k| < +\infty \right\}$$

implies that

$$u_n \to u \quad \text{strongly in } l_1 \quad \Leftrightarrow \quad |u_n - u|_1 = \sum_{k=1}^{+\infty} |x_{nk} - x_k| \to +0 \quad \text{as } n \to +\infty.$$

Therefore, the identity operator id : $l_1 \to l_1$ is totally continuous but, obviously, is not compact.

However, under the additional condition of the reflexivity of the Banach space $\mathbb{B}_1$[1] the total continuity implies the complete continuity even for a *nonlinear operator*

$$L : \mathbb{B}_1 \to \mathbb{B}_2.$$

Namely, the following assertion is valid.

Theorem 8.2. *Let* $K : \mathbb{B}_1 \to \mathbb{B}_2$ *be a totally continuous operator and the Banach space* $\mathbb{B}_1$ *is reflexive. Then the operator* K *is completely continuous.*

Proof. 1. First, we prove the continuity of the operator K. Indeed, let

$$u_n \to u \quad \text{strongly in } \mathbb{B}_1 \text{ as } n \to +\infty.$$

Then, obviously,

$$u_n \rightharpoonup u \quad \text{weakly in } \mathbb{B}_1 \text{ as } n \to +\infty.$$

Due to the total continuity of the operator K, we conclude that

$$K(u_n) \to K(u) \quad \text{strongly in } \mathbb{B}_2 \text{ as } n \to +\infty.$$

Thus, the continuity of the operator K is proved.

---

[1] Note that the Banach space l_1 is not reflexive.

2. Now we prove the complete continuity of the operator K. Let $D \subset \mathbb{B}_1$ be a bounded set and let $\{u_n\} \subset D$. Then (see Theorem 2.1), owing to the reflexivity of $\mathbb{B}_1$, one can extract from this sequence a subsequence $\{u_{n_k}\} \subset \{u_n\}$ such that

$$u_{n_k} \rightharpoonup u \quad \text{weakly in } \mathbb{B}_1 \text{ as } n_k \to +\infty.$$

Therefore, due to the total continuity of the operator K, we conclude that

$$K(u_{n_k}) \to K(u) \quad \text{strongly in } \mathbb{B}_2 \text{ as } n_k \to +\infty.$$

Thus, the complete continuity of the operator K is proved. $\qquad\qquad\qquad \square$

Note that the converse assertion in general is invalid.

Example 8.2. Let $\mathbb{B}_1 = L^2(0,1)$ and $\mathbb{B}_2 = \mathbb{R}_1$. Prove that the nonlinear operator

$$K(u) := \int_0^1 u^2(s)\,ds = \|u\|_2^2$$

is completely continuous. First, we prove its continuity. Let

$$u_n \to u \quad \text{strongly in } L^2(0,1).$$

Due to the obvious inequality

$$\big|\|u_n\|_2 - \|u\|_2\big| \le \|u_n - u\|_2$$

we conclude that

$$\|u_n\|_2 \to \|u\|_2 \quad \text{as } n \to +\infty;$$

therefore,

$$K(u_n) \to K(u) \quad \text{as } n \to +\infty.$$

Now we prove the complete continuity of the operator K. Let $D \subset L^2(0,1)$ be an arbitrary bounded set. We prove that $\overline{K(D)}$ is compact in $\mathbb{R}^1$. For this, it suffices to prove that $K(D)$ is a bounded set. Due to the boundedness of D in $L^2(0,1)$, for all $u \in D$ we have the inequality

$$\|u\|_2 \le c,$$

where $c > 0$ is independent of u. Then $0 < K(u) \le c^2 < +\infty$. The complete continuity of the operator K is proved.

However, the operator K is not totally continuous. Indeed, consider the following sequence $\{u_n\} \subset L^2(0,1)$:

$$u_n(s) := \sin(\pi n s), \quad s \in (0,1), \quad n \in \mathbb{N}.$$

By the Riemann–Lebesgue theorem, for any fixed function $v(s) \in L^2(0,1)$ we have

$$\int_0^1 v(s)\sin(\pi n s)\,ds \to 0 \quad \text{as } n \to +\infty,$$

i.e.,

$$u_n \rightharpoonup 0 \quad \text{weakly in } L^2(0,1) \text{ as } n \to +\infty.$$

However,

$$K(u_n) = \int_0^1 u_n^2(s)\,ds = \frac{1}{2} \not\to 0 = K(0) \quad \text{as } n \to +\infty.$$

8.2 Compact Sets: A Reminder

We recall the Hausdorff precompactness criterion (without a proof) based on the notion of an ε-net.

Let $\mathbb{B}$ be a Banach space, and $M \subset \mathbb{B}$, and $\varepsilon > 0$.

Definition 8.6. A set $M_\varepsilon \subset \mathbb{B}$ is called an ε-net of the set M if for any point $x \in M$, there exists a point $x_\varepsilon \in M_\varepsilon$ such that $\|x - x_\varepsilon\| < \varepsilon$.

The notion of an ε-net of a set M admits the following geometric interpretation. Let M_ε be an ε-net of M and $x_\varepsilon \in M_\varepsilon$. Take the ball

$$O(x_\varepsilon, \varepsilon) := \{x \in \mathbb{B} : \|x - x_\varepsilon\| < \varepsilon\}.$$

Then

$$M \subset \bigcup_{x_\varepsilon \in M_\varepsilon} O(x_\varepsilon, \varepsilon),$$

i.e., M is contained in the union of balls of radius $\varepsilon > 0$ centered at $x_\varepsilon \in M_\varepsilon$. In other words, the collection of these balls covers M.

We say that an ε-net is finite if M_ε is a finite set, i.e., consists of a finite number of elements.

Definition 8.7. A set $B \subset \mathbb{B}$ of a Banach space $\mathbb{B}$ is said to be precompact (or *relatively compact*) if its completion is compact.

Theorem 8.3 (Hausdorff precompact criterion). *A set M is a Banach space $\mathbb{B}$. It is precompact if and only if for any $\varepsilon > 0$, there exists a finite ε-net M_ε in $\mathbb{B}$.*

Lemma 8.1. *A subset $K \subset \mathbb{B}$ is precompact if for any $\varepsilon > 0$, there exists a precompact set $K_\varepsilon \subset \mathbb{B}$ such that for any $u \in K$, there exists $u_\varepsilon \in K_\varepsilon$ such that*

$$\|u - u_\varepsilon\| < \varepsilon.$$

Proof. Let $\varepsilon > 0$ be an arbitrary fixed number. By the condition of the lemma, there exists a precompact set $K_{\varepsilon/2} \subset \mathbb{B}$ such that the condition of the lemma also holds for the number $\varepsilon/2$. Due to the Hausdorff precompactness criterion, this means that there exist points

$$u_\varepsilon^k \in \mathbb{B}, \quad k = \overline{1, n}, \quad n = n(\varepsilon) \in \mathbb{N},$$

such that

$$K_{\varepsilon/2} \subset \bigcup_{k=1}^{n} O(u_\varepsilon^k, \varepsilon/2), \tag{8.2}$$

where

$$O(u_\varepsilon^k, \varepsilon/2) := \left\{u \in \mathbb{B} : \|u - u_\varepsilon^k\| < \frac{\varepsilon}{2}\right\}.$$

On the one hand, by the condition of the lemma, for each $u \in K$ there exists $u_{\varepsilon/2} \in K_{\varepsilon/2}$ such that

$$\|u - u_{\varepsilon/2}\| < \frac{\varepsilon}{2}. \tag{8.3}$$

On the other hand, due to (8.2), for $u_{\varepsilon/2}$ there exists $k_0 \in \overline{1,n}$ such that

$$\left\|u_{\varepsilon/2} - u_\varepsilon^{k_0}\right\| < \frac{\varepsilon}{2}.$$

This and (8.3) imply that

$$\|u - u_\varepsilon^{k_0}\| \le \left\|u_{\varepsilon/2} - u_\varepsilon^{k_0}\right\| + \|u - u_{\varepsilon/2}\| < \frac{\varepsilon}{2} + \frac{\varepsilon}{2} = \varepsilon.$$

This means that

$$K \subset \bigcup_{k=1}^{n} O(u_\varepsilon^k, \varepsilon),$$

i.e., the set K is precompact by the Hausdorff criterion. $\qquad\square$

8.3 Completely Continuous Operators

Definition 8.8. A mapping $F : \mathbb{B}_1 \to \mathbb{B}_2$ is said to be bounded if the image of any bounded subset of the space $\mathbb{B}_1$ is a bounded subset of the space $\mathbb{B}_2$.

Theorem 8.4. *Let $\mathbb{B}_1$ and $\mathbb{B}_2$ be Banach space, $D \subset \mathbb{B}_1$ be a bounded set, and*

$$F : D \to \mathbb{B}_2$$

be a certain mapping. Then the following two conditions are equivalent:

(i) *F is a completely continuous mapping;*

(ii) *for any $\varepsilon > 0$, there exists a bounded continuous mapping*

$$F_\varepsilon : D \to \mathbb{B}_2,$$

such that $\dim(\operatorname{span} F_\varepsilon(D)) < +\infty$ and

$$\|F(u) - F_\varepsilon(u)\|_2 < \varepsilon \quad \text{for all } u \in D. \tag{8.4}$$

Proof. 1. First, we prove that (i) implies (ii). Indeed, let F be a completely continuous mapping. Then due to the boundedness of $D \subset \mathbb{B}_1$, the set $F(D)$ is precompact in $\mathbb{B}_2$. Obviously, then the set $\overline{F(D)}$ is also precompact since

$$\overline{\overline{F(D)}} = \overline{F(D)}.$$

Therefore, for each $\varepsilon > 0$, there exist points $v_\varepsilon^k \in \mathbb{B}_2$, $k = \overline{1,n}$, such that

$$\overline{F(D)} \subset \bigcup_{k=1}^{n} O(v_\varepsilon^k, \varepsilon), \tag{8.5}$$

where

$$O(v_\varepsilon^k, \varepsilon) := \left\{ v \in \mathbb{B}_2 : \|v - v_\varepsilon^k\|_2 < \varepsilon \right\}.$$

Introduce the following functions:

$$f_k(v) := \max\left\{\varepsilon - \|v - v_\varepsilon^k\|_2,\ 0\right\}.$$

Consider the function

$$\overline{f}_m(v) := \begin{cases} f_m(v) \Big/ \sum\limits_{k=1}^{n} f_k(v), & f_m(v) \neq 0; \\ 0, & f_m(v) = 0, \end{cases} \tag{8.6}$$

for $m \in \overline{1,n}$ and for all $v \in \overline{F(D)}$. Introduce the mapping $F_\varepsilon(u)$ as follows:

$$F_\varepsilon(u) := \sum_{m=1}^{n} \overline{f}_m(F(u)) v_\varepsilon^m \quad \text{for all } u \in D.$$

The boundedness of this mapping for each fixed $\varepsilon > 0$ is obvious. Prove the continuity. By the construction of the function (8.6), the function

$$\overline{f}_m = \overline{f}_m(f_1, \ldots, f_n) \quad \text{for } m = \overline{1,n}$$

is jointly continuous by the real variables $f_k \in \mathbb{R}^1$ on the set $\{(f_1, \ldots, f_n) \in \mathbb{R}^n \mid f_k \geq 0, \sum\limits_{k=1}^{n} f_k > 0\}$ and the function $f_k = f_k(v)$ is continuous for all $v \in \overline{F(D)}$. Indeed, on the one hand, the function of the real variable $x^+ := \max\{x, 0\}$ is continuous for all $x \in \mathbb{R}^1$. On the other hand, the function

$$g_k(v) := \varepsilon - \|v - v_k^\varepsilon\|_2$$

is continuous as a function $\mathbb{B}_2 \to \mathbb{R}^1$ since if a sequence $\{v_l\}_{l=1}^{+\infty} \subset \mathbb{B}_2$ is such that

$$v_l \to v \quad \text{strongly in } \mathbb{B}_2 \text{ as } l \to +\infty,$$

then the following limit property holds:

$$|g_k(v_l) - g_k(v)| = \big|\|v_l - v_k^\varepsilon\|_2 - \|v - v_k^\varepsilon\|_2\big| \leq \|v_l - v\| \to +0$$

as $l \to +\infty$. The function $f_k(v)$ is continuous on $\overline{F(D)}$ as the composition of continuous function.

Now we note that for any $v \in \overline{F(D)}$, there exists $k_0 \in \overline{1,n}$ such that $v \in O(v_\varepsilon^{k_0}, \varepsilon)$, i.e., if

$$\|v - v_\varepsilon^{k_0}\| < \varepsilon,$$

then $f_{k_0}(v) > 0$. Therefore,

$$\sum_{k=1}^{n} f_k(v) > 0 \quad \text{for all } v \in \overline{F(D)}.$$

The function $\overline{f}_m = \overline{f}_m(f_1, \ldots, f_n)$ is continuous at each point where $\sum\limits_{k=1}^{n} f_k > 0$.

Finally, by the condition of the lemma, the operator F is continuous on $D \subset \mathbb{B}_1$. Therefore, by the theorem on the composition of continuous mappings, the operator $F_\varepsilon(u)$ is continuous. Moreover,

$$\operatorname{span} F_\varepsilon(D) \subset \operatorname{span}\{v_\varepsilon^1, \ldots, v_\varepsilon^n\},$$

i.e., $F_\varepsilon(u)$ is a finite-dimensional operator. By (8.5) and (8.6), we have

$$\|F(u) - F_\varepsilon(u)\|_2 = \left\| \sum_{m=1}^{n} \overline{f}_m(F(u))F(u) - \sum_{m=1}^{n} \overline{f}_m(F(u))v_\varepsilon^m \right\|_2$$

$$\leq \sum_{m=1}^{n} \overline{f}_m(F(u))\|F(u) - v_\varepsilon^m\|_2.$$

Note that if

$$\|F(u) - v_\varepsilon^m\|_2 \geq \varepsilon,$$

then

$$f_m(F(u)) = \max\{\varepsilon - \|F(u) - v_\varepsilon^m\|_2, 0\} = 0.$$

Therefore, the following inequality holds:

$$\sum_{m=1}^{n} \overline{f}_m(F(u))\|F(u) - v_\varepsilon^m\|_2 < \frac{1}{\sum\limits_{m=1}^{n} f_m(F(u))} \sum_{m=1}^{n} \varepsilon f_m(F(u)) = \varepsilon.$$

Thus, the following inequality holds:

$$\|F(u) - F_\varepsilon(u)\|_2 < \varepsilon \quad \text{for all } u \in D.$$

2. Now we prove that (ii) implies (i).

On the one hand, as $\varepsilon \to +0$, the uniform limit of F_ε is a mapping F, which is *continuous* and *bounded*. Indeed, for any $\varepsilon > 0$, due to the continuity of the mapping $F_{\varepsilon/3}$, there exists $\delta = \delta(\varepsilon) > 0$ such that the inequality

$$\|F_{\varepsilon/3}(u_1) - F_{\varepsilon/3}(u_2)\|_2 < \frac{\varepsilon}{3}$$

holds for all

$$\|u_1 - u_2\|_1 < \delta, \quad u_1, u_2 \in D.$$

Thus, due to (8.4), we arrive at the inequality

$$\|F(u_1) - F(u_2)\|_2$$
$$= \|F(u_1) - F_{\varepsilon/3}(u_1) + F_{\varepsilon/3}(u_1) - F_{\varepsilon/3}(u_2) + F_{\varepsilon/3}(u_2) - F(u_2)\|_2$$
$$\leq \|F(u_1) - F_{\varepsilon/3}(u_1)\|_2 + \|F_{\varepsilon/3}(u_1) - F_{\varepsilon/3}(u_2)\|_2$$
$$+ \|F(u_2) - F_{\varepsilon/3}(u_2)\|_2 < \frac{\varepsilon}{3} + \frac{\varepsilon}{3} + \frac{\varepsilon}{3} = \varepsilon.$$

On the other hand, due to (ii), the following inequality holds:

$$\|F(u) - F_\varepsilon(u)\|_2 < \varepsilon \quad \text{for all } u \in D.$$

The set $F_\varepsilon(D)$ is precompact as a bounded[2] set of finite dimension; therefore, by Lemma 8.1, we see that $F(D)$ is precompact in $\mathbb{B}_2$. Therefore, F is a *compact* mapping.

Finally, we conclude that the mapping F is *completely continuous*. $\qquad\square$

8.4   Bibliographical Notes

The contents of this lecture is taken from [14, 23, 24].

---

[2] Due to the boundedness of the set D and the mapping F_ε.

Local Invertibility Theorem

9.1 Space $\text{Isom}(X; Y)$

Definition 9.1. We denote by $\text{Isom}(X; Y)$ the set of all linear continuous isomorphisms of a Banach space X onto a Banach space Y.

Remark 9.1. The set $\text{Isom}(X; Y)$ is empty if Banach spaces X and Y are not isomorphic. Obviously, $\text{Isom}(X; Y) \subset \mathscr{L}(X, Y)$.

Lemma 9.1 (Lemma on the Neumann series). *Let X be a Banach space and let $A \in \mathscr{L}(X, X)$ be an operator such that $\|A\| < 1$. Then the operator $\text{id} - A$ is invertible, $(\text{id} - A)^{-1} \in \mathscr{L}(X, X)$, and*

$$(\text{id} - A)^{-1} = \sum_{n=0}^{+\infty} A^n. \tag{9.1}$$

Proof. 1. *Convergence of the series.* If X is a Banach space, then $\mathscr{L}(X, X)$ is also a Banach space with respect to the norm

$$\|A\| := \sup_{\|x\| \leq 1} \|A(x)\|. \tag{9.2}$$

The definition (9.2) of the operator norm immediately implies that for any $A, B \in \mathscr{L}(X, X)$, the inequality

$$\|A \cdot B\| \leq \|A\| \|B\| \tag{9.3}$$

holds. In particular, from (9.3) we obtain

$$\|A^n\| \leq \|A\|^n \quad \text{for any } n \in \mathbb{N}. \tag{9.4}$$

Consider the partial sum

$$S_N := \sum_{n=0}^{N} A^n. \tag{9.5}$$

Then due to (9.4), introducing the notation

$$S_{N+p} - S_N := \sum_{n=N+1}^{N+p} A^n$$

we see that

$$\|S_{N+p} - S_N\| \le \sum_{n=N+1}^{N+p} \|A^n\| \le \sum_{n=N+1}^{N+p} \|A\|^n \to +0 \quad \text{as } N, p \to +\infty, \qquad (9.6)$$

since $\|A\| < 1$ and hence the numerical series

$$\sum_{n=0}^{+\infty} \|A\|^n$$

converges. Therefore, $\{S_N\} \subset \mathscr{L}(X, X)$ is a Cauchy sequence and, due to the completeness of $\mathscr{L}(X, X)$ with respect to the operator norm (9.2), we obtain the limit relation

$$S_N := \sum_{n=0}^{N} A^n \to S_\infty := \sum_{n=0}^{+\infty} A_n \in \mathscr{L}(X, X) \quad \text{strongly in } \mathscr{L}(X, X) \qquad (9.7)$$

as $N \to +\infty$.

2. Note that the following equalities hold:

$$(\mathrm{id} - A)S_N = S_N(\mathrm{id} - A) = \mathrm{id} - A^{N+1}. \qquad (9.8)$$

Thus, we obtain the following estimates:

$$\|(\mathrm{id} - A)S_N - \mathrm{id}\| \le \|A\|^{N+1}, \quad \|S_N(\mathrm{id} - A) - \mathrm{id}\| \le \|A\|^{N+1}. \qquad (9.9)$$

Since $\|A\| < 1$, we obtain from (9.9) the following limit relations:

$$(\mathrm{id} - A)S_N \to \mathrm{id} \quad \text{strongly in } \mathscr{L}(X, X), \qquad (9.10)$$
$$S_N(\mathrm{id} - A) \to \mathrm{id} \quad \text{strongly in } \mathscr{L}(X, X) \qquad (9.11)$$

as $N \to +\infty$. On the other hand, due to (9.7) we have

$$(\mathrm{id} - A)S_N \to (\mathrm{id} - A)S_\infty \quad \text{strongly in } \mathscr{L}(X, X), \qquad (9.12)$$
$$S_N(\mathrm{id} - A) \to S_\infty(\mathrm{id} - A) \quad \text{strongly in } \mathscr{L}(X, X) \qquad (9.13)$$

as $N \to +\infty$. Using (9.10)–(9.13), we conclude that

$$(\mathrm{id} - A)S_\infty = S_\infty(\mathrm{id} - A) = \mathrm{id} \quad \Rightarrow \quad (\mathrm{id} - A)^{-1} = \sum_{n=0}^{+\infty} A^n. \qquad (9.14)$$

Lemma 9.1 is proved.   □

This lemma implies the following theorem.

Theorem 9.1. *Let X and Y be Banach spaces. Then the following assertions are valid:*

(a) *the set $\mathrm{Isom}(X; Y)$ is open in $\mathscr{L}(X, Y)$;*

(b) *the mapping $A \to A^{-1}$ of the set $\mathrm{Isom}(X; Y)$ into the space $\mathscr{L}(Y, X)$ is continuous.*

Proof. Let $\mathrm{Isom}(X;Y) \neq \varnothing$ (otherwise there is nothing to prove). Let $A_0 \in \mathrm{Isom}(X;Y)$.

(a) Prove that any operator $A \in \mathcal{L}(X;Y)$, which is close to A_0 with respect to the operator norm, belongs to $\mathrm{Isom}(X;Y)$.

Note that $A : X \to Y$ is an isomorphism if and only if

$$A_0^{-1}A : X \to X$$

is an isomorphism. By Lemma 9.1, the operator $A_0^{-1}A$ is an isomorphism if there exists an operator $B \in \mathcal{L}(X,X)$ such that

$$A_0^{-1}A = \mathrm{id} - B, \quad \|B\| < 1. \tag{9.15}$$

Then

$$B = A_0^{-1}(A_0 - A) \quad \Rightarrow \quad \|B\| \leq \|A_0^{-1}\| \|A_0 - A\|. \tag{9.16}$$

Thus,

$$\|A - A_0\| < \frac{1}{\|A_0^{-1}\|} \quad \Rightarrow \quad \|B\| < 1. \tag{9.17}$$

Therefore, under the condition

$$A \in O(A_0, \varepsilon) \in \mathcal{L}(X,Y), \quad \varepsilon = \frac{1}{\|A_0^{-1}\|}, \tag{9.18}$$

we have $A_0^{-1}A \in \mathrm{Isom}(X;X)$ and hence $A \in \mathrm{Isom}(X;Y)$.

(b) Let the inequality (9.17) be fulfilled; then due to (9.15), the following equalities holds:

$$A = A_0(\mathrm{id} - B) \quad \Rightarrow \quad A^{-1} = (\mathrm{id} - B)^{-1} A_0^{-1}$$
$$\Rightarrow \quad A^{-1} - A_0^{-1} = \left[(\mathrm{id} - B)^{-1} - \mathrm{id} \right] A_0^{-1}. \tag{9.19}$$

Since $\|B\| < 1$, by Lemma 9.1 we have

$$\|(\mathrm{id} - B)^{-1} - \mathrm{id}\| \leq \sum_{n=1}^{+\infty} \|B\|^n = \frac{\|B\|}{1 - \|B\|}. \tag{9.20}$$

Thus, from (9.19) and (9.20) we conclude that

$$\|A^{-1} - A_0^{-1}\| \leq \|A_0^{-1}\| \frac{\|B\|}{1 - \|B\|} \to +0 \tag{9.21}$$

as $\|A - A_0\| \to +0$ due to the inequality (9.16). $\qquad\square$

9.2 Auxiliary Results

Theorem 9.2. *Let X and Y be Banach spaces, $U \subset X$ and $V \subset Y$ be open sets, and $F : U \to V$ be a homeomorphism Fréchet-differentiable at a point $a \in U$. The mapping $G = F^{-1} : V \to U$ is Fréchet-differentiable at the point $b = F(a)$ if and only if $F_f'(a) \in \mathrm{Isom}(X;Y)$. Moreover, the following equality holds:*

$$G_f'(b) = (F_f'(a))^{-1}. \tag{9.22}$$

Proof. 1. *Necessity.* Let the mapping G be Fréchet-differentiable at the point $b = F(a) \in V$. The following equalities hold:

$$G \cdot F = \text{id}_X, \quad F \cdot G = \text{id}_Y. \tag{9.23}$$

Using the chain rule for Fréchet derivatives, we obtain

$$G'_f(b) \cdot F'_f(a) = \text{id}_X, \quad F'_f(a) \cdot G'_f(b) = \text{id}_Y, \tag{9.24}$$

i.e., $F'_f(a) \in \text{Isom}(X; Y)$ and Eq. (9.22) holds.

2. *Sufficiency.* Let $F'_f(a) \in \text{Isom}(X; Y)$. We prove that the mapping $G = F^{-1}$ is Fréchet-differentiable at the point $b = F(a)$.

Since the mapping F is differentiable at the point a, setting $y = F(x)$, we obtain the following formula (see Definition 5.2 of the Fréchet differentiability):

$$y - b = F(x) - F(a) = F'_f(a)(x - a) + \omega(a, x - a), \tag{9.25}$$

$$\omega(a, x - a) = \|x - a\| \varphi(x - a), \quad \lim_{\|x - a\| \to 0} \|\varphi(x - a)\| = 0. \tag{9.26}$$

Applying the operator $(F'_f(a))^{-1}$ to both sides of Eq. (9.25), we obtain the following equality:

$$x - a = (F'_f(a))^{-1}(y - b) - \|x - a\| \psi(x - a), \tag{9.27}$$

$$\psi(x - a) := (F'_f(a))^{-1} \varphi(x - a). \tag{9.28}$$

To complete the proof, we must verify that

$$\|\psi(x - a)\| = o(\|y - b\|). \tag{9.29}$$

Indeed, due to (9.26) we have

$$\lim_{\|x - a\| \to 0} \|\psi(x - a)\| = 0. \tag{9.30}$$

Equality (9.27) implies the inequality

$$\|x - a\| \leq \|(F'_f(a))^{-1}(y - b)\| + \|x - a\| \|\psi(x - a)\|, \tag{9.31}$$

and, consequently,

$$\|(F'_f(a))^{-1}(y - b)\| \geq \|x - a\|(1 - \|\psi(x - a)\|). \tag{9.32}$$

By (9.30), for sufficiently small $\|x - a\|$, we have the inequality

$$\|\psi(x - a)\| < 1. \tag{9.33}$$

Thus, there exists a neighborhood $O(a, \varepsilon) \subset X$, $\varepsilon > 0$, such that for all $x \in O(a, \varepsilon)$, the following inequality holds:

$$\|x - a\| \leq \|y - b\| \frac{\|(F'_f(a))^{-1}\|}{1 - \|\psi(x - a)\|} \quad \text{for all } x \in O(a, \varepsilon). \tag{9.34}$$

On the one hand, this implies that

$$\|x - a\| \|\psi(x - a)\| \leq \|y - b\| \frac{\|(F'_f(a))^{-1}\|}{1 - \|\psi(x - a)\|} \|\psi(x - a)\| \tag{9.35}$$

for all $x \in O(a, \varepsilon)$. On the other hand, from (9.34) and (9.30) we conclude that

$$\lim_{\|y-b\|\to 0} \|\psi(x-a)\| = 0. \tag{9.36}$$

Therefore, Eq. (9.29) is fulfilled.

The formulas (9.27) and (9.29) imply the following equality:

$$G(y) - G(b) = (F_f'(a))^{-1}(y-b) + \omega(b, y-b), \tag{9.37}$$

$$\lim_{\|y-b\|\to +0} \frac{\|\omega(b, y-b)\|}{\|y-b\|} = 0. \tag{9.38}$$

By the definition of the Fréchet derivative, we obtain Eq. (9.22). □

Definition 9.2. A mapping $F : U \subset X \to Y$ is called a *mapping of class C^1* if the following conditions hold:

(1) the mapping F is Fréchet-different on U;
(2) the mapping $F_f'(\cdot) : U \to \mathscr{L}(X, Y)$ is continuous at each point of U.

Theorem 9.3. *Let $F : U \to V$ be a homeomorphism of class C^1, where U is an open set in the Banach space X and V be an open set in Y. The mapping F is a diffeomorphism of class C^1 if and only if $F_f'(u) \in \mathrm{Isom}(X; Y)$ for any $u \in U$.*

Proof. The necessity is obvious. Prove the sufficiency. Let $F_f'(u) \in \mathrm{Isom}(X; Y)$ for any $u \in U$. By Theorem 9.2, for each $v = F(u) \in V$, there exists the Fréchet derivative $G_f'(v)$ and the following equality holds:

$$G_f'(v) = \big(F_f'(G(v))\big)^{-1}. \tag{9.39}$$

It remains to verify that the mapping G belongs to the class C^1, i.e., the mapping $G_f'(\cdot) : V \to \mathscr{L}(Y, X)$ is continuous at each point $v \in V$.

Indeed, Eq. (9.39) implies that the mapping $G_f'(\cdot)$ is the composition of the following three *continuous mappings*:

1° the mapping $v \to G(v)$ from the set $V \subset Y$ into the set $U \subset X$, which is continuous since $F : U \to V$ is a homeomorphism;
2° the mapping $u \to F_f'(u)$ from the set $U \subset X$ into the set $\mathrm{Isom}(X; Y)$, which is continuous since, by assumption, the operator F belongs to the class C^1;
3° the mapping $u \to u^{-1}$ from the space $\mathrm{Isom}(X; Y)$ into the space $\mathscr{L}(Y; X)$, which is continuous by Theorem 9.1. □

In the sequel, we need the following auxiliary assertion.

Lemma 9.2. *If a mapping $F : U \subset X \to Y$ is Fréchet differentiable of an open set U and the Fréchet derivative*

$$F_f'(\cdot) : U \to \mathscr{L}(X, Y)$$

is continuous at a point $x = a \in U$, then for any number $\varepsilon > 0$, there exists a number $r > 0$ such that the mapping

$$G(x) := F(x) - F(a) - F'_f(a)(x - a) \tag{9.40}$$

satisfies the Lipschitz inequality

$$\|G(x) - G(y)\| \le \varepsilon \|x - y\| \quad \text{for all } x, y \in O(a, r) \subset U, \tag{9.41}$$

where $O(a, r) := \{x \in X : \|x - a\| < r\}$ is the open ball in X of radius $r > 0$ centered at a point a.

Proof. 1. Consider the segment $[x, y] \subset U$. The following inequality holds:

$$\|G(x) - G(y)\| \le \|x - y\| \sup_{t \in [0,1]} \|G'_f(tx + (1-t)y)\|. \tag{9.42}$$

Indeed, the mapping $G(x)$ is Fréchet-differentiable in U. Then the following equalities hold:

$$G(x) - G(y) = \int_0^1 dt \frac{d}{dt} G(tx + (1-t)y) = \int_0^1 dt G'_f(tx + (1-t)y)(x - y). \tag{9.43}$$

2. The Fréchet derivative of $G(x)$ has the following form:

$$G'_f(x) = F'_f(x) - F'_f(a). \tag{9.44}$$

By the condition, the Fréchet derivative $F'_f(x)$ is continuous at the point $x = a$; therefore,

$$\lim_{\|x-a\| \to 0} \|G'_f(x)\| = 0. \tag{9.45}$$

Hence, for any $\varepsilon > 0$, there exists $r > 0$ such that

$$\|G'_f(x)\| \le \varepsilon \quad \text{for all } x \in O(a, r) \subset U. \tag{9.46}$$

It remains to apply the inequality (9.42).   □

Theorem 9.4. *If a mapping $F : O(a, r) \subset X \to X$ is continuous and the mapping $G(x) := x - F(x)$ satisfies the Lipschitz inequality*

$$\|G(x) - G(y)\| \le k\|x - y\| \quad \text{for all } x, y \in O(a, r) \tag{9.47}$$

for $k \in (0, 1)$, then there exists a neighborhood $a \in U \subset O(a, r)$ in the ball $O(a, r)$ such that F is a homeomorphism from U onto an open ball $O(b, (1 - k)r) \subset X$, $b = F(a)$. The inverse mapping

$$F^{-1} : O(b, (1 - k)r) \to O(a, r) \tag{9.48}$$

satisfies the Lipschitz condition with constant $1/(1 - k)$.

Proof. 1. Let $x, x' \in O(a, r)$. Then

$$F(x) - F(x') = (x - x') - (G(x) - G(x')), \tag{9.49}$$

and, using (9.47), we obtain the chain of inequalities

$$\|F(x) - F(x')\| \geq \|x - x'\| - \|G(x) - G(x')\|$$
$$\geq \|x - x'\| - k\|x - x'\| = (1 - k)\|x - x'\|, \quad k \in (0, 1). \tag{9.50}$$

2. Prove the following auxiliary lemma.

Lemma 9.3. *For any* $y \in O(b, (1 - k)r)$, *there exists a unique point* $x \in O(a, r)$ *such that* $y = F(x)$.

Proof. Let $y \in O(b, (1 - k)r)$. First, we prove the uniqueness. Indeed, if $y = F(x_1) = F(x_2)$, then due to the inequality (9.50) we obtain the equality $x_1 = x_2$. Now we prove the existence of the point $x \in O(a, r)$. Consider the following sequence $\{x_n\}$:

$$x_0 = a, \ x_1 = y + G(x_0), \ \ldots, \ x_{n+1} = y + G(x_n), \ \ldots \tag{9.51}$$

The following inequality holds:

$$\|x_n - a\| \leq \frac{1 - k^n}{1 - k}\|y - b\| \quad \text{for all } n \in \mathbb{N}. \tag{9.52}$$

Indeed, for $n = 1$ from (9.51) we obtain the equalities

$$x_1 - a = y + G(a) - a = y - F(a) = y - b \quad \Rightarrow \quad \|x_1 - a\| = \|y - b\|. \tag{9.53}$$

Assume that the inequality (9.52) holds for $n \in \mathbb{N}$. Prove that it also holds for $n + 1$. From (9.51) we obtain the equality

$$x_{n+1} - x_n = G(x_n) - G(x_{n-1}). \tag{9.54}$$

On the one hand, this immediately implies the following chain of inequalities:

$$\|x_{n+1} - x_n\| = \|G(x_n) - G(x_{n-1})\| \leq k\|x_n - x_{n-1}\|$$
$$= k\|G(x_{n-1}) - G(x_{n-2})\| \leq \cdots \leq k^n\|x_1 - x_0\| = k^n\|y - b\|. \tag{9.55}$$

On the other hand, we have

$$\|x_{n+1} - a\| \leq \|x_n - a\| + \|x_{n+1} - x_n\|$$
$$\leq \frac{1 - k^n}{1 - k}\|y - b\| + k^n\|y - b\| = \frac{1 - k^{n+1}}{1 - k}\|y - b\|. \tag{9.56}$$

First, the inequality (9.52) implies that for $y \in O(b, (1-k)r)$ the following inequality holds:

$$\|x_n - a\| < r \quad \Rightarrow \quad \{x_n\} \subset O(a, r).$$

Moreover, from (9.55) we conclude that the sequence $\{x_n\}$ is a Cauchy sequence in the Banach space X. Indeed, the following equality holds:

$$x_N - x_M = \pm \sum_{n=\min\{M,N\}}^{\max\{M,N\}-1} (x_{n+1} - x_n), \tag{9.57}$$

where one must choose "+" if $N > M$ and "−" if $N < M$; hence we obtain the estimate

$$\|x_N - x_M\| \le \sum_{n=\min\{M,N\}}^{\max\{M,N\}-1} \|x_{n+1} - x_n\|$$

$$\le \sum_{n=\min\{M,N\}}^{\max\{M,N\}-1} k^n \|y - b\| \to +0 \quad \text{as } \min\{M,N\} \to +\infty \qquad (9.58)$$

since $k \in (0,1)$.

Since X is a Banach space, we conclude that

$$x_n \to x \quad \text{strongly in } X \text{ as } n \to +\infty. \qquad (9.59)$$

Passing to the limit as $n \to +\infty$ in the inequality (9.52) and Eq. (9.51), we obtain the following inequality:

$$\|x - a\| \le \frac{1}{1-k}\|y - b\|, \quad x = y + G(x). \qquad (9.60)$$

By the condition, $y \in O(b,(1-k)r)$; therefore, (9.60) implies that

$$y = F(x), \quad x \in O(a,r). \qquad (9.61)$$

Lemma 9.3 is proved.   □

3. By Lemma 9.3, for any $y \in O(b,(1-k)r)$, there exists unique $x \in O(a,r)$ such that $y = F(x)$. Thus, there exists the inverse mapping

$$x = \Phi(y) : O(b,(1-k)r) \to O(a,r). \qquad (9.62)$$

From the inequality (9.50) we obtain the inequality

$$\|\Phi(y) - \Phi(y')\| \le \frac{1}{1-k}\|y - y'\| \qquad (9.63)$$

for all $y, y' \in O(b,(1-k)r)$. In particular, Φ is continuous on $O(b,(1-k)r)$.

Let $U := \Phi(O(b,(1-k)r)) \subset O(a,r)$. The set $U = F^{-1}(O(b,(1-k)r))$ is the preimage of the open set $O(b,(1-k)r)$. Since F is continuous, the set $U \subset O(a,r)$ is open in X due to the theorem on open mappings.   □

9.3   Local Invertibility Theorem

In this section, we eliminate from Theorem 9.3 the condition that the mapping $F : U \to V$ is a homeomorphism.

Theorem 9.5 (Local invertibility theorem).   *Let U be an open set in a Banach space X, $F : U \to Y$ be a mapping of class C^1, and Y be a Banach space. Assume that at a point $a \in U$ the mapping $F'_f(a)$ belongs to $\mathrm{Isom}(X;Y)$. Then there exists a neighborhood $U_1 \subset U$ of the point a and an open neighborhood V of the point $b = F(a)$ such that F is a C^1-diffeomorphism from U_1 on V and for $G := F^{-1}$, the following equality holds:*

$$G'_f(y) = \left(F'_f(G(y))\right)^{-1}. \qquad (9.64)$$

Proof. We partition the proof of Theorem 9.5 into several lemmas.

Lemma 9.4. *Let $U \subset X$ be an open set in a Banach space X and $F : U \to Y$ be a continuous mapping into a Banach space Y. Assume that the mapping F satisfies the following conditions:*

(i) *$F(x)$ is Fréchet-differentiable in U;*
(ii) *the Fréchet derivative $F'_f(x)$ is continuous at the point $x = a \in U$;*
(iii) *$F'_f(a) \in \mathrm{Isom}(X; Y)$.*

Then there exists an open neighborhoods $U' \subset U$ and V' of the points a and $b = F(a)$, respectively, such that F maps U' to V' homeomorphically.

Proof. 1. Consider the composition of the mappings

$$F_1(x) := (F'_f(a))^{-1} \cdot F(x) : \ U \subset X \to X. \tag{9.65}$$

Since $F(x)$ is Fréchet differentiable in U and the operator $(F'_f(a))^{-1}$ is linear, the mapping $F_1(x)$ is also Fréchet differentiable. The Fréchet derivative of the operator $F_1(x)$ has the following form:

$$F'_{1f}(x) = (F'_f(a))^{-1} F'_f(x) : \ U \to \mathscr{L}(Y, X), \quad F'_{1f}(a) = \mathrm{id}_X; \tag{9.66}$$

since $(F'_f(a))^{-1} \in \mathscr{L}(Y, X)$ and the operator $F'_f(x)$ is continuous at the point $x = a \in U$, the operator $F'_{1f}(x)$ is continuous at the point $x = a$. Therefore, by Lemma 9.2, for any $k \in (0, 1)$ there exists $r > 0$ such that the mapping

$$G_1(x) := F_1(x) - F_1(a) - F'_{1f}(a)(x - a) = F_1(x) - F_1(a) - x + a \tag{9.67}$$

satisfies the Lipschitz inequality

$$\|G_1(x) - G_1(y)\| \le k\|x - y\| \quad \text{for all } x, y \in O(a, r) \subset U. \tag{9.68}$$

Comparing the mapping $G(x) := x - F_1(x)$ with (9.67), we obtain from (9.68) the inequality

$$\|G(x) - G(y)\| \le k\|x - y\| \quad \text{for all } x, y \in O(a, r) \subset U \tag{9.69}$$

for $k \in (0, 1)$.

2. Thus, all the conditions of Theorem 9.4 for the mapping $F_1(x)$ are fulfilled, and hence there exists an open neighborhood $a \in U' \subset O(a, r)$ such that the mapping $F_1(x)$ homeomorphically maps the neighborhood U' onto a certain neighborhood of $b_1 = F_1(a) \in V_1$. The derivative $F'_f(a)$ lies in $\mathrm{Isom}(X; Y)$ and hence it is a homeomorphism from the Banach space X onto the Banach space Y. Therefore, the mapping

$$F(x) = F'_f(a) \cdot F_1(x)$$

is a homeomorphic mapping from the neighborhood $a \in U' \subset O(a, r) \subset U$ on the neighborhood $b = F(a) \in V' \subset Y$, where $V' := F'_f(a)V_1$. $\qquad\square$

(A) By the condition of the theorem, the mapping $F : U \subset X \to Y$ is a mapping of class C^1. Therefore, at each point $x \in U$, the mapping F has the Fréchet derivative

$$F'_f(\cdot) : U \to \mathscr{L}(X, Y),$$

which is continuous on U.

(B) By condition, $F'_f(a) \in \mathrm{Isom}(X; Y)$. By Theorem 9.1, the set $\mathrm{Isom}(X; Y)$ is open in $\mathscr{L}(X, Y)$. Then by (A) there exists a small open neighborhood $a \in U'' \subset U'$ such that

$$F'_f(x) \in \mathrm{Isom}(X; Y) \quad \text{for all } x \in U''. \tag{9.70}$$

(C) Let $V'' := F(U'') \subset Y$. Since F is a continuous mapping, the set V'' is open and by Lemma 9.4 the mapping $F : U'' \to V''$ is a homeomorphism.

(D) Due to (B) and (C) and Theorem 9.3, we conclude that $F : U'' \to V''$ is a diffeomorphism of class C^1. □

9.4 Two Examples of Application of the Local Invertibility Theorem

First, we consider a problem whose classical statement is as follows:

$$\Delta u + u^2 = f(x, t) \quad (x, t) \in D = \Omega \otimes [0, T], \tag{9.71}$$
$$u(x, t) = 0, \quad\quad (x, t) \in S = \partial\Omega \otimes [0, T], \tag{9.72}$$

where $\Omega \subset \mathbb{R}^3$ is a bounded domain with sufficiently smooth boundary $\partial\Omega$. The right-hand side $f(x, t)$ depends on the parameter $t \in [0, T]$.

Definition 9.3. A *weak solution* of the problem (9.71)–(9.72) is a function $u(x, t) \in H_0^1(\Omega)$, $t \in [0, T]$, satisfying the equation

$$\langle \Delta u + u^2 - f(t), \varphi \rangle = 0 \quad \text{for all } \varphi(x) \in H_0^1(\Omega), \, t \in [0, T], \tag{9.73}$$

where $f(t) \in C^{(1)}([0, T]; H^{-1}(\Omega))$ and $\langle \cdot, \cdot \rangle$ is the duality bracket between $H_0^1(\Omega)$ and $H^{-1}(\Omega)$.

Below, we will use the following notation:

$$O_1(\theta, r_1) := \{ v \in H_0^1(\Omega) : \|v\|_{H_0^1(\Omega)} < r_1 \},$$
$$O_2(\theta, r_2) := \{ v \in H^{-1}(\Omega) : \|v\|_{H^{-1}(\Omega)} < r_2 \}.$$

Theorem 9.6. *For any function $f(t) \in C^{(1)}([0, T]; H^{-1}(\Omega) \cap O_2(\theta, r_2))$ and sufficiently small $r_2 > 0$, there exists a unique weak solution $u(x, t) \in C^{(1)}([0, T]; H_0^1(\Omega) \cap O_1(\theta, r_1))$ in the sense of Definition 9.3 for some small $r_1 > 0$.*

Proof. We apply Theorem 9.5 on the local invertibility in the case where $X = H_0^1(\Omega)$, $Y = H^{-1}(\Omega)$, and $F(u) = \Delta u + u^2$.

1. Prove the validity of Definition 9.3. Since $\Omega \subset \mathbb{R}^3$ is a bounded domain, the following continuous and dense embeddings hold:

$$H_0^1(\Omega) \overset{ds}{\subset} L^3(\Omega) \overset{ds}{\subset} L^{3/2}(\Omega) \overset{ds}{\subset} H^{-1}(\Omega) \tag{9.74}$$

with the nonlinear mapping

$$u^2 : L^3(\Omega) \to L^{3/2}(\Omega). \tag{9.75}$$

Therefore, taking into account (9.74) and (9.75), we have

$$u^2 : H_0^1(\Omega) \to H^{-1}(\Omega). \tag{9.76}$$

Therefore,

$$F(u) = \Delta u + u^2 : H_0^1(\Omega) \to H^{-1}(\Omega) \tag{9.77}$$

and Definition 9.3 is valid.

2. Prove the continuous differentiability of the operator F. The Fréchet derivative of the operator F has the following form:

$$F_f'(u) \cdot h = \Delta h + 2uh, \quad F_f'(\cdot) : H_0^1(\Omega) \to \mathscr{L}(H_0^1(\Omega), H^{-1}(\Omega)). \tag{9.78}$$

We prove the continuity of the Fréchet derivative (9.78) on $H_0^1(\Omega)$. Indeed, the following chain of relations holds:

$$\|F_f'(u) - F_f'(v)\|_{\mathscr{L}(H_0^1(\Omega), H^{-1}(\Omega))} = 2 \sup_{\|h\|_{H_0^1(\Omega)} = 1} \|(u-v)h\|_{H^{-1}(\Omega)}$$

$$= 2 \sup_{\|h\|_{H_0^1(\Omega)} = 1} \sup_{\|\varphi\|_{H_0^1(\Omega)} = 1} \left| \int_\Omega (u(x) - v(x))h(x)\varphi(x)\,dx \right|$$

$$\leq 2 \sup_{\|h\|_{H_0^1(\Omega)} = 1} \sup_{\|\varphi\|_{H_0^1(\Omega)} = 1} \|u - v\|_3 \|h\|_3 \|\varphi\|_3 \leq 2c_1^3 \|u - v\|_{H_0^1(\Omega)}, \tag{9.79}$$

$$\|w\|_3 := \left(\int_\Omega |w(x)|^3\,dx \right)^{1/3},$$

$$\|w\|_3 \leq c_1 \|w\|_{H_0^1(\Omega)} \quad \text{for all } w(x) \in H_0^1(\Omega).$$

Thus, the mapping $F : H_0^1(\Omega) \to H^{-1}(\Omega)$ is a mapping of class C^1 on $H_0^1(\Omega)$.

3. Prove that $F_f'(\theta) \in \mathrm{Isom}(H_0^1(\Omega); H^{-1}(\Omega))$. Since $F_f'(\theta) \cdot h = \Delta h$ and $\Delta \in \mathrm{Isom}(H_0^1(\Omega); H^{-1}(\Omega))$, we have, obviously, $F_f'(\theta) \in \mathrm{Isom}(H_0^1(\Omega); H^{-1}(\Omega))$.

4. Prove the local invertibility. Since $F(\theta) = \theta$, all the conditions of Theorem 9.5 on local invertibility holds for $X = H_0^1(\Omega)$, $Y = H^{-1}(\Omega)$, $F(u) = \Delta u + u^2$, $a = \theta$, and $b = \theta$, and there exist small $r_1 > 0$ and $r_2 > 0$ such that

$$F : O_1(\theta, r_1) \to O_2(\theta, r_2) \tag{9.80}$$

and F is a C^1-diffeomorphism from $O_1(\theta, r_1)$ onto $O_2(\theta, r_2)$. Moreover,

$$G_f'(y) = \left(F_f'(G(y)) \right)^{-1}, \quad G := F^{-1}. \tag{9.81}$$

5. From the result of step 4 we conclude that for each $t \in [0,T]$, there exists a weak solution of Eq. (9.73) in the ball $u(x,t) \in O_1(\theta, r_1) \subset H_0^1(\Omega)$ for sufficiently small $r_1 > 0$. Note that r_1 is independent of $t \in [0,T]$. This local equation has the following form:

$$u(t) = G(f(t)) \in C^1([0,T]; O_1(\theta, r_1)), \qquad (9.82)$$

since $G = F^{-1}$ is continuously differentiable in the Fréchet sense on the ball $O_2(\theta, r_2)$ and the function $f(t) \in C^1([0,T]; H^{-1}(\Omega))$ takes values in the ball $O_2(\theta, r_2)$:

$$\sup_{t \in [0,T]} \|f(t)\|_{H^{-1}(\Omega)} < r_2.$$

It remain to apply the chain rule for compositions of Fréchet-differentiable mappings. $\qquad \square$

Now we consider the problem whose classical statement is as follows[1]:

$$-\Delta u + u^3 = f(x,t), \quad (x,t) \in D = \Omega \otimes [0,T], \qquad (9.83)$$

$$u(x,t) = 0 \qquad\qquad (x,t) \in S = \partial\Omega \otimes [0,T], \qquad (9.84)$$

where $\Omega \subset \mathbb{R}^3$ is a bounded domain with sufficiently smooth boundary $\partial\Omega$. The right-hand side $f(x,t)$ depends on the parameter $t \in [0,T]$.

Definition 9.4. A *weak solution* of the problem (9.83)–(9.84) is a function $u(x,t) \in H_0^1(\Omega)$, $t \in [0,T]$, satisfying the equation

$$\langle -\Delta u + u^3 - f(t), \varphi \rangle = 0 \quad \text{for all } \varphi(x) \in H_0^1(\Omega), \, t \in [0,T], \qquad (9.85)$$

where $f(t) \in C^{(1)}([0,T]; H^{-1}(\Omega))$ and $\langle \cdot, \cdot \rangle$ is the duality bracket between $H_0^1(\Omega)$ and $H^{-1}(\Omega)$.

Theorem 9.7. *For any function* $f(t) \in C^{(1)}([0,T]; H^{-1}(\Omega))$, *there exists a unique weak solution* $u(t) \in C^{(1)}([0,T]; H_0^1(\Omega))$ *of the problem* (9.85).

Proof. 1. Prove the validity of Definition 9.4. Since $\Omega \subset \mathbb{R}^3$ is a bounded domain, the following continuous and dense embeddings hold:

$$H_0^1(\Omega) \overset{ds}{\subset} L^4(\Omega) \overset{ds}{\subset} L^{4/3}(\Omega) \overset{ds}{\subset} H^{-1}(\Omega) \qquad (9.86)$$

with the nonlinear mapping

$$u^3 : L^4(\Omega) \to L^{4/3}(\Omega). \qquad (9.87)$$

Therefore, taking into account (9.86) and (9.87), we have

$$u^3 : H_0^1(\Omega) \to H^{-1}(\Omega). \qquad (9.88)$$

Therefore,

$$F(u) = -\Delta u + u^3 : H_0^1(\Omega) \to H^{-1}(\Omega) \qquad (9.89)$$

---

[1]This examples requires the knowledge of the Browder–Minty theory and can be omitted at the first reading.

and Definition 9.4 is valid.

2. Prove the continuous differentiability of the operator F. The Fréchet derivative of the operator F has the following form:

$$F'_f(u) \cdot h = -\Delta h + 3u^2 h, \quad F'_f(\cdot) : H^1_0(\Omega) \to \mathscr{L}(H^1_0(\Omega), H^{-1}(\Omega)). \tag{9.90}$$

We prove the continuity of the Fréchet derivative (9.78) on $H^1_0(\Omega)$. Indeed, the following chain of relations holds:

$$\|F'_f(u) - F'_f(v)\|_{\mathscr{L}(H^1_0(\Omega), H^{-1}(\Omega))} = 3 \sup_{\|h\|_{H^1_0(\Omega)}=1} \|(u^2 - v^2)h\|_{H^{-1}(\Omega)}$$

$$= 3 \sup_{\|h\|_{H^1_0(\Omega)}=1} \sup_{\|\varphi\|_{H^1_0(\Omega)}=1} \left| \int_\Omega (u(x) - v(x))(u(x) + v(x))h(x)\varphi(x)\,dx \right|$$

$$\leq 3 \sup_{\|h\|_{H^1_0(\Omega)}=1} \sup_{\|\varphi\|_{H^1_0(\Omega)}=1} \|u - v\|_4 \|u + v\|_4 \|h\|_4 \|\varphi\|_4$$

$$\leq 3c_2^4 \|u - v\|_{H^1_0(\Omega)} \|u + v\|_{H^1_0(\Omega)}$$

$$\leq 6c_2^4 \max\left\{ \|u\|_{H^1_0(\Omega)}, \|v\|_{H^1_0(\Omega)} \right\} \|u - v\|_{H^1_0(\Omega)}, \tag{9.91}$$

$$\|w\|_4 := \left(\int_\Omega |w(x)|^4\,dx \right)^{1/4},$$

$$\|w\|_4 \leq c_2 \|w\|_{H^1_0(\Omega)} \quad \text{for all } w(x) \in H^1_0(\Omega).$$

Thus, the mapping $F : H^1_0(\Omega) \to H^{-1}(\Omega)$ is a mapping of class C^1 on $H^1_0(\Omega)$.

3. Prove that $F'_f(u) \in \text{Isom}(H^1_0(\Omega); H^{-1}(\Omega))$ for all $u \in H^1_0(\Omega)$. We use results of Lecture 21 devoted to the theory of monotonic Browder–Minty operators. Let $u_0 \in H^1_0(\Omega)$ be an arbitrary fixed function. We note the following facts:

   (i) the operator $F'_f(\cdot)$ is continuous;
   (ii) the operator $F'_f(u_0)$ is strongly monotonic (see Definition 1.2(iii));
   (iii) the operator $F'_f(u_0)$ is coercive (see Definition 1.3).

The continuity follows from the result of step 1. Prove the strong monotonicity:

$$\langle F'_f(u_0)h_1 - F'_f(u_0)h_2, h_1 - h_2 \rangle$$

$$= -\langle \Delta h_1 - \Delta h_2, h_1 - h_2 \rangle + 3\langle u_0^2(h_1 - h_2), h_1 - h_2 \rangle$$

$$= \int_\Omega |D_x(h_1(x) - h_2(x))|^2\,dx + 3\int_\Omega u_0^2(x)(h_1(x) - h_2(x))^2\,dx$$

$$\geq \|h_1 - h_2\|^2_{H^1_0(\Omega)}. \tag{9.92}$$

To prove the coercivity, we set in the inequality (9.92) $h_1(x) = h(x) \in H^1_0(\Omega)$ and $h_2(x) = \theta \in H^1_0(\Omega)$ and obtain the following inequality:

$$\langle F'_f(u_0)h, h \rangle \geq \|h\|^2_{H^1_0(\Omega)}. \tag{9.93}$$

By the Browder–Minty theorem (see Theorem 20.2), we conclude that $F'_f(u_0) \in$ $\mathrm{Isom}(H_0^1(\Omega); H^{-1}(\Omega))$ for any $u_0 \in H_0^1(\Omega)$.

4. Therefore, by Theorem 9.5, the mapping $F : H_0^1(\Omega) \to H^{-1}(\Omega)$ is a C^1-diffeomorphism. Therefore, for any function $f(t) \in C^{(1)}([0, T]; H^{-1}(\Omega))$, there exists a unique solution $u \in H_0^1(\Omega)$ of the form

$$u(t) = G(f(t)), \quad G = F^{-1}, \tag{9.94}$$

where the mapping $G = F^{-1}$ is continuous differentiable in the Fréchet sense on $H^{-1}(\Omega)$. Using the chain rule for the composition of Fréchet-differentiable mappings, we conclude from (9.94) that $u(t) \in C^{(1)}([0, T]; H_0^1(\Omega))$. $\qquad \square$

9.5 Bibliographical Notes

The contents of this lecture is taken from [6, 7].

PART 3
Variational Methods of the Study of Nonlinear Operator Equations

Lecture 10

Potential Operators

In this lecture, we start to study various variational methods os examining nonlinear operator equations. Basically, these methods are used in the study of boundary-value problems for nonlinear elliptic equations, but they are also useful in the study of stability of stationary solutions of various evolutionary nonlinear equations, for example, the Korteweg–de Vries equation, nonlinear Schrödinger equation, and nonlinear wave equation.

10.1 Basic Definitions

Before applying a variational method to a nonlinear operator problem, we must establish whether this problem has a variational statement, i.e., whether it is equivalent to the problem of search for a minimum or a maximum of a certain functional on a subset of a Banach space.

Let $\mathbb{B}$ be a Banach with norm $\|\cdot\|$ and the duality bracket $\langle \cdot, \cdot \rangle$ between $\mathbb{B}$ and its dual space $\mathbb{B}^*$. Consider a (nonlinear) functional $\psi : \mathbb{B} \to \mathbb{R}^1$ defined on this Banach space. As in the previous lecture, we denote by $\psi'_g(u)$ and $\psi'_f(u)$ the Gâteaux and Fréchet derivatives of the functional ψ, respectively.

Definition 10.1. An operator $F : \mathbb{B} \to \mathbb{B}^*$ is said to be *strongly potential* or *potential* if there exists a Fréchet-differentiable functional $\psi(u) : \mathbb{B} \to \mathbb{R}^1$ such that

$$F(u) = \psi'_f(u). \tag{10.1}$$

Definition 10.2. An operator $F : \mathbb{B} \to \mathbb{B}^*$ is said to be *weakly potential* if there exists a Gâteaux-differentiable functional $\psi(u) : \mathbb{B} \to \mathbb{R}^1$ such that

$$F(u) = \psi'_g(u). \tag{10.2}$$

Naturally, the question on sufficient conditions of the potentiality of a given operator $F : \mathbb{B} \to \mathbb{B}^*$ arises. To answer this question, we need the notion of the *bounded continuity* in the Lipschitz sense.

Definition 10.3. An operator $F : \mathbb{B}_1 \to \mathbb{B}_2$, where $\mathbb{B}_1$ and $\mathbb{B}_2$ are Banach spaces with the norms $\| \cdot \|_1$ and $\| \cdot \|_2$, respectively, is said to be *boundedly Lipschitz* *continuous*[1] if there exists a function $\mu : [0, +\infty) \to [0, +\infty)$ bounded on each bounded subset of the half-line $[0, +\infty)$ such that

$$\|F(u_1) - F(u_2)\|_2 \leq \mu(R)\|u_1 - u_2\|_1 \tag{10.3}$$

for all $u_1, u_2 \in \mathbb{B}_1$ such that $\|u_k\|_1 \leq R$, $k = 1, 2$.

Theorem 10.1. *An operator $F : \mathbb{B} \to \mathbb{B}^*$ satisfying the condition of bounded Lipschitz continuity is potential if and only if for all $u, v \in \mathbb{B}$, the following equality holds:*

$$\int_0^1 \langle F(tu), u \rangle \, dt - \int_0^1 \langle F(tv), v \rangle \, dt = \int_0^1 \langle F(tu + (1-t)v), u - v \rangle \, dt. \tag{10.4}$$

If the condition (10.4) is fulfilled, then the functional

$$\psi(u) = \psi(\theta) + \int_0^1 \langle F(tu), u \rangle \, dt \quad \text{for all } u \in \mathbb{B}, \tag{10.5}$$

where $\theta \in \mathbb{B}$ is the zero element, is called the strong potential *of the operator F (below we usually omit the word "strong").*

Proof. 1. Let an operator F be strongly potential; then there exists a Fréchet-differentiable functional $\psi(u) : \mathbb{B} \to \mathbb{R}^1$ such that $F(u) = \psi'_f(u)$. In this case, the following formula holds[2]:

$$\psi(u) - \psi(v) = \int_0^1 \frac{d}{dt}\psi(tu + (1-t)v) \, dt = \int_0^1 \langle \psi'_f(tu + (1-t)v), u - v \rangle \, dt$$

$$= \int_0^1 \langle F(tu + (1-t)v), u - v \rangle \, dt. \tag{10.6}$$

Setting $v = \theta \in \mathbb{B}$ in Eq. (10.6), we obtain the following equality:

$$\psi(u) = \psi(\theta) + \int_0^1 \langle F(tu), u \rangle \, dt. \tag{10.7}$$

Now setting $u = \theta$ in Eq. (10.6), we obtain

$$\psi(v) = \psi(\theta) + \int_0^1 \langle F(tv), v \rangle \, dt. \tag{10.8}$$

---

[1] Note that there also exists the term "locally Lipschitz continuous operator." Do not mix these different notions!

[2] Here we used the formula for the Fréchet derivative of composite mappings.

Taking into account Eqs. (10.7) and (10.8), we arrive at the following relation:

$$\psi(u) - \psi(v) = \int\limits_0^1 \langle F(tu), u \rangle \, dt - \int\limits_0^1 \langle F(tv), v \rangle \, dt.$$

This and (10.6) imply (10.4).

2. Now assume that for the operator F, Eq. (10.4) is fulfilled. Introduce the functional

$$\psi(u) := \psi(\theta) + \int\limits_0^1 \langle F(tu), u \rangle \, dt \qquad (10.9)$$

and prove that it is Fréchet differentiable and its Fréchet derivative is equal to $F(u)$. Indeed, due to (10.4), we have

$$\psi(u + h) - \psi(u) = \int\limits_0^1 \langle F(t(u + h)), u + h \rangle \, dt - \int\limits_0^1 \langle F(tu), u \rangle \, dt$$

$$= \int\limits_0^1 \langle F(t(u + h) + (1 - t)u), h \rangle \, dt. \qquad (10.10)$$

Introduce the notation

$$\omega(u, h) \overset{\text{def}}{=\!=} \psi(u + h) - \psi(u) - \langle F(u), h \rangle.$$

Then for $\omega(u, h)$, the following inequality holds:

$$|\omega(u, h)| \leq \int\limits_0^1 |\langle F(t(u + h) + (1 - t)u) - F(u), h \rangle| \, dt$$

$$\leq \int\limits_0^1 \|F(t(u + h) + (1 - t)u) - F(u)\|_* \|h\| \, dt$$

$$\leq c(2R) \int\limits_0^1 \|t(u + h) + (1 - t)u - u\| \|h\| \, dt = c(2R)\|h\|^2 \frac{1}{2}$$

for all $u, h \in \mathbb{B}$ such that $\|u\| \leq R$ and $\|h\| \leq R$. Therefore, we conclude that

$$\lim_{\|h\| \to 0} \frac{|\omega(u, h)|}{\|h\|} = 0.$$

Thus, the functional $\psi(u)$ is Fréchet differentiable on $\mathbb{B}$ and its Fréchet derivative is $\psi'_f(u) = F(u)$. $\qquad \square$

Examples of potential and nonpotential operators

Example 10.1. Consider the functional

$$\psi(u) = \frac{1}{p} \int_\Omega |u|^p \, dx : \ L^p(\Omega) \to \mathbb{R}^1, \quad p > 1.$$

As was proved in Lemma 5.1, this functional is Fréchet differentiable and its Fréchet derivative is equal to

$$F(u) := \psi'_f(u) = |u|^{p-2} u : L^p(\Omega) \to L^{p'}(\Omega), \quad p' = \frac{p}{p-1}.$$

Therefore, the operator $F(u)$ is potential (strongly potential).

Example 10.2. Consider the nonlinear operator

$$F : u \mapsto \frac{d(u^2)}{dx}.$$

First, we prove that it is a bounded operator acting as follows:

$$F : \mathbb{B} = W_0^{1,4}(0,1) \to W^{-1,4/3}(0,1) = \mathbb{B}^*.$$

Indeed,

$$F(u) = 2uu', \quad u(x) \in W_0^{1,4}(0,1) \subset L^4(0,1), \quad u' \in L^4(0,1);$$

therefore, $u(x)u'(x) \in L^2(0,1)$. Note that the following chain of continuous and dense embeddings hold:

$$W_0^{1,4}(0,1) \overset{ds}{\subset} W_0^{1,2}(0,1) \overset{ds}{\subset} L^2(0,1) \overset{ds}{\subset} W^{-1,2}(0,1) \overset{ds}{\subset} W^{-1,4/3}(0,1).$$

Thus, $u(x)u'(x) \in L^2(0,1) \overset{ds}{\subset} W^{-1,4/3}(0,1)$.

Assume that the operator $F(u)$ is potential. Then its potential is defined by the formula

$$\psi(u) = \psi(\theta) + \int_0^1 \langle F(tu), u \rangle \, dt.$$

The following chain of equalities holds:

$$\langle F(tu), u \rangle = \int_0^1 \frac{d}{dx}(tu(x))^2 \, u(x) \, dx = t^2 \frac{2}{3} \int_0^1 \frac{d}{dx} u^3(x) \, dx = 0;$$

however, then $\psi(u) = \text{constant}$ and its Fréchet derivative is equal to zero. Therefore, the operator $F(u)$ is not potential.

10.2   Taylor's Formula

Find a solution of the following operator equation:

$$F(u) = \theta \in \mathbb{B}^*, \quad u \in \mathbb{B}. \tag{10.11}$$

Assume that the operator F is potential and a functional $\psi(u)$ is the potential of F.

Definition 10.4. Let $M \subset \mathbb{B}$ be a nonempty, closed subset of a Banach space $\mathbb{B}$. A point $\hat{u} \in M$ is called an *extremal point* of the functional $\psi(u)$ on M if

$$\inf_{u \in M} \psi(u) = \psi(\hat{u}) \quad \text{or} \quad \sup_{u \in M} \psi(u) = \psi(\hat{u}).$$

Consider the function

$$\varphi(t) = \psi(\hat{u} + th), \quad t \in (-1, 1),$$

where $\hat{u}$ is an extremal point of the functional $\psi(\cdot)$ on a set $M = \mathbb{B}$. Then the function $\varphi(t)$ reaches an extremum at the point $t = 0$. Due to the Fréchet differentiability of the functional $\psi(u)$ at the point $\hat{u} \in \mathbb{B}$, we conclude that $\varphi(t)$ is differentiable at the point $t = 0$. Thus, we obtain a necessary condition of extremum:

$$\varphi'(t) = \langle \psi'_f(\hat{u} + th), h \rangle, \ \varphi'(0) = 0$$
$$\Rightarrow \quad \langle \psi'_f(\hat{u}), h \rangle = 0 \ \forall h \in \mathbb{B} \quad \Rightarrow \quad \psi'_f(\hat{u}) = \theta \in \mathbb{B}^* \quad \Rightarrow \quad F(\hat{u}) = \theta.$$

Therefore, on the one hand, each extremal point of a Fréchet differentiable functional $\psi(u)$ is necessarily a solution of the operator equation (10.11). On the other hand, clearly, not every solution of the operator equation (10.11) is an extremal point of the functional $\psi(u)$ since Eq. (10.11) is only a necessary, but not sufficient condition.

To find sufficient conditions of the existence of an extremal of the functional $\psi(u)$, we need an analog of Taylor's formula for functionals that are twice differentiable in the Fréchet sense.

Consider a functional $\psi(v) \in C^{(2)}(\overline{O(u, \varepsilon)}; \mathbb{R}^1)$; let

$$O(u, \varepsilon) = \{ v \in \mathbb{B} : \ \|v - u\| < \varepsilon \}.$$

Lemma 10.1. *The equality*

$$\psi(u + h) = \psi(u) + \langle \psi'_f(u), h \rangle + \frac{1}{2} \langle \psi''_{ff}(u)h, h \rangle + \omega_2(u, h) \tag{10.12}$$

holds for all $h \in \mathbb{B}$ such that $\|h\| < \varepsilon$, where for $\omega_2(u, h)$, the following limit equality is fulfilled:

$$\lim_{\|h\| \to 0} \frac{|\omega_2(u, h)|}{\|h\|^2} = 0. \tag{10.13}$$

Proof. Let $\psi(v) \in C^{(2)}(\overline{O(u, \varepsilon)}; \mathbb{R}^1)$. Note that for the Fréchet derivative $\psi'_f(v)$, due its Fréchet differentiability in $\overline{O(u, \varepsilon)}$, the following equality holds:

$$\psi'_f(u + h) = \psi'_f(u) + \psi''_{ff}(u)h + \omega_1(u, h), \quad \|h\| < \varepsilon,$$

where

$$\lim_{\|h\| \to 0} \frac{\|\omega_1(u, h)\|_*}{\|h\|} = 0.$$

Therefore, the following chain of equalities holds:

$$\psi(u + h) - \psi(u) = \int_0^1 \frac{d}{dt} \psi(u + th)\, dt = \int_0^1 \langle \psi_f'(u + th), h \rangle\, dt$$

$$= \int_0^1 \langle \psi_f'(u) + t\psi_{ff}''(u)h, h \rangle\, dt + \omega_2(u, h),$$

where

$$\omega_2(u, h) = \int_0^1 \langle \omega_1(u, th), h \rangle\, dt.$$

Thus, we obtain the equality

$$\psi(u + h) - \psi(u) = \langle \psi_f'(u), h \rangle + \langle \psi_{ff}''(u)h, h \rangle \int_0^1 t\, dt + \omega_2(u, h)$$

$$= \langle \psi_f'(u), h \rangle + \frac{1}{2} \langle \psi_{ff}''(u)h, h \rangle + \omega_2(u, h),$$

where $\omega_2(u, h)$ has the following representation:

$$\omega_2(u, h) = \int_0^1 \langle \omega_1(u, th), h \rangle\, dt.$$

We arrive at the inequality

$$|\omega_2(u, h)| \le \int_0^1 \|\omega_1(u, th)\|_* \|h\|\, dt.$$

Therefore, the following limit relation holds:

$$\lim_{\|h\| \to 0} \frac{|\omega_2(u, h)|}{\|h\|^2} \le \lim_{\|h\| \to 0} \int_0^1 \frac{\|\omega_1(u, th)\|_*}{\|h\|}\, dt = 0.$$

The formulas (10.12) and (10.13) are proved.   □

10.3 Extremum Conditions for Functionals

Now we clarify the result on a necessary condition of an extremum of a functional. Let $\psi(u) \in C^{(2)}(\overline{O(\hat{u}, \varepsilon)}; \mathbb{R}^1)$, where $\varepsilon > 0$.

Lemma 10.2. *The following conditions are necessary for a minimum of the functional $\psi(u)$ at a point $\hat{u}$:*

$$\psi'_f(\hat{u}) = 0 \quad and \quad \langle \psi''_{ff}(\hat{u})h, h \rangle \geq 0 \quad for\ all\ \|h\| < \varepsilon. \tag{10.14}$$

Proof. Consider the following decomposition of the functional $\psi(u)$ in a neighborhood $O(\hat{u}, \varepsilon)$ of the extremal point $\hat{u} \in \mathbb{B}$:

$$\psi(\hat{u} + h) = \psi(\hat{u}) + \langle \psi'_f(\hat{u}), h \rangle + \frac{1}{2} \langle \psi''_{ff}(\hat{u})h, h \rangle + \omega_2(\hat{u}, h),$$

where $\|h\| < \varepsilon$. As was proved above, at the extremal point $\hat{u}$, the equality $\psi'_f(\hat{u}) = 0$ holds; therefore, we obtain the following equality:

$$\psi(\hat{u} + h) - \psi(\hat{u}) = \frac{1}{2} \langle \psi''_{ff}(\hat{u})h, h \rangle + \omega_2(\hat{u}, h). \tag{10.15}$$

Assume that $\hat{u}$ is a point of local minimum and for some $h_1 \in O(\theta, \varepsilon) \subset \mathbb{B}$, $h_1 \neq \theta$, the following inequality holds:

$$\langle \psi''_{ff}(\hat{u})h_1, h_1 \rangle < 0.$$

Then for $h = \varepsilon_1 h_1$, $\varepsilon_1 \in (0, 1)$, the following relation holds:

$$\langle \psi''_{ff}(\hat{u})h, h \rangle = \varepsilon_1^2 \langle \psi''_{ff}(\hat{u})h_1, h_1 \rangle < 0. \tag{10.16}$$

Now, choosing $\varepsilon_1 > 0$ sufficiently small, we obtain the inequality

$$|\omega_2(\hat{u}, h)| = |\omega_2(\hat{u}, \varepsilon_1 h_1)| \leq \varepsilon_1^2 \|h_1\|^2 \frac{|c_1|}{4\|h_1\|^2} = \varepsilon_1^2 \frac{|c_1|}{4}, \tag{10.17}$$

where $c_1 = \langle \psi''_{ff}(\hat{u})h_1, h_1 \rangle < 0$. From (10.16) and (10.17) we obtain the following inequality:

$$\frac{1}{2} \langle \psi''_{ff}(\hat{u})h, h \rangle + \omega_2(\hat{u}, h) \leq \frac{\varepsilon_1^2}{2} \left(c_1 + \frac{1}{2}|c_1| \right) = -\frac{\varepsilon_1^2}{4}|c_1| < 0. \tag{10.18}$$

Due to Eq. (10.15), from the inequality (10.18) we obtain the inequality

$$\psi(\hat{u} + h) - \psi(\hat{u}) < 0,$$

i.e., there is no a minimum at the point $\hat{u} \in \mathbb{B}$. Therefore, the necessary condition of a minimum at a point $\hat{u} \in \mathbb{B}$ has the following form:

$$\langle \psi''_{ff}(\hat{u})h, h \rangle \geq 0 \quad for\ all\ \|h\| < \varepsilon.$$

Lemma 10.2 is proved. $\qquad\square$

Corollary 10.1. *The following conditions are necessary for a maximum of the functional $\psi(u)$ at a point $\hat{u} \in \mathbb{B}$:*

$$\psi'_f(\hat{u}) = 0 \quad and \quad \langle \psi''_f(\hat{u})h, h \rangle \leq 0 \quad for\ all\ \|h\| < \varepsilon. \tag{10.19}$$

Remark 10.1. Note that, in contrast to a finite-dimensional Banach space $\mathbb{B}$, the condition

$$\langle \psi_{ff}''(\hat{u})h, h \rangle \geq 0 \ (\leq 0) \quad \text{for all } h \in \mathbb{B}$$

is not a sufficient condition of a minimum (maximum). This can be confirmed by the following example.

Example 10.3. On the Banach space $C[0,1]$ with the standard sup-norm, we consider the functional

$$\psi(u) = \int_0^1 u^2(x)(x - u(x)) \, dx.$$

The following chain of equalities holds:

$$\psi(u+h) = \int_0^1 (u+h)^2 (x - u - h) \, dx$$

$$= \int_0^1 u^2(x-u) \, dx + \int_0^1 (2ux - 3u^2) \, h \, dx + \int_0^1 (x - 3u)h^2 \, dx - \int_0^1 h^3 \, dx.$$

Hence we conclude that $\psi_f'(u) = 0$ on the following two functions from $C[0,1]$: $u(x) = 0$ and $u(x) = 2x/3$. Further, for $u = 0$ we have

$$\psi_{ff}''(u) = 2x - 6u(x) \quad \Rightarrow \quad \psi_{ff}''(0) = 2x.$$

Therefore,

$$\langle \psi_{ff}''(0)h, h \rangle = 2 \int_0^1 h^2(x)x \, dx \geq 0 \quad \text{for all } h(x) \in C[0,1];$$

moreover, $\psi(0) = 0$, i.e., all necessary condition of a local minimum are fulfilled for the function $u(x) = 0$, but the function does not attain a local minimum at this function.

Indeed, consider the following one-parameter family of functions

$$u_\varepsilon(x) = \begin{cases} \varepsilon - x, & x \in [0, \varepsilon], \\ 0, & x \geq \varepsilon. \end{cases}$$

Obviously, $u_\varepsilon \in C[0,1]$ for all $\varepsilon \in (0,1)$. Calculate the norm of this function:

$$\sup_{x \in [0,1]} |u_\varepsilon(x)| = \varepsilon \to 0 \quad \text{as } \varepsilon \to 0,$$

i.e., any neighborhood of the function $u(x) = 0 \in C[0,1]$ contains a function $u_\varepsilon(x) \in C[0,1]$ for certain $\varepsilon > 0$. Now we calculate the value of the functional ψ at the function u_ε:

$$\psi(u_\varepsilon) = \int_0^1 u_\varepsilon^2(x) \, (x - u_\varepsilon(x)) \, dx$$

$$= \int_0^\varepsilon (\varepsilon - x)^2 (2x - \varepsilon) \, dx = \frac{\varepsilon^4}{3} - \frac{\varepsilon^4}{2} = -\frac{\varepsilon^4}{6} < 0 = \psi(0).$$

Thus, the functional $\psi(u)$ does not attain a minimum at the function $u(x) = 0$.

However, the following theorem is valid.

Theorem 10.2. *Let* $\psi(u) \in C^{(2)}\left(\overline{O(\hat{u}, \varepsilon)}; \mathbb{R}^1\right)$, *where* $\varepsilon > 0$. *Then the following conditions are sufficient for a minimum (respectively, maximum) of the functional* $\psi(u)$ *at a point* $\hat{u} \in \mathbb{B}$:

(I) $\psi_f'(\hat{u}) = 0$;
(II) $\langle \psi_{ff}''(\hat{u})h, h\rangle \geq c\|h\|^2$ *(respectively,* $\leq -c\|h\|^2$*) for all* $h \in \mathbb{B}$ *for certain* $c = c(\hat{u}) > 0$.

Proof. We prove the theorem for minimum conditions of the functional $\psi(u)$ at a point $\hat{u}$. Due to the conditions of the theorem, we have the following representation in a neighborhood $O(\hat{u}, \varepsilon)$ of the point $\hat{u} \in \mathbb{B}$:

$$\psi(\hat{u} + h) - \psi(\hat{u}) = \langle \psi_f'(\hat{u}), h\rangle + \frac{1}{2}\langle \psi_{ff}''(\hat{u})h, h\rangle + \omega_2(\hat{u} + h, h). \tag{10.20}$$

Moreover, since the limit equality

$$\lim_{\|h\| \to 0} \frac{|\omega_2(u, h)|}{\|h\|^2} = 0$$

is valid, we see that for sufficiently small $\|h\|$ and for $c = c(\hat{u}) > 0$ stated in the condition of the theorem, the following inequality holds:

$$|\omega_2(u, h)| < \frac{c}{4}\|h\|^2.$$

Then from (10.20) we obtain the inequality

$$\psi(\hat{u} + h) - \psi(\hat{u}) > \frac{1}{2}\langle \psi_{ff}''(\hat{u})h, h\rangle - \frac{c}{4}\|h\|^2 \geq \frac{c}{2}\|h\|^2 - \frac{c}{4}\|h\|^2 = \frac{c}{4}\|h\|^2$$

for such $h \in \mathbb{B}$, i.e., the functional ψ attains a minimum at the point $\hat{u} \in \mathbb{B}$. $\square$

Remark 10.2. Under the conditions of Theorem 10.2, each extremal point of the functional $\psi(u) : \mathbb{B} \to \mathbb{R}^1$ is a solution of the operator equation $\psi_f'(u) = 0$. In the general case, the converse assertions is invalid.

10.4 Bibliographical Notes

The contents of this lecture is taken from $[14, 20, 23, 24, 35, 56, 67]$.

Semicontinuous and Weakly Coercive Functionals

11.1 Introduction

The sufficient extremum condition (II) (naturally, together with the condition (I)) obtained in Theorem 10.2 is very restrictive: in practice, the existence of the continuous second Fréchet derivative of a functional is not expected; moreover, if this derivative exists, then the requirement of the strong positivity (or negativity) of $\psi''_{ff}(\hat{u})$ is not fulfilled in practice. In particular, if a functional $\psi(u) : \mathbb{B} \to \mathbb{R}^1$ is the potential of a certain operator $F(u) = \psi'_f(u) : \mathbb{B} \to \mathbb{B}^*$, then this requirement means the existence of the continuous Fréchet derivative of this operator such that

$$\langle F'_f(\hat{u})h, h \rangle \geq c\|h\|^2 \ (\leq -c\|h\|^2) \quad \text{for all } h \in \mathbb{B}$$

for $c = c(\hat{u}) > 0$. Hence, in this section we weaken the condition (II) of Theorem 10.2.

11.2 Semicontinuous Functionals

Definition 11.1. A set $M \subset \mathbb{B}$ is said to be *weakly closed* if for any sequence $\{u_n\} \subset M$ such that

$$u_n \rightharpoonup u \quad \text{weakly in } \mathbb{B} \text{ as } n \to +\infty,$$

we have $u \in M$.

Definition 11.2. A functional $\psi(u) : \mathbb{B} \to \mathbb{R}^1$ is said to be *weakly sequentially lower semicontinuous at a point* $u_0 \in M \subset \mathbb{B}$ with respect to a weakly closed set $M \subset \mathbb{B}$ if for any sequence $\{u_n\} \subset M$ such that

$$u_n \rightharpoonup u_0 \quad \text{weakly in } \mathbb{B} \text{ as } n \to +\infty,$$

the following relation holds:

$$\psi(u_0) \leq \liminf_{n \to +\infty} \psi(u_n) \quad \text{as } n \to +\infty.$$

Definition 11.3. We say that a functional $\psi(u) : \mathbb{B} \to \mathbb{R}^1$ is *weakly lower semicontinuous on a weakly closed set* $M \subset \mathbb{B}$ if it is weakly lower semicontinuous at each point $u \in M$.

Definition 11.4. A functional $\psi(u) : \mathbb{B} \to \mathbb{R}^1$ satisfying the condition

$$\lim_{\|u\| \to +\infty} \psi(u) = +\infty$$

is said to be *weakly coercive*.

The following important theorem is valid.

Theorem 11.1. *Let $\mathbb{B}$ be a reflexive Banach space. Assume that the following conditions hold:*

(i) *$M \subset \mathbb{B}$ is weakly closed in $\mathbb{B}$;*
(ii) *the functional $\psi(u) : \mathbb{B} \to \mathbb{R}^1$ is weakly coercive on $\mathbb{B}$;*
(iii) *the functional $\psi(u)$ is weakly lower semicontinuous on the set M.*

Then the functional $\psi(u)$ is lower bounded and attains a minimum on M:

$$\psi(u_0) = \inf_{u \in M} \psi(u).$$

Proof. 1. Let $d > \inf_{u \in M} \psi(u)$. Since the functional $\psi(u)$ is weakly coercive on $\mathbb{B}$, there exists $R > 0$ such that

$$\psi(u) \geq d \quad \text{for all } u \in \mathbb{B} \backslash B_R, \quad B_R = \{u \in \mathbb{B} : \|u\| \leq R\}.$$

Therefore,

$$\inf_{u \in M} \psi(u) = \inf_{u \in M \cap B_R} \psi(u).$$

2. Let $\alpha_0 = \inf_{u \in M} \psi(u)$ and $\{u_n\} \subset M$ be a minimizing sequence:

$$\psi(u_n) \to \alpha_0 \quad \text{as } n \to +\infty,$$

whose terms, obviously, belong to the set $B_R \cap M$ starting from a certain number $n_0 \in \mathbb{N}$. This means that

$$\|u_n\| \leq R \quad \text{for all } n \geq n_0.$$

Then, due to the reflexivity of $\mathbb{B}$, there exists a subsequence $\{u_{n_k}\} \subset \{u_n\}$ such that

$$u_{n_k} \rightharpoonup u_0 \in B_R \quad \text{weakly in } \mathbb{B} \text{ as } n_k \to +\infty;$$

moreover, due to the weak closedness of M, we obtain $u_0 \in M$. Due to the weak sequentially lower semicontinuity of the functional ψ on M, we conclude that

$$\psi(u_0) \leq \liminf_{k \to +\infty} \psi(u_{n_k}).$$

3. Thus, we obtain the following chain of inequalities:

$$\inf_{u \in M} \psi(u) \leq \psi(u_0) \leq \liminf_{k \to +\infty} \psi(u_{n_k}) = \lim_{n \to +\infty} \psi(u_n) = \inf_{u \in M} \psi(u),$$

which implies

$$\inf_{u \in M} \psi(u) = \psi(u_0).$$

Theorem 11.1 is proved. $\square$

11.3 Compact and Weakly Compact Sets

Definition 11.5. A set $M \in \mathbb{B}$ is said to be *compact* if from each sequence $\{u_m\} \subset M$ one can extract a strongly converging subsequence $\{u_{n_k}\} \subset \{u_n\}$:

$$u_{n_k} \to u_0 \in M \quad \text{strongly in } \mathbb{B} \text{ as } n_k \to +\infty. \tag{11.1}$$

Definition 11.6. A set $M \in \mathbb{B}$ is said to be *weakly compact* if from each sequence $\{u_m\} \subset M$ one can extract a weakly converging subsequence $\{u_{n_k}\} \subset \{u_n\}$:

$$u_{n_k} \rightharpoonup u_0 \in M \quad \text{weakly in } \mathbb{B} \text{ as } n_k \to +\infty. \tag{11.2}$$

The following important assertions hold.

Theorem 11.2. *If a functional ψ is continuous on a compact set $M \subset \mathbb{B}$, then it attains a minimum and a maximum on M.*

Proof. We prove the theorem for a minimum. Let

$$A := \inf_{u \in M} \psi(u). \tag{11.3}$$

Let $\{u_m\} \subset M$ be a minimizing sequence

$$\psi(u_m) \to A \quad \text{as } m \to +\infty. \tag{11.4}$$

Since M is a compact set, there exists a subsequence $\{u_{m_k}\} \subset \{u_n\}$ such that

$$u_{m_k} \to u_0 \in M \quad \text{strongly ing } \mathbb{B} \text{ as } m_k \to +\infty. \tag{11.5}$$

Due to the continuity of the functional ψ on M, we obtain from (11.5) the limit relation

$$\psi(u_{m_k}) \to \psi(u_0) \quad \text{as } m_k \to +\infty. \tag{11.6}$$

From (11.4) and (11.6) we obtain the equality

$$\inf_{u \in M} \psi(u) = A = \psi(u_0), \quad u_0 \in M.$$

For a maximum, the proof is similar: instead of a minimizing sequence, one must consider a maximizing sequence. □

Theorem 11.3. *If a functional ψ is weakly continuous on a weakly compact set $M \subset \mathbb{B}$, then it attains a minimum and a maximum on M.*

Proof. Let $B := \inf_{u \in M} \psi(u)$. Consider a minimizing sequence $\{u_m\} \subset M$:

$$\psi(u_m) \to B \quad \text{as } m \to +\infty. \tag{11.7}$$

Since the set M is weakly compact, there exists a subsequence $\{u_{m_k}\} \subset \{u_m\}$ such that

$$u_{m_k} \rightharpoonup u_0 \in M \quad \text{weakly in } \mathbb{B} \text{ as } m_k \to +\infty. \tag{11.8}$$

Due to the weak continuity of the functional ψ on M, we arrive at the following limit relation:

$$\psi(u_{m_k}) \to \psi(u_0) \quad \text{as } m_k \to +\infty. \tag{11.9}$$

From (11.7) and (11.9) we obtain

$$\inf_{u \in M} \psi(u) = B = \psi(u_0), \quad u_0 \in M. \tag{11.10}$$

The proof for a maximum is the same, but instead of a minimizing sequence, one must consider a maximizing sequence. □

Problem 11.1. Prove that Theorem 11.3 for a minimum remains valid if the condition of the weak continuity of the functional is replaced by the weak lower semicontinuity.

11.4 Example

Consider the following nonlinear boundary-value problem:

$$\Delta_p u = f(x) \quad \text{in } \Omega, \tag{11.11}$$

$$u(x) = 0 \qquad \text{on } \partial\Omega, \tag{11.12}$$

where $\Omega \subset \mathbb{R}^N$ is a bounded domain with smooth boundary $\partial\Omega \in C^{2,\delta}$, $\delta \in (0,1)$, and $\Delta_p u$ is the following nonlinear (for $p \neq 2$) operator:

$$\Delta_p u(x) \stackrel{\text{def}}{=} \operatorname{div}(|D_x u(x)|^{p-2} D_x u(x)).$$

Definition 11.7. A *weak solution* of the problem (11.11) is a function of class $u(x) \in W_0^{1,p}(\Omega)$ satisfying the equality

$$\langle -\Delta_p u(x) + f(x), \varphi(x) \rangle = 0 \quad \text{for all } \varphi \in W_0^{1,p}(\Omega), \tag{11.13}$$

where $\langle \cdot, \cdot \rangle$ is the duality bracket between the Banach spaces $W_0^{1,p}(\Omega)$ and $W^{-1,p'}(\Omega)$ and $f(x) \in W^{-1,p'}(\Omega)$.

Before examining the corresponding variational problem, we consider the operator Δ_p. We prove that

$$\Delta_p : W_0^{1,p}(\Omega) \to W^{-1,p'}(\Omega), \quad p > 1. \tag{11.14}$$

Indeed, as the domain of the operator we take $\operatorname{dom} \Delta_p = W_0^{1,p}(\Omega)$; then the weak gradient acts as follows:

$$D_x : W_0^{1,p}(\Omega) \to \underbrace{L^p(\Omega) \otimes \cdots \otimes L^p(\Omega)}_{N}.$$

Consider the nonlinear vector-valued function

$$\eta(x) = (\eta_1(x), \dots, \eta_N(x))$$
$$= |\xi(x)|^{p-2}\xi(x) : \underbrace{L^p(\Omega) \otimes \cdots \otimes L^p(\Omega)}_{N} \to \underbrace{L^{p'}(\Omega) \otimes \cdots \otimes L^{p'}(\Omega)}_{N},$$

where $\xi(x) = (\xi_1(x), \ldots, \xi_N(x))$. Now we introduce the functional $\mathrm{div}(\eta(x)) \in W^{-1,p'}(\Omega)$ on the space $W_0^{1,p}(\Omega)$ as follows:

$$\langle \mathrm{div}(\eta(x)), \varphi(x) \rangle_* \overset{\mathrm{def}}{=} -\sum_{j=1}^N \int_\Omega \eta_j(x) \frac{\partial \varphi(x)}{\partial x_j} \, dx \tag{11.15}$$

for all $\varphi(x) \in W_0^{1,p}(\Omega)$ and a given function

$$\eta(x) = (\eta_1(x), \ldots, \eta_N(x)) \in \underbrace{L^{p'}(\Omega) \otimes \cdots \otimes L^{p'}(\Omega)}_{N}.$$

Then we represent the operator Δ_p as follows:

$$\Delta_p u \overset{\mathrm{def}}{=} \mathrm{div}(\eta(x)), \quad \eta(x) = |\xi(x)|^{p-2}\xi(x), \quad \xi(x) = D_x u(x),$$

for functions $u(x) \in W_0^{1,p}(\Omega)$. Therefore,

$$\Delta_p : W_0^{1,p}(\Omega) \to W^{-1,p'}(\Omega).$$

To the problem (11.11), we assign the following functional:

$$\psi(u) = \psi_1(u) + \psi_2(u) = \frac{1}{p}\int_\Omega |D_x u(x)|^p \, dx + \langle f, u \rangle. \tag{11.16}$$

Find the Fréchet derivative of this functional. The Fréchet derivative of the second term is calculated easily:

$$\psi_2(u+h) - \psi_2(u) = \langle f, u+h \rangle - \langle f, u \rangle = \langle f, h \rangle,$$

i.e.,

$$\psi'_{2f}(u) = f(x) \in W^{-1,p'}(\Omega).$$

Therefore, the Fréchet derivative of the functional $\psi(u)$ is equal to

$$\psi'_f(u) = -\Delta_p u + f,$$

i.e., the operator

$$F(u) := -\Delta_p u + f : W_0^{1,p}(\Omega) \to W^{-1,p'}(\Omega)$$

is potential.

We prove that the functional $\psi(u)$ satisfies all the conditions of Theorem 11.1.

1. The Banach space $\mathbb{B} = W_0^{1,p}(\Omega)$ is weakly closed for $p > 1$ since it is reflexive.

2. Note that due to the condition $f \in W^{-1,p'}(\Omega)$, we have the inequality

$$|\langle f, u \rangle| \le \|f\|_{-1,p'}\|\,|D_x u|\,\|_p.$$

Therefore, for the functional (11.16), the following lower estimate holds:

$$\psi(u) \ge \frac{1}{p}\|\,|D_x u|\,\|_p^p - \|f\|_{-1,p'}\|\,|D_x u|\,\|_p.$$

Introduce the notation $c := \|f\|_{-1,p'}$; then we have

$$\psi(u) \ge \frac{1}{p}\|\,|D_x u|\,\|_p^p - c\|\,|D_x u|\,\|_p. \tag{11.17}$$

Let $\varepsilon \in (0,1)$. Using the three-parameter Young inequality

$$ab \leq \varepsilon a^p + \frac{1}{p'(\varepsilon p)^{p'/p}} b^{p'}, \quad \varepsilon > 0, \quad \frac{1}{p} + \frac{1}{p'} = 1, \quad p > 1, \quad a, b \geq 0, \qquad (11.18)$$

we obtain the following chain of inequalities:

$$c \big\| |D_x u| \big\|_p = \frac{c}{\varepsilon^{1/p}} \varepsilon^{1/p} \big\| |D_x u| \big\|_p \leq \frac{1}{p'} \left(\frac{c}{\varepsilon^{1/p}} \right)^{p'} + \frac{\varepsilon}{p} \big\| |D_x u| \big\|_p^p.$$

From the inequality (11.17) we further have

$$\psi(u) \geq \frac{1 - \varepsilon}{p} \big\| |D_x u| \big\|_p^p - c_1, \quad c_1 = \frac{1}{p'} \left(\frac{c}{\varepsilon^{1/p}} \right)^{p'}.$$

Therefore,

$$\psi(u) \to +\infty \quad \text{as} \quad \big\| |D_x u| \big\|_p \to +\infty$$

and hence the functional (11.16) is weakly coercive.

3. Now we verify the weak lower semicontinuity of the functional $\psi(u)$ on $W_0^{1,p}(\Omega)$. Let

$$u_n \rightharpoonup u_0 \quad \text{weakly in } W_0^{1,p}(\Omega) \text{ as } n \to +\infty;$$

then, due to the weak lower semicontinuity of the norm of the Banach space $W_0^{1,p}(\Omega)$, we conclude that

$$\liminf_{n \to +\infty} \int_\Omega |D_x u_n|^p \, dx \geq \int_\Omega |D_x u_0|^p \, dx.$$

Moreover, since $f \in W^{-1,p'}(\Omega)$, we obtain the following limit equality:

$$\langle f, u_n \rangle \to \langle f, u_0 \rangle \quad \text{as } n \to +\infty.$$

Therefore,

$$\psi(u_0) \leq \liminf_{n \to +\infty} \psi(u_n) \quad \text{as } n \to +\infty.$$

Now we can apply Theorem 11.1, where we set $M = W_0^{1,p}(\Omega)$, $p > 1$. Thus, there exists a minimum point $\hat{u}(x) \in W_0^{1,p}(\Omega)$ of the functional $\psi(u)$ and hence a weak solution of the equation

$$\psi_f'(\hat{u}(x)) = \theta \quad \Longleftrightarrow \quad -\Delta_p u(x) + f(x) = \theta.$$

For completeness, we prove the uniqueness of a weak solution of the boundary-value problem considered.

Let $u_1(x), u_2(x) \in W_0^{1,p}(\Omega)$ be two solutions of the problem (11.11). Then by the definition of a weak solution (Definition 11.7), the following two equalities are valid:

$$\langle -\Delta_p u_k(x) + f(x), \varphi(x) \rangle = 0 \quad \text{for all } \varphi \in W_0^{1,p}(\Omega), \ k = 1, 2.$$

Subtracting one equality from the other, we obtain

$$\langle -\Delta_p u_1 + \Delta_p u_2, \varphi \rangle = 0 \quad \text{for all } \varphi \in W_0^{1,p}(\Omega).$$

Now we take $\varphi(x) = u_1(x) - u_2(x) \in W_0^{1,p}(\Omega)$; then we obtain the equality

$$\langle -\Delta_p u_1 + \Delta_p u_2, u_1 - u_2 \rangle = 0.$$

Then, due to the definition (11.15) of the operator $\operatorname{div}(\cdot)$, we obtain the following equality:

$$\int\limits_{\Omega} \Big(|D_x u_1(x)|^{p-2} D_x u_1(x) - |D_x u_2(x)|^{p-2} D_x u_2(x), D_x u_1(x) - D_x u_2(x) \Big) \, dx = 0.$$

Note that for arbitrary vectors $a, b \in \mathbb{R}^N$, the following chain of inequalities holds:

$$(|b|^{p-2} b - |a|^{p-2} a, b - a) \geq c(p) \, (|b| + |a|)^{p-2} \, |b - a|^2, \quad p > 1.$$

Therefore, we arrive at the inequality

$$0 = \int\limits_{\Omega} \Big(|D_x u_1(x)|^{p-2} D_x u_1(x) - |D_x u_2(x)|^{p-2} D_x u_2(x), D_x u_1(x) - D_x u_2(x) \Big) \, dx$$

$$\geq c(p) \int\limits_{\Omega} (|D_x u_1(x)| + |D_x u_2(x)|)^{p-2} \, |D_x u_1(x) - D_x u_2(x)|^2 \, dx.$$

This easily implies that $u_1(x) = u_2(x)$ almost everywhere on Ω.

11.5 Bibliographical Notes

The contents of this lecture is taken from $[13, 14, 23, 44, 56, 65, 67]$.

L. A. Lyusternik's Theorem
on Conditional Extrema

In many cases, the functional that immediately corresponds to a nonlinear boundary-value problem, in unbounded and hence does not possess extremal points on the given Banach space. However, to the boundary-value problem considered, one can assign a conditional variational problem such that all conditions of Theorem 4.1 are fulfilled and hence extremals of this variational problem satisfy the Lagrange equation, which will be obtained in this lecture.

12.1 L. A. Lyusternik Theorem: A Particular Case

Let

$$\varphi : \mathbb{B} \to \mathbb{R}^1, \quad \psi : \mathbb{B} \to \mathbb{R}^1$$

be functionals defined on a Banach space $\mathbb{B}$. Consider the manifold in $\mathbb{B}$ determined by the equation

$$V_c \overset{\text{def}}{=\!=} \{u \in \mathbb{B} : \varphi(u) = c\}, \quad c \in \mathbb{R}^1.$$

Definition 12.1. A point $u_0 \in V_c$ is called a minimum (maximum) point of the functional ψ with respect to the manifold V_c if there exists a neighborhood

$$O(u_0, r) := \{u \in \mathbb{B} : \|u - u_0\| < r\}, \quad r > 0,$$

such that

$$\psi(u) \geq \psi(u_0) \quad (\leq \psi(u_0)) \quad \text{for all } u \in V_c \cap O(u_0, r).$$

In the sequel, we consider the important case where the functionals ψ and φ are Fréchet-differentiable at the point $u_0 \in V_c$.

Definition 12.2. A point $u_0 \in V_c$ is called an *ordinary* point of the manifold V_c if

$$\left\| \varphi'_f(u_0) \right\|_* > 0.$$

Definition 12.3. A point $u_0 \in V_c$ is called a *conditionally critical* point of a functional ψ with respect to the manifold V_c if there exists a number $\mu \in \mathbb{R}^1$ such that

$$\psi'_f(u_0) = \mu \varphi'_f(u_0).$$

The following theorem is the basic result of this lecture.

Theorem 12.1. *Let functionals φ and ψ be Fréchet-differentiable at a point $u_0 \in$
$\mathbb{B}$, which is a ordinary point of the manifold $\varphi(u) = \varphi(u_0)$:*

$$\left\| \varphi_f'(u_0) \right\|_* > 0.$$

*If the point $u_0 \in \mathbb{B}$ is a point of conditional extremum of the functional ψ with
respect to the manifold*

$$V_{c_0} \overset{\text{def}}{=} \left\{ u \in \mathbb{B} : \; \varphi(u) = \varphi(u_0) = c_0 \right\},$$

then the point $u_0 \in V_{c_0}$ is conditionally critical, i.e., there exists $\mu \in \mathbb{R}^1$ such that

$$\psi_f'(u_0) = \mu \varphi_f'(u_0).$$

Proof. The general proof of this theorem will be presented in the following lecture
in connection with the Lyusternik–Schnirelmann category; now we prove it in one
important case where $\mathbb{B} = \mathbb{H}$ is a real Hilbert space with the scalar product $(\cdot, \cdot)$,
the functional $\varphi(u)$ has the form $\varphi(u) = (u, u)$, and the point u_0 lies on the sphere

$$\mathbb{S}_a \overset{\text{def}}{=} \left\{ u \in \mathbb{H} : \varphi(u) = (u, u) = a^2 \right\}, \quad a > 0.$$

1. First, we prove that each point of the sphere $\mathbb{S}_a$ is an ordinary point. Introduce the isometric Riesz–Fréchet operator

$$J : \mathbb{H}^* \to \mathbb{H},$$

which exists due to the well-known Riesz–Fréchet theorem on representation of a
linear continuous functional on a Hilbert space. The following equality holds:

$$\langle f, u \rangle = (Jf, u) \quad \text{for all } f \in \mathbb{H}^*, \, u \in \mathbb{H}.$$

We have the following chain of equalities:

$$\varphi(u + h) - \varphi(u) = (u + h, u + h) - (u, u)$$
$$= 2(u, h) + (h, h) = 2\langle J^{-1}u, h \rangle + \|h\|^2,$$

i.e.,

$$\varphi_f'(u) = 2J^{-1}u \quad \Rightarrow \quad \left\| \varphi_f'(u) \right\|_* = 2\|J^{-1}u\|_* = 2\|u\| = 2a > 0.$$

2. Assume that a point $u_0 \in \mathbb{H}$ is a point of conditional extremum of a functional
$\psi : \mathbb{H} \to \mathbb{R}^1$ with respect to the manifold $\mathbb{S}_a$. We prove that in this case, there exists
a number $\mu \in \mathbb{R}^1$ such that

$$\psi_f'(u_0) = \frac{\mu}{2} \varphi_f'(u_0) = \mu J^{-1} u_0.$$

Introduce the gradient operator

$$\operatorname{grad} \psi(u_0) := J \psi_f'(u_0).$$

We must prove the following equivalent relation:

$$\operatorname{grad} \psi(u_0) = \mu u_0. \tag{12.1}$$

Consider the one-dimensional subspace

$$H_1 \overset{\text{def}}{=} \{\lambda u_0 : \lambda \in \mathbb{R}^1\} \subset \mathbb{H}.$$

Let H_2 be the orthogonal completion to H_1 in $\mathbb{H}$, i.e., $\mathbb{H}$ possesses the orthogonal decomposition $\mathbb{H} = H_1 \oplus H_2$. Moreover, let $h \in H_2$ be an arbitrary vector belonging to the sphere $\mathbb{S}_a$, i.e., $\|h\| = a > 0$. Now we consider the vector

$$u = (1 + \alpha\varepsilon)u_0 + \varepsilon h, \tag{12.2}$$

where the value $\alpha = \alpha(\varepsilon)$ is defined by the condition that this vector lies on the sphere $\mathbb{S}_a$:

$$\|u\|^2 = a^2 \quad \Rightarrow \quad (1 + \varepsilon\alpha)^2 a^2 + \varepsilon^2 a^2 = a^2,$$

where we used the fact that $u_0 \perp h$. Thus, we arrive at the following equation:

$$(1 + \varepsilon\alpha)^2 + \varepsilon^2 = 1 \quad \Rightarrow \quad \varepsilon\alpha^2 + 2\alpha + \varepsilon = 0$$

whose roots are

$$\alpha_{1,2} = \frac{-1 \pm \sqrt{1 - \varepsilon^2}}{\varepsilon}.$$

From these two roots we choose

$$\alpha = \frac{-1 + \sqrt{1 - \varepsilon^2}}{\varepsilon}.$$

Note that

$$\alpha = -\frac{1}{2}\varepsilon + o(\varepsilon) \quad \text{as } \varepsilon \to 0.$$

3. Since, by the condition, the functional

$$\psi(u) : \mathbb{H} \to \mathbb{R}^1$$

is Fréchet-differentiable at the point $u_0 \in \mathbb{S}_a$, we conclude that for all $u \in \mathbb{S}_a$ of the form (12.2), the following representation holds:

$$\psi(u) - \psi(u_0) = \langle \psi'_f(u_0), u - u_0 \rangle + w(u_0, u - u_0)$$
$$= \big(\operatorname{grad} \psi(u_0), u - u_0 \big) + w(u_0, u - u_0). \tag{12.3}$$

Due to (12.2), we obtain the equality

$$\psi(u) - \psi(u_0) = \big(\operatorname{grad} \psi(u_0), \alpha\varepsilon u_0 + \varepsilon h \big) + w\big(u_0, \alpha\varepsilon u_0 + \varepsilon h\big);$$

moreover,

$$\alpha\varepsilon = o(\varepsilon), \quad w\big(u_0, \alpha\varepsilon u_0 + \varepsilon h\big) = o(\varepsilon).$$

Thus, we arrive at the relation

$$\psi(u) - \psi(u_0) = \varepsilon\big(\operatorname{grad} \psi(u_0), h \big) + o(\varepsilon).$$

By the condition of the theorem, $u = u_0 \in \mathbb{S}_a$ is a conditional extremal point of the functional ψ with respect to the sphere $\mathbb{S}_a$; therefore, for sufficiently small $\varepsilon > 0$, the sign of the left-hand side is preserved. This is possible only under the condition

$$\big(\operatorname{grad} \psi(u_0), h \big) = 0 \quad \text{for all } h \in H_2,$$

that is,

$$\operatorname{grad} \psi(u_0) \in H_1 \quad \Rightarrow \quad \operatorname{grad} \psi(u_0) = \mu u_0 \quad \text{for some } \mu \in \mathbb{R}^1. \qquad \square$$

12.2 Lyusternik Theorem: The General Case

Let $\mathbb{B}$ be a real separable Banach space and $\mathbb{B}^*$ be the dual space. Moreover, let

$$\varphi \in C^{(2)}(\mathbb{B}; \mathbb{R}^1),$$

i.e., a real functional φ is twice Fréchet-differentiable and its second derivative

$$\varphi''_{ff}(u) : \mathbb{B} \to \mathscr{L}(\mathbb{B}; \mathbb{B}^*)$$

is a continuous mapping

$$\varphi''_{ff}(u) \in C(\mathbb{B}; \mathscr{L}(\mathbb{B}; \mathbb{B}^*)).$$

Consider the following set:

$$\mathscr{V} \overset{\text{def}}{=} \{v \in \mathbb{B} : \varphi(v) = 1\}; \tag{12.4}$$

we assume that

$$\left\| \varphi'_f(v) \right\|_* > 0 \quad \text{for all } v \in \mathscr{V}. \tag{12.5}$$

Due to the last condition, the set $\mathscr{V} \subset \mathbb{B}$ is an ordinary manifold. From the theory of smooth manifolds we conclude that the manifold $\mathscr{V}$ is a C^2-manifold; moreover, the norm of the Banach space $\mathbb{B}$ induces the standard metric on this manifold, and the manifold $\mathscr{V}$ is a metric space with respect to this metric satisfying the ANR-property.

Now we introduce the *tangent space* at a point $v \in \mathscr{V}$:

$$T_v \mathscr{V} \overset{\text{def}}{=} \left\{ u \in \mathbb{B} : \langle \varphi'_f(v), u \rangle = 0 \right\}. \tag{12.6}$$

Remark 12.1. We can propose the following motivation of the notion of the tangent space $T_v \mathscr{V}$. Consider an arbitrary smooth curve $\gamma(t) \in \mathscr{V}$, $t \in (-\varepsilon, \varepsilon)$, such that $\gamma(0) = v \in \mathscr{V}$. It suffices to take $\gamma(t) \in C^{(1)}((-\varepsilon, \varepsilon); \mathscr{V})$. By the definition (12.4) of the manifold $\mathscr{V}$,

$$\varphi(\gamma(t)) = 0 \quad \text{for } t \in (-\varepsilon, \varepsilon). \tag{12.7}$$

On the one hand, since the functional $\varphi(u)$ is Fréchet-differentiable, we have

$$\frac{d}{dt} \varphi(\gamma(t)) \Big|_{t=0} = \langle \varphi'_f(\gamma(0)), \dot{\gamma}(0) \rangle = \langle \varphi'_f(v), u \rangle, \quad u := \dot{\gamma}(0) \in \mathbb{B}. \tag{12.8}$$

The sense of u is the following: it is a tangent vector of the manifold $\mathscr{V}$ at the point $v \in \mathscr{V}$. Moreover, for any tangent vector u, there exists a smooth curve $\gamma(t)$ such that $u = \dot{\gamma}(0)$. On the other hand, by (12.7), we obtain the equality

$$\frac{d}{dt} \varphi(\gamma(t)) \Big|_{t=0} = 0. \tag{12.9}$$

Thus, we conclude from (12.8) and (12.9) that the whole set of tangent vectors u at the point $v \in \mathscr{V}$ is described by Eq. (12.6).

We state the following properties of the tangent space.

Theorem 12.2. *Note that $T_v\mathscr{V}$ is a linear subspace in $\mathbb{B}$, which is weakly closed in $\mathbb{B}$. Moreover, $T_v\mathscr{V}$ is a Banach space with respect to the norm $\|\cdot\|$ of the Banach space $\mathbb{B}$.*

Proof. Indeed, let $\{u_n\} \subset T_v\mathscr{V}$ and

$$\|u_n - u_m\| \to +0 \quad \text{as } n, m \to +\infty.$$

Due to the completeness of the space $\mathbb{B}$ we have

$$u_n \to u \quad \text{strongly in } \mathbb{B} \text{ as } n \to +\infty.$$

Then, obviously,

$$u_n \rightharpoonup u \quad \text{weakly in } \mathbb{B} \text{ as } n \to +\infty.$$

Therefore, $u \in T_v\mathscr{V}$. $\qquad\square$

Theorem 12.3. *The dual space $(T_v\mathscr{V})^*$ is a Banach space with respect to the norm*

$$\|f^*\|_* (T_v\mathscr{V}) := \sup_{\substack{\|u\| \leq 1, \\ u \in T_v\mathscr{V}}} |\langle f^*, u \rangle|,$$

where $\langle \cdot, \cdot \rangle$ is the duality bracket between $(T_v\mathscr{V})^*$ and $T_v\mathscr{V}$, which coincide with the duality bracket between the Banach spaces $\mathbb{B}^*$ and $\mathbb{B}$ in the case where $f^* \in \mathbb{B}^*$.

Remark 12.2. Note that the tangent space is actually nondegenerate since, due to (12.5), at each point $v \in \mathscr{V}$, we have

$$\varphi'_f(v) \neq \theta^* \quad \Rightarrow \quad T_v\mathscr{V} \neq \mathbb{B}.$$

In the sequel, we consider a conditional extremum of a functional ψ on the manifold $\mathscr{V}$ generated by a functional φ.

Introduce a functional

$$\psi : \mathbb{B} \to \mathbb{R}^1$$

of the class $C^{(1)}(\mathbb{B}; \mathbb{R}^1)$, i.e., ψ is Fréchet-differentiable on $\mathbb{B}$ and its Fréchet derivative is a continuous mapping:

$$\psi'_f(u) \in C(\mathbb{B}; \mathbb{B}^*) : \ u \in \mathbb{B} \mapsto \psi'_f(u) \in \mathbb{B}^*.$$

Lemma 12.1. *If a functional $\psi(u) \in C^{(1)}(\mathbb{B}; \mathbb{R}^1)$ attains an extremum at a point $v \in \mathscr{V}$, then the following equality holds:*

$$\langle \psi'_f(v), u \rangle = 0 \quad \text{for all } u \in T_v\mathscr{V}. \tag{12.10}$$

Proof. By the condition, for any smooth curve $\gamma(t) \in C((-\varepsilon, \varepsilon); \mathscr{V})$ such that $\gamma(0) = v$ and $\dot{\gamma}(0) = u$, the necessary condition of extremum holds:

$$0 = \frac{d}{dt}\psi(\gamma(t))\Big|_{t=0} = 0 \quad \Leftrightarrow \quad \langle \psi'_f(\gamma(0)), \dot{\gamma}(0) \rangle = \langle \psi'_f(v), u \rangle$$

for any $u \in \{u \in \mathbb{B} : \langle \varphi'_f(v), u \rangle = 0\}$ (see Remark 12.1). $\qquad\square$

Consider the norm of the Fréchet derivative $\psi'_f(v)$ restricted to the tangent space $T_v \mathscr{V}$:

$$\left\| \psi'_f(v) \right\|_* (T_v \mathscr{V}) \overset{\text{def}}{=} \sup_{\substack{\|u\| \leq 1, \\ u \in T_v \mathscr{V}}} \left| \langle \psi'_f(v), u \rangle \right|, \tag{12.11}$$

where $v \in \mathscr{V}$.

Remark 12.3. The following inequality holds:

$$\left\| \psi'_f(v) \right\|_* (T_v \mathscr{V}) \leq \left\| \psi'_f(v) \right\|_*, \tag{12.12}$$

since in the definition of the norm with restriction, the supremum is taken over by a narrower set.

The following inequality will be used below:

$$\langle f^*, w \rangle \leq \|f^*\|_* (T_v \mathscr{V}) \|w\| \quad \text{for all } w \in T_v \mathscr{V}, \, f^* \in \mathbb{B}^*,$$

which can be proved as follows. Due to (12.11), we have the inequality

$$\left\langle f^*, \frac{w}{\|w\|} \right\rangle \leq \|f^*\|_* (T_v \mathscr{V}) \quad \text{for } w \neq \theta, \, w \in T_v \mathscr{V}, \, f^* \in \mathbb{B}^*,$$

since the inequality considered is valid for $w = \theta$.

Definition 12.4. A point $v \in \mathscr{V} \subset \mathbb{B}$ is called a *critical point* of a functional $\psi \in C^1(\mathbb{B}; \mathbb{R}^1)$ relative to a manifold $\mathscr{V}$ if the following equality holds:

$$\left\| \psi'_f(v) \right\|_* (T_v \mathscr{V}) = 0. \tag{12.13}$$

Remark 12.4. Note that the manifold $\mathscr{V}$ is "curved" and, in general, it is not Banach space, but only a metric space. Therefore, a necessary condition of extremum of a functional $\psi(u)$ is changed: instead of the condition

$$\psi'_f(u_0) = \theta^* \in \mathbb{B}^*, \quad \ker(\theta^*) = \mathbb{B},$$

we now have the condition

$$\psi'_f(u_0) = \theta^*_{u_0} \in \mathbb{B}^*, \quad \ker(\theta^*_{u_0}) = \ker(\varphi'_f(u_0)) = T_{u_0} \mathscr{V} \subset \mathbb{B}.$$

Lemma 12.2. *If a functional $\psi(u) \in C^{(1)}(\mathbb{B}; \mathbb{R}^1)$ attains a local extremum at a point $v \in \mathscr{V}$, then*

$$\left\| \psi'_f(v) \right\|_* (T_v \mathscr{V}) = 0.$$

Proof. Lemma 12.2 immediately follows from Lemma 12.1 and the definition (12.11) of the norm with restriction on the tangent space $T_v \mathscr{V}$. $\qquad\square$

Lemma 12.3 (duality lemma). *Let $f, g \in \mathbb{B}^*$. Then the following equality holds:*

$$\sup_{\substack{\|v\| \leq 1, \\ \langle g, v \rangle = 0}} |\langle f, v \rangle| = \min_{\lambda \in \mathbb{R}^1} \|f - \lambda g\|_*. \tag{12.14}$$

Proof. Indeed, on the one hand,

$$\|f\|_* (\ker(g)) := \sup_{\substack{\|v\|\leq 1, \\ \langle g,v\rangle=0}} |\langle f,v\rangle| = \sup_{\substack{\|v\|\leq 1, \\ \langle g,v\rangle=0}} |\langle f - \lambda g, v\rangle|. \tag{12.15}$$

On the other hand, by the Hahn–Banach theorem, there exists an extension $\overline{f} \in \mathbb{B}^*$ of the functional f such that

$$\langle \overline{f}, v\rangle = \langle f, v\rangle \quad \text{for all} \quad v \in \ker(g) := \big\{v \in \mathbb{B} : \langle g, v\rangle = 0\big\} \tag{12.16}$$

and the following equality holds:

$$\|\overline{f}\|_* = \|f\|_* (\ker(g)). \tag{12.17}$$

Moreover, due to (12.16), we have the embedding

$$\ker(g) \subset \ker(\overline{f} - f),$$

which implies the existence of $\lambda_0 \in \mathbb{R}^1$ such that

$$\overline{f} - f = \lambda_0 g \tag{12.18}$$

(see [62, Lemma 3.9]); in particular, $\ker(\overline{f} - f) = \ker(\lambda_0 g)$. From (12.15), (12.17), and (12.18), we obtain the following chain of inequalities:

$$\|f - \lambda_0 g\|_* = \|\overline{f}\|_* = \|f\|_* (\ker(g))$$
$$= \sup_{\substack{\|v\|\leq 1, \\ \langle g,v\rangle=0}} |\langle f - \lambda g, v\rangle| \leq \|f - \lambda g\|_* \quad \text{for all } \lambda \in \mathbb{R}^1.$$

Therefore,

$$\|f\|_* (\ker(g)) = \min_{\lambda \in \mathbb{R}^1} \|f - \lambda g\|_*. \qquad \square$$

Remark 12.5. For completeness, we prove the following auxiliary assertion, which is needed for the proof of Lemma 12.3 (see [62]).

Theorem 12.4. *Let Λ and Λ_i, $i = \overline{1, N}$, be linear functionals on a vector space X and let $\langle \Lambda, x\rangle = 0$ for all $x \in \mathcal{N}$,*

$$\mathcal{N} \overset{\text{def}}{=\!=} \big\{x \in X : \langle \Lambda_1, x\rangle = 0, \dots, \langle \Lambda_n, x\rangle = 0\big\}.$$

Then

$$\Lambda = \alpha^1 \Lambda_1 + \cdots + \alpha^n \Lambda_n$$

for some $\alpha^1, \dots, \alpha^n \in K$, where K is a field of scalars ($K = \mathbb{C}$ or $K = \mathbb{R}$).

Proof. Introduce the mapping

$$\pi(x) : X \to K^n \overset{\text{def}}{=\!=} K \otimes \cdots \otimes K$$

as follows:

$$\pi(x) \overset{\text{def}}{=\!=} \big(\langle \Lambda_1, x\rangle, \dots, \langle \Lambda_n, x\rangle\big).$$

Note that if $\pi(x) = \pi(x')$, then the following chain of implications holds:

$$\pi(x) = \pi(x') \quad \Rightarrow \quad (\langle \Lambda_1, x - x' \rangle, \dots, \langle \Lambda_n, x - x' \rangle) = (0, \dots, 0) \in K^n$$
$$\Rightarrow \quad x - x' \in N \quad \Rightarrow \quad \langle \Lambda, x - x' \rangle = 0 \quad \Rightarrow \quad \langle \Lambda, x \rangle = \langle \Lambda, x' \rangle.$$

Thus on K^n there exists a *single-valued* function $F\circ$ defined as follows:

$$\langle \Lambda, x \rangle = F \circ \pi(x).$$

We prove that this function is *linear*. Note that

$$\alpha^1 \pi(x^1) + \alpha^2 \pi(x^2) = \pi(\alpha^1 x^1 + \alpha^2 x^2) \quad \text{for all } \alpha^1, \alpha^2 \in K, \ x^1, x^2 \in X;$$

therefore, we have

$$F(\pi(\alpha^1 x^1 + \alpha^2 x^2)) = \langle \Lambda, \alpha^1 x^1 + \alpha^2 x^2 \rangle$$
$$= \alpha^1 \langle \Lambda, x^1 \rangle + \alpha^2 \langle \Lambda, x^2 \rangle = \alpha^1 F(\pi(x^1)) + \alpha^2 F(\pi(x^2)).$$

Therefore, the function F is linear on K^n; hence it has the form

$$F(u_1, \dots, u_n) = \alpha^1 u_1 + \cdots + \alpha^n u_n,$$

where $\alpha^k \in K$ are constants, $k = \overline{1, n}$. This implies the following chain of equalities:

$$\langle \Lambda, x \rangle = F(\pi(x)) = F(\langle \Lambda_1, x \rangle, \dots, \langle \Lambda_n, x \rangle) = \sum_{i=1}^{n} \alpha^i \langle \Lambda_i, x \rangle = \left\langle \sum_{i=1}^{n} \alpha^i \Lambda_i, x \right\rangle$$

for all $x \in X$. Thus,

$$\Lambda = \sum_{i=1}^{n} \alpha^i \Lambda_i.$$

The theorem is proved. $\qquad\qquad\qquad\qquad\qquad\qquad\qquad\qquad\qquad\qquad\square$

Lemma 12.3 immediately implies the following important assertion.

Theorem 12.5. *Let* $\varphi \in C^{(1)}(\mathbb{B}; \mathbb{R}^1)$ $\psi \in C^{(1)}(\mathbb{B}; \mathbb{R}^1)$, *and* $u \in \mathcal{V}$, *where the manifold* $\mathcal{V}$ *is defined by the formula* (12.4). *Then the following equality holds:*

$$\left\| \psi'_f(u) \right\|_* (T_u \mathcal{V}) = \min_{\lambda \in \mathbb{R}^1} \left\| \psi'_f(u) - \lambda \varphi'_f(u) \right\|_*. \tag{12.19}$$

In particular, if $u \in \mathcal{V}$ *is a critical point of the functional* ψ *with respect to the manifold* $\mathcal{V}$, *then there exists* $\mu \in \mathbb{R}^1$ *such that*

$$\psi'_f(u) - \mu \varphi'_f(u) = \theta \in \mathbb{B}^*. \tag{12.20}$$

Proof. It suffices to set $f = \psi'_f(u) \in \mathbb{B}^*$ and $g = \varphi'_f(u) \in \mathbb{B}^*$ in Lemma 13.1, where $u \in \mathbb{B}$ is fixed and $\mu = \lambda_0$. $\qquad\qquad\qquad\qquad\qquad\qquad\qquad\qquad\qquad\square$

Remark 12.6. The proof of Theorem 12.3 is the promised proof of Theorem 11.1 in the general case for functionals $\varphi, \psi \in C^{(1)}(\mathbb{B}; \mathbb{R}^1)$.

12.3 Example

Consider the following nonlinear boundary-value problem:

$$-\Delta u + \lambda u = |u|^{p-2}u \quad \text{in } \Omega, \tag{12.21}$$

$$u(x) = 0 \qquad\qquad \text{on } \partial\Omega. \tag{12.22}$$

Assume that $\Omega \subset \mathbb{R}^N$, $N \geq 3$, is a bounded domain with smooth boundary $\partial\Omega \in C^{(2,\delta)}$, $\delta \in (0,1]$. We also assume that

$$2 < p < \frac{2N}{N-2}, \quad N \geq 3. \tag{12.23}$$

Then due to the Sobolev embedding theorem (see [1]), we have the completely continuous embedding

$$H_0^1(\Omega) \hookrightarrow\hookrightarrow L^p(\Omega), \quad N \geq 3.$$

Note that the Euler functional of this problem has the form

$$E(u) := \frac{1}{2}\int_\Omega |D_x u|^2 \, dx + \frac{\lambda}{2}\int_\Omega |u|^2 \, dx - \frac{1}{p}\int_\Omega |u|^p \, dx.$$

The Fréchet derivative $E'_f(u)$ satisfies the equation

$$\langle E'_f(u), v \rangle = 0 \quad \text{for all } v \in H_0^1(\Omega),$$

where

$$E'_f(u) = -\Delta u + \lambda u - |u|^{p-2}u.$$

This is the weak statement of the problem (12.21). This functional is unbounded from below and from above and hence it does not attain extrema on $H_0^1(\Omega)$. However, this nonlinear boundary-valued problem admits the variational statement as a *conditional extremal problem*.

Definition 12.5. A function $u \in H_0^1(\Omega)$ satisfying the condition

$$\langle -\Delta u + \lambda u - |u|^{p-2}u, v \rangle = 0 \quad \text{for all } v \in H_0^1(\Omega), \tag{12.24}$$

where $\langle \cdot, \cdot \rangle$ is the duality bracket between the Hilbert spaces $H_0^1(\Omega)$ and $H^{-1}(\Omega)$, is called a *weak solution* of the problem (12.21).

Earlier, we proved that

$$\Delta_p : W_0^{1,p}(\Omega) \to W^{-1,p'}(\Omega), \quad p \geq 2.$$

For $p = 2$, we have $\Delta_p = \Delta$; therefore,

$$\Delta : H_0^1(\Omega) \overset{\text{def}}{=\!=} W_0^{1,2}(\Omega) \to H^{-1}(\Omega) \overset{\text{def}}{=\!=} W^{-1,2}(\Omega). \tag{12.25}$$

The following chain of dense and continuous embedding holds:

$$H_0^1(\Omega) \overset{ds}{\subset} L^p(\Omega), \quad L^{p'}(\Omega) \overset{ds}{\subset} H^{-1}(\Omega), \quad p \in [2, 2^*). \tag{12.26}$$

Indeed, under the conditions specified, $H_0^1(\Omega)$ and $L^p(\Omega)$ are reflexive Banach spaces and the following dense embeddings hold:

$$H_0^1(\Omega) \overset{ds}{\subset} L^p(\Omega) \quad \Rightarrow \quad L^{p'}(\Omega) \overset{ds}{\subset} H^{-1}(\Omega).$$

Moreover, we have the equalities

$$\langle f, u \rangle = \int_\Omega f(x)u(x)\,dx \quad \text{for all } u(x) \in H_0^1(\Omega), \qquad f(x) \in L^{p'}(\Omega) \overset{ds}{\subset} H^{-1}(\Omega),$$

$$\langle f, u \rangle = \int_\Omega f(x)u(x)\,dx \quad \text{for all } u(x) \in H_0^1(\Omega), \qquad f(x) \in L^2(\Omega) \overset{ds}{\subset} H^{-1}(\Omega).$$

Note that

$$|u|^{p-2}u : L^p(\Omega) \to L^{p'}(\Omega), \quad p > 2, \quad p' = \frac{p}{p-1}.$$

Indeed,

$$\int_\Omega \left| |u(x)|^{p-2}u(x) \right|^{p'}\,dx = \int_\Omega |u(x)|^{(p-1)p'}\,dx = \int_\Omega |u(x)|^p\,dx.$$

Therefore,

$$|u|^{p-2}u : H_0^1(\Omega) \overset{ds}{\subset} L^p(\Omega) \to L^{p'}(\Omega) \overset{ds}{\subset} H^{-1}(\Omega).$$

Finally, we have for the identity operator

$$Iu : L^2(\Omega) \to L^2(\Omega),$$

and due to the chain of embeddings (12.26) we conclude that

$$Iu : H_0^1(\Omega) \overset{ds}{\subset} L^2(\Omega) \to L^2(\Omega) \overset{ds}{\subset} H^{-1}(\Omega).$$

Thus we see that under the condition (12.23), the nonlinear operator acts as follows:

$$-\Delta u + \lambda u - |u|^{p-2}u : H_0^1(\Omega) \to H^{-1}(\Omega).$$

Therefore, the definition of a weak solution (Definition 12.5) is valid.

Now, to the boundary-value problem (12.24) treated in the weak sense (12.24), we assign the following conditional variational problem. Consider the functional

$$\psi(u) := \frac{1}{2} \int_\Omega \left(|D_x u(x)|^2 + \lambda |u(x)|^2 \right) dx \tag{12.27}$$

on the Hilbert space $H_0^1(\Omega)$ and the manifold

$$V \overset{\text{def}}{=} \left\{ u \in H_0^1(\Omega) : \ \varphi(u) := \int_\Omega |u(x)|^p\,dx = 1 \right\}. \tag{12.28}$$

1. First, we verify that the functional $\psi(u)$ is weakly lower semicontinuous on V. Let

$$\{u_n\} \subset V \subset H_0^1(\Omega)$$

and

$$u_n \rightharpoonup u \in V \quad \text{weakly in } H_0^1(\Omega) \text{ as } n \to +\infty.$$

Then

$$\liminf_{n \to +\infty} \psi(u_n) \geq \psi(u).$$

Indeed, this follows from the fact that

$$\liminf_{n \to +\infty} \int_{\Omega} |D_x u_n(x)|^2 \, dx \geq \int_{\Omega} |D_x u(x)|^2 \, dx,$$

since

$$\left(\int_{\Omega} |D_x u(x)|^2 \, dx \right)^{1/2}$$

is the norm on $H_0^1(\Omega)$. Moreover, due to the Sobolev embedding theorem, we have the totally continuous embedding

$$H_0^1(\Omega) \hookrightarrow\hookrightarrow L^2(\Omega).$$

Therefore,

$$u_n \to u \quad \text{strongly in } L^2(\Omega) \text{ as } n \to +\infty.$$

Hence

$$\lambda \int_{\Omega} |u_n(x)|^2 \, dx \to \lambda \int_{\Omega} |u(x)|^2 \, dx \quad \text{as } n \to +\infty.$$

The weak lower semicontinuity of the functional ψ is proved.

2. Now we prove that the set V is weakly closed. Let $\{u_n\} \subset V$ and

$$u_n \rightharpoonup u \quad \text{weakly in } H_0^1(\Omega).$$

Due to the assumption (12.23), we have the following totally continuous embedding:

$$H_0^1(\Omega) \hookrightarrow\hookrightarrow L^p(\Omega);$$

therefore,

$$u_n \to u \quad \text{strongly in } L^p(\Omega) \text{ as } n \to +\infty.$$

Then

$$1 = \int_{\Omega} |u_n(x)|^p \, dx \to \int_{\Omega} |u(x)|^p \, dx.$$

Hence

$$\int_{\Omega} |u(x)|^p \, dx = 1,$$

i.e., $u \in V$. The weak closedness is proved.

3. Prove the weak coerciveness of the functional $\psi(u)$ on $H_0^1(\Omega)$. Due to the Friedrichs inequality, we have the inequality

$$\int_\Omega |D_x u(x)|^2 \, dx \geq \lambda_1 \int_\Omega |u(x)|^2 \, dx, \quad u \in H_0^1(\Omega),$$

where $0 < \lambda_1$ is the first eigenvalue of the operator $-\Delta$ with the homogeneous Dirichlet condition. Due to the Friedrichs inequality, for $\lambda < 0$ we have the following lower estimate:

$$\frac{\lambda}{2} \int_\Omega |u(x)|^2 \, dx = -\frac{|\lambda|}{2} \int_\Omega |u(x)|^2 \, dx$$

$$\geq -\frac{|\lambda|}{2\lambda_1} \int_\Omega |D_x u(x)|^2 \, dx = \frac{\lambda}{2\lambda_1} \int_\Omega |D_x u(x)|^2 \, dx.$$

Therefore, the functional $\psi(u)$ satisfies the following lower estimate for $\lambda \in (-\lambda_1, 0)$:

$$\psi(u) \geq \frac{1}{2} \int_\Omega |D_x u(x)|^2 \, dx + \frac{\lambda}{2\lambda_1} \int_\Omega |D_x u(x)|^2 \, dx = \frac{1}{2}\left(1 + \frac{\lambda}{\lambda_1}\right) \int_\Omega |D_x u(x)|^2 \, dx.$$

In the case where $\lambda \geq 0$, the coerciveness of this functional is obvious. Therefore, the functional $\psi(u)$ is weakly coercive under the condition $\lambda > -\lambda_1$.

4. It remains to verify that all points of the manifold V are ordinary. Consider the functional

$$\varphi(u) := \int_\Omega |u(x)|^p \, dx : \ H_0^1(\Omega) \subset L^p(\Omega) \to \mathbb{R}^1, \quad 2 < p < \frac{2N}{N-2}, \quad N \geq 3.$$

Its Fréchet derivative has the form

$$\varphi_f'(u) = p|u|^{p-2} u \in L^{p'}(\Omega).$$

On the one hand,

$$\left\|\varphi_f'(u)\right\|_* = \sup_{\|D_x v\|_2 \leq 1} \left|\langle \varphi_f'(u), v \rangle\right|.$$

On the other hand, note that

$$\langle \varphi_f'(u), u \rangle = p \int_\Omega |u(x)|^p \, dx = p > 0 \quad \text{on } V.$$

Therefore,

$$\left\|\varphi_f'(u)\right\|_* = \sup_{\|D_x v\|_2 \leq 1} \left|\langle \varphi_f'(u), v \rangle\right| \geq \frac{\langle \varphi_f'(u), u \rangle}{\||D_x u\||_2} = \frac{p}{\||D_x u\||_2} > 0 \quad \text{for all } u \in V.$$

Thus, all the conditions of Theorem 4.1 are fulfilled. Therefore, there exists a point $u_0 \in V$ at which the functional ψ attains a minimum. Moreover, this implies the fulfillment of all conditions of Theorem 12.1. Hence, there exists a number $\mu \in \mathbb{R}^1$ such that the following equality holds:

$$\langle \psi_f'(u_0) - \mu \varphi_f'(u_0), v \rangle = 0 \quad \text{for all } v \in H_0^1(\Omega),$$

which is equivalent to the equality

$$\langle -\Delta u_0 + \lambda u_0 - \mu p |u_0|^{p-2} u_0, \ v \rangle = 0 \quad \text{for all } v \in H_0^1(\Omega). \tag{12.29}$$

5. Now we prove that $\mu > 0$. Setting $v = u_0 \in H_0^1(\Omega)$ in Eq. (12.29) and integrating by parts, we obtain the equality

$$2\psi(u_0) = \int_\Omega \left[|D_x u_0(x)|^2 + \lambda |u_0(x)|^2 \right] dx = \mu \int_\Omega |u_0(x)|^p \, dx = \mu$$

since $u_0 \in V$. As was proved above, $\psi(u_0) > 0$; therefore, $\mu > 0$. Performing the substitution

$$u_0 = c_1 u, \quad c_1 = \left(\frac{1}{\mu p} \right)^{1/(p-2)},$$

we arrive at Eq. (12.24).

Thus, the nonlinear boundary-value problem (12.21) is solvable in the weak sense for $\lambda > -\lambda_1$.

12.4 Bibliographical Notes

The contents of this lecture is taken from [14, 20, 23, 24, 35, 62, 65, 67].

Lecture 13

S. I. Pokhozhaev's Method
of Spherical Fibering

In 1948, L. V. Ovsyannikov in his Ph.D. thesis (see [57]) obtained the nonlinear equation

$$\Delta u = u^2, \quad u = u(x, y, z), \tag{13.1}$$

which describes transonic flows of an ideal polytropic gas. In this lecture, we present S. I. Pokhozhaev's method of spherical fibering for the boundary-value problem for Ovsyannikov's equation, and, in particular, prove the existence of solutions for two nonlinear solutions.

13.1 Ovsyannikov's Equation

L. V. Ovsyannikov obtained the following nonlinear equation describing three-dimensional transonic flows of an ideal polytropic gas:

$$\frac{\partial^2 u}{\partial x^2} + \frac{\partial^2 u}{\partial y^2} = \frac{\partial^2 (u-1)^2}{\partial z^2}. \tag{13.2}$$

An exact solution of this nonlinear equation has the following form:

$$u(x, y, z) = 1 + u_0(x, y) + u_1(x, y)z + \frac{1}{12}u_2(x, y)z^2, \tag{13.3}$$

where the functions $u_2(x, y)$, $u_1(x, y)$, and $u_0(x, y)$ are solutions of the following equations:

$$\Delta u_2 - u_2^2 = 0, \tag{13.4}$$

$$\Delta u_1 - u_2 u_1 = 0, \tag{13.5}$$

$$\Delta u_0 - \frac{1}{3}u_2 u_0 = 2u_1^2. \tag{13.6}$$

If u_2 is given, then Eq. (13.5) is linear with respect to u_1, and if u_1 and u_2 are given, then Eq. (13.6) is linear with respect to u_0. Therefore, we first must examine the nonlinear equation (13.4).

13.2   Statement of the Problem

Below, we consider the three-dimensional version of Eq. (13.4). Let $\Omega \subset \mathbb{R}^3$ be a bounded domain with smooth boundary $\partial\Omega$. Consider the following boundary-value problem:

$$\Delta\Phi(x) + \Phi^2(x) = 0, \quad x \in \Omega, \tag{13.7}$$

$$\Phi(x) = h(x) \qquad\qquad x \in \partial\Omega. \tag{13.8}$$

Assume that $h(x) \in H^{1/2}(\partial\Omega)$.[1] Let $H(x) \in H^1(\Omega)$ be a unique weak solution of the problem

$$\Delta H(x) = 0, \quad x \in \Omega, \tag{13.9}$$

$$H(x) = h(x), \quad x \in \partial\Omega. \tag{13.10}$$

Consider the function $u(x) = \Phi(x) - H(x)$; owing to (13.9) and (13.10) we obtain the following boundary-value problem for this function:

$$\Delta u(x) + (u(x) + H(x))^2 = 0, \quad x \in \Omega, \tag{13.11}$$

$$u(x) = 0, \qquad\qquad x \in \partial\Omega, \tag{13.12}$$

where $H(x) \in H^1(\Omega)$. The remaining part of this lecture is devoted to the study of the boundary-value problem (13.11)–(13.12).

13.3   Variational Statement of the Problem

Since $\Omega \subset \mathbb{R}^3$ is a bounded domain, we have the following continuous and dense embeddings:

$$H_0^1(\Omega) \overset{ds}{\subset} L^4(\Omega) \overset{ds}{\subset} L^{4/3}(\Omega) \overset{ds}{\subset} H^{-1}(\Omega). \tag{13.13}$$

First, we consider the Euler functional, which, due to (13.13), is defined on the Hilbert space $H_0^1(\Omega)$:

$$E(u) = F(u) + G(u), \tag{13.14}$$

where

$$F(u) := \frac{1}{2} \int_\Omega |D_x u|^2 \, dx, \quad G(u) := -\frac{1}{3} \int_\Omega \left[(u(x) + H(x))^3 - H^3(x) \right] dx. \tag{13.15}$$

The Fréchet derivatives of the functionals $F(u)$ and $G(u)$ have the form

$$F_f'(u) := -\Delta u(x), \quad G_f'(u) = -(u(x) + H(x))^2. \tag{13.16}$$

Indeed, the first equality in (13.16) was proved in Lecture 5. Prove the second relation. Applying the theorem on the necessary and sufficient condition of the potentiality of an operator (Theorem 10.1) to the nonlinear operator

$$g(u) = -\big(u(x) + H(x)\big)^2 : L^4(\Omega) \to L^{4/3}(\Omega),$$

---

[1] We have not introduced the space $H^{1/2}(\partial\Omega)$ above. The need of considering this space is related to the existence of a unique solution $H(x)$ of the class $H^1(\Omega)$ to the problem (13.9), (13.10) stated below.

we obtain that its potential has the form

$$\psi(u) := -\int_0^1 \int_\Omega (tu(x) + H(x))^2 u(x)\, dx$$

$$= -\int_\Omega \int_0^1 (tu(x) + H(x))^2 u(x)\, dt\, dx = -\int_\Omega \int_{H(x)}^{u(x)+H(x)} z^2\, dz\, dx$$

$$= -\frac{1}{3}\int_\Omega \left[(u(x) + H(x))^3 - H^3(x) \right] dx, \quad \psi(\theta) = 0. \tag{13.17}$$

Now we immediately obtain that the Fréchet derivative of the functional $\psi(u)$ is equal to $g(u)$. Thus, we have the following assertion.

Theorem 13.1. *Each weak solution $u(x) \in H_0^1(\Omega)$ of the boundary-value problem* (13.11)–(13.12) *is a critical point of the functional $E(u) = F(u) + G(u)$, and conversely, each critical point of the functional $E(u)$ is a weak solution of the boundary-value problem* (13.11), (13.12).

For convenience, instead of the functional $E(u)$, we consider the functional

$$E_1(u) = 2E(u) = \int_\Omega |D_x u|^2\, dx + \frac{2}{3}\int_\Omega \left(-(u(x) + H(x))^3 + H^3(x) \right) dx. \tag{13.18}$$

We also introduce the functional

$$\psi_1(t, v) := E_1(t \cdot v) : \mathbb{R}^1 \otimes H_0^1(\Omega) \to \mathbb{R}^1. \tag{13.19}$$

Obviously, the vector space $\mathbb{R}^1 \otimes H_0^1(\Omega)$ is a Banach space with respect to the norm

$$\|(t, v)\| := |t| + \|v\|, \quad \|v\| := \left(\int_\Omega |D_x v(x)|^2\, dx \right)^{1/2}. \tag{13.20}$$

We search for conditionally critical points of the functional $\psi_1(t, v)$ with respect to the manifold

$$\mathscr{V} := \mathbb{R}^1 \backslash \{0\} \otimes S, \tag{13.21}$$

where

$$S := \{ v \in H_0^1(\Omega) : \varphi(t, v) = 1 \}, \quad \varphi(t, v) := \int_\Omega |D_x v|^2\, dx. \tag{13.22}$$

We need an explicit expression for the Fréchet derivative of the functional $\psi_1(t, v)$. For any $(\tau, h) \in \mathbb{R}^1 \otimes H_0^1(\Omega)$, we have the equality

$$\psi_1(t + \tau, v + h) = \psi_1(t, v) + \frac{\partial \psi_1}{\partial t}(t, v) + \langle \psi_{1v}'(t, v), h \rangle + \omega((t, v), (\tau, h)); \tag{13.23}$$

moreover, due to the Fréchet differentiability of the functional $E(u)$, the following limit relation holds:

$$\lim_{|\tau|+\|h\|\to+0} \frac{\omega((t,v),(\tau,h))}{|\tau|+\|h\|} = 0. \tag{13.24}$$

Therefore, the Fréchet derivative has the following form:

$$\psi'_{1w}(t,v) = \left(\frac{\partial \psi_1(t,v)}{\partial t}, \psi'_{1v}(t,v)\right). \tag{13.25}$$

Now we prove that the manifold $\mathcal{V}$ is ordinary. Indeed, assume that let there exists a point $w_0 = (t_0, v_0) \in \mathcal{V}$ such that

$$\varphi'_w(w_0) = (0, -2\Delta v_0) = 0 \quad \Rightarrow \quad \Delta v_0 = \theta \quad \Rightarrow \quad \|v_0\|^2 = \langle -\Delta v_0, v_0 \rangle = 0. \tag{13.26}$$

However, $v_0 \in S$ and hence $\|v_0\|^2 = 1$, a contradiction.

Now we examine the equality

$$\|\psi'_{1w}(w)\|_* (T_w \mathcal{V}) = 0, \quad w = (t,v) \in \mathcal{V}. \tag{13.27}$$

Note that

$$T_w \mathcal{V} = \Big\{ z = (\tau, h) \in \mathbb{R}^1 \otimes H_0^1(\Omega) :$$

$$\langle\!\langle \varphi'_w(w), z \rangle\!\rangle = \frac{\partial \varphi(w)}{\partial t}\tau + \langle \varphi'_v(w), h \rangle = 0 \Big\}. \tag{13.28}$$

Since

$$\varphi(w) := \int_{\Omega} |D_x v|^2 \, dx$$

we obtain

$$\frac{\partial \varphi(w)}{\partial t} = 0, \quad \varphi'_v(w) = -\Delta v \tag{13.29}$$

for all $w = (t, w) \in \mathcal{V}$. Therefore, we arrive at the following equality:

$$T_w \mathcal{V} = \mathbb{R} \otimes T_v S, \quad T_v S = \{ h \in H_0^1(\Omega) : \langle \varphi'_v(v), h \rangle = 0 \}. \tag{13.30}$$

The following chain of equalities holds:

$$\|\psi'_{1w}(w)\|_* (T_w \mathcal{V}) = \sup_{\substack{\tau \in \mathbb{R}, \, h \in T_v S, \\ |\tau|+\|h\| \leq 1}} |\langle\!\langle \psi'_{1w}(w), z \rangle\!\rangle|$$

$$= \sup_{\substack{\tau \in \mathbb{R}, \, h \in T_v S, \\ |\tau|+\|h\| \leq 1}} \left| \frac{\partial \psi_1(w)}{\partial t}\tau + \langle \psi'_{1v}(w), h \rangle \right|. \tag{13.31}$$

Therefore, due to (13.31), Eq. (13.27) is equivalent to the following two equalities:

$$\frac{\partial \psi_1(w)}{\partial t}\Big|_{v \in S} = 0, \quad \|\psi'_{1v}(w)\|_* (T_v S) = 0. \tag{13.32}$$

Indeed, from (13.27) and (13.31) we obtain the following equalities:

$$0 = \sup_{\substack{\tau \in \mathbb{R}, \, h \in T_v S \\ |\tau| + \|h\| \leq 1}} \left| \frac{\partial \psi_1(w)}{\partial t} \tau + \langle \psi'_{1v}(w), h \rangle \right|$$

$$\geq \sup_{|\tau| \leq 1} \left| \frac{\partial \psi_1(w)}{\partial t} \tau \right| \quad \Rightarrow \quad \frac{\partial \psi_1(w)}{\partial t} = 0, \tag{13.33}$$

which follows from (13.31) if $h = \theta \in T_v S$. Moreover,

$$0 = \sup_{\substack{\tau \in \mathbb{R}, \, h \in T_v S \\ |\tau| + \|h\| \leq 1}} \left| \frac{\partial \psi_1(w)}{\partial t} \tau + \langle \psi'_{1v}(w), h \rangle \right|$$

$$\geq \sup_{\substack{\|h\| \leq 1, \\ h \in T_v S}} |\langle \psi'_{1v}(w), h \rangle| \quad \Rightarrow \quad \psi'_{1v}(w) = \theta, \tag{13.34}$$

which is obtained if $\tau = 0$.

Thus, from (13.17) and (13.19) we obtain the following equality:

$$\psi_1(t, v) = t^2 \int_\Omega |D_x v(x)|^2 \, dx + \frac{2}{3} \int_\Omega \left(-(tv(x) + H(x))^3 + H^3(x) \right) \, dx$$

$$= t^2 + \frac{2}{3} \int_\Omega \left(-(tv(x) + H(x))^3 + H^3(x) \right) \, dx, \tag{13.35}$$

since $w = (t, v) \in \mathscr{V} = \left(\mathbb{R}^1 \backslash \{0\} \right) \otimes S$ and, therefore,

$$\int_\Omega |D_x v(x)|^2 \, dx = 1.$$

Due to (13.32), the following equalities hold:

$$\frac{\partial \psi_1(t, v)}{\partial t} = 2t - 2 \int_\Omega (tv(x) + H(x))^2 v(x) \, dx = 0, \tag{13.36}$$

$$\psi'_{1v}(t, v) = -2(tv(x) + H(x))^2 t, \tag{13.37}$$

$$\|\psi'_{1v}(t, v)\|_* (T_v S) = 0. \tag{13.38}$$

Equality (13.38) is equivalent to the existence of $\mu \in \mathbb{R}$ such that the following relation holds:

$$-2(tv(x) + H(x))^2 t = 2\mu \Delta v(x), \quad v(x) \in S, \tag{13.39}$$

where $t = t(v)$ is a solution of the quadratic equation (13.36).

Setting $\lambda = -2\mu$, we reduce Eq. (13.39) to the form

$$2(tv(x) + H(x))^2 t = \lambda \Delta v(x), \quad v(x) \in S. \tag{13.40}$$

Solving the quadratic equation (13.36), we obtain the following two solutions:

$$t_j = t_j(v) = \Bigg\{ 1 - 2 \int_\Omega v^2(x) H(x)\, dx$$

$$+ \nu_j \Bigg[\Big(1 - 2 \int_\Omega v^2(x) H(x)\, dx \Big)^2 - 4 \int_\Omega H^2(x) v(x)\, dx \int_\Omega v^3(x)\, dx \Bigg]^{1/2} \Bigg\}$$

$$\times \Big(2 \int_\Omega v^3(x)\, dx \Big)^{-1}, \quad j = 1, 2, \quad \nu_1 = -1, \quad \nu_2 = 1. \tag{13.41}$$

Introduce the following functionals:

$$J_1 := \int_\Omega H^2(x) v(x)\, dx, \quad J_2 := \int_\Omega H(x) v^2(x)\, dx, \quad J_3 := \int_\Omega v^3(x)\, dx. \tag{13.42}$$

Then the expression for $t_j(v)$ can be rewritten in the following form:

$$t_j = \frac{1}{2 J_3} \Big(1 - 2 J_2 + \nu_j \big((1 - 2 J_2)^2 - 4 J_1 J_3 \big)^{1/2} \Big). \tag{13.43}$$

Note that the following inequalities holds for all $v(x) \in B_1$:

$$|J_1| \le \Big(\int_\Omega H^4(x)\, dx \Big)^{1/2} \Big(\int_\Omega v^2(x)\, dx \Big)^{1/2}$$

$$\le \Big(\int_\Omega H^4(x)\, dx \Big)^{1/2} c_1 \Big(\int_\Omega |D_x v(x)|^2\, dx \Big)^{1/2} \le c_1 \| H \|_4^2, \tag{13.44}$$

$$|J_2| \le \Big(\int_\Omega H^2(x)\, dx \Big)^{1/2} \Big(\int_\Omega v^4(x)\, dx \Big)^{1/2}$$

$$\le \Big(\int_\Omega H^2(x)\, dx \Big)^{1/2} c_2^2 \int_\Omega |D_x v(x)|^2\, dx \le c_2^2 \| H \|_2^2, \tag{13.45}$$

$$|J_3| \le \int_\Omega |v(x)|^3\, dx \le c_3^3 \Big(\int_\Omega |D_x v(x)|\, dx \Big)^{3/2} \le c_3^3. \tag{13.46}$$

Substituting the values $t_1 = t_1(v)$ and $t_2 = t_2(v)$ found above into the functional $\psi_1(t, v)$ defined by Eq. (13.19), we reduce the variational problem considered to the equivalent problem of the search for conditional critical points of the functionals

$$F_1(v) = E_1(t_1(v) v), \quad F_2(v) = E_1(t_2(v) v) \tag{13.47}$$

on the unit sphere

$$S = \Big\{ v \in H_0^1(\Omega) : \int_\Omega |D_x v(x)|^2\, dx = 1 \Big\}.$$

These functionals have the following explicit form:

$$F_j(v) = -t_j^3(v)\frac{2}{3}\int\limits_\Omega v^3(x)\,dx$$

$$+ t_j^2(v)\left(1 - 2\int\limits_\Omega H(x)v^2(x)\,dx\right) - 2t_j(v)\int\limits_\Omega H^2(x)v(x)\,dx. \qquad (13.48)$$

Note that Eq. (13.36) implies a quadratic equation, which yields the following equality after multiplying by $t_j(v)$:

$$t_j^3(v)\int\limits_\Omega v^3(x)\,dx = \left(1 - 2\int\limits_\Omega H(x)v^2(x)\,dx\right)t_j^2(v) - \int\limits_\Omega H^2(x)v(x)\,dx\,t_j(v).$$

$$(13.49)$$

Substituting (13.49) into (13.48), we obtain the following equality:

$$F_j(v) = \frac{1}{3}\left(1 - 2\int\limits_\Omega H(x)v^2(x)\,dx\right)t_j^2(v) - \frac{4}{3}\int\limits_\Omega H^2(x)u(x)\,dx\,t_j(v). \qquad (13.50)$$

Now we assume that

$$1 - 2\int\limits_\Omega H(x)v^2(x)\,dx \geq c_0 > 0 \quad \text{for all } v(x) \in B_1; \qquad (13.51)$$

obviously, this assumption holds for the function $H(x) \leq 0$.

We examine the behavior of the functionals $t_1 = t_1(v)$ and $t_2 = t_2(v)$ defined by Eq. (13.41) in the closed unit ball $B_1 := \{u \in H_0^1(\Omega) : \|u\| \leq 1\}$ under the assumption (13.51). Note that $t_1 = t_1(v)$ is defined everywhere in the ball B_1 and $t_2 = t_2(v)$ has a singularity if

$$\int\limits_\Omega v^3(x)\,dx = 0.$$

We prove that $t_1 = t_1(v)$ has no singularities if

$$\int\limits_\Omega v^3(x)\,dx \to 0.$$

Indeed, under the assumption (13.51), the numerator in Eq. (13.41) has the following asymptotic expansion for $j = 1$:

$$1 - 2\int\limits_\Omega H(x)v^2(x)\,dx$$

$$-\left[\left(1 - 2\int\limits_\Omega H(x)v^2(x)\,dx\right)^2 - 4\int\limits_\Omega v^3(x)\,dx\int\limits_\Omega H^2(x)v(x)\,dx\right]^{1/2}$$

$$= 2 \frac{1}{\left(1 - 2 \int\limits_{\Omega} H(x) v^2(x) \, dx\right)} \int\limits_{\Omega} v^3(x) \, dx \int\limits_{\Omega} H^2(x) v(x) \, dx$$

$$+ o \left(\int\limits_{\Omega} v^3(x) \, dx\right). \tag{13.52}$$

Thus, the functional $F_1(v)$ is defined everywhere in the ball $B_1 = \{w \in H_0^1(\Omega) : \|w\| \leq 1\}$, whereas the functional $F_2(v)$ is defined on the set

$$B_1 \setminus \left\{\int\limits_{\Omega} w^3(x) \, dx = 0\right\}.$$

Now we assume that there exists a constant $c_1 > 0$ such that the inequality

$$\left(1 - 2 \int\limits_{\Omega} H(x) w^2(x) \, dx\right)^2 - 4 \int\limits_{\Omega} H^2(x) w(x) \, dx \int\limits_{\Omega} w^3(x) \, dx \geq c_1 > 0 \tag{13.53}$$

holds for all $v(x) \in B_1$ and for any sufficiently small function $h(x) \in H^{1/2}(\partial\Omega)$. Under this condition, both roots $t_1 = t_1(v)$ and $t_2 = t_2(v)$ are real. The following assertion holds.

Lemma 13.1. *The functional $F_1(v)$ attains a minimum on the closed unit ball B_1.*

Proof. As is well known, a closed ball in a reflexive Banach space is weakly compact. The space $H_0^1(\Omega)$ is a reflexive Banach space since it is a Hilbert space.

Now we prove that the functional $F_1(v)$ is weakly continuous on the closed unit ball. Namely, we prove that the following functionals are weakly continuous on $H_0^1(\Omega)$:

$$J_1(v) := \int\limits_{\Omega} H(x) v^2(x) \, dx, \quad J_2(v) := \int\limits_{\Omega} H^2(x) v(x) \, dx, \quad J_3(v) := \int\limits_{\Omega} v^3(x) \, dx. \tag{13.54}$$

The following inequalities hold:

$$\left|J_1(v_1) - J_1(v_2)\right|$$
$$\leq \int\limits_{\Omega} |H(x)| \cdot |v_1(x) + v_2(x)| \cdot |v_1(x) - v_2(x)| \leq \mu(R) \|v_1 - v_2\|_3, \tag{13.55}$$

$$\mu(R) = 2\|H(x)\|_3 R, \quad \max\{\|v_1\|_3, \|v_2\|_3\} \leq R,$$
$$|J_2(v_1) - J_2(v_2)| \leq \|H(x)\|_4^2 \|v_1 - v_2\|_2,$$

$$|J_3(v_1) - J_3(v_2)| \leq \int\limits_{\Omega} |v_1^3(x) - v_2^3(x)| \, dx$$

$$\leq 3 \int\limits_{\Omega} \max\{v_1^2(x), v_2^2(x)\} |v_1(x) - v_2(x)| \, dx$$

$$\leq 3 \max\{\|v_1\|_3^2, \|v_2\|_3^2\} \|v_1 - v_2\|_3. \tag{13.56}$$

Note that for $N = 3$, the embedding $H_0^1(\Omega) \hookrightarrow L^p(\Omega)$ is completely continuous for $1 < p < 6$ and hence is totally continuous. Therefore, in particular, the embeddings $H_0^1(\Omega) \hookrightarrow L^2(\Omega)$ and $H_0^1(\Omega) \hookrightarrow L^3(\Omega)$ are totally continuous. Therefore, due to the inequalities (13.55)–(13.56), the functionals $J_1(v)$, $J_2(v)$, and $J_3(v)$ are weakly continuous on $H_0^1(\Omega)$.

From the explicit form of the functional $t_1 = t_1(J_1, J_2, J_3)$ we obtain that it continuously depends on the numbers J_1, J_2, and J_3 and hence it is a weakly continuous functional on B_1 as a composite function. Therefore, $F_1 = F_1(v)$ is a weakly continuous functional on B_1.

Since the ball B_1 is a weakly compact set, we conclude that, due to Theorem 11.3, the weakly continuous functional $F_1 = F_1(v)$ on B_1 attains a minimum. □

A similar assertion is valid for the functional $F_2(v)$.

Lemma 13.2. *The functional $F_2 = F_2(v)$ attains its minimum on the closed unit ball B_1.*

Proof. 1. Let $\varepsilon > 0$ be a sufficiently mall number. Consider the set

$$B_1^\varepsilon := B_1 \cap \left\{ w(x) \in H_0^1(\Omega) : \left| \int_\Omega w^3(x)\, dx \right| \geq \varepsilon \right\}. \tag{13.57}$$

We prove that the set B_1^ε is weakly compact. Indeed, let $\{u_m\} \subset B_1^\varepsilon$ be an arbitrary sequence. Since the set B_1^ε is bounded in $H_0^1(\Omega)$, there exists a subsequence $\{u_{m_k}\} \subset \{u_m\}$ such that

$$u_{m_k} \rightharpoonup u_0 \quad \text{weakly in } H_0^1(\Omega) \text{ as } m_k \to +\infty. \tag{13.58}$$

Since $B_1^\varepsilon \subset B_1$ and the set B_1 is weakly closed, we have $u_0(x) \in B_1$. Moreover, the following completely continuous embedding holds:

$$H_0^1(\Omega) \hookrightarrow L^3(\Omega);$$

therefore, the embedding operator is totally continuous. Due to (13.58) we have

$$u_{m_k} \to u_0 \quad \text{strongly in } L^3(\Omega) \text{ as } m_k \to +\infty. \tag{13.59}$$

Thus,

$$\int_\Omega u_{m_k}^3(x)\, dx \to \int_\Omega u_0^3(x)\, dx \quad \text{as } m_k \to +\infty. \tag{13.60}$$

Consequently,

$$\left| \int_\Omega u_{m_k}^3(x)\, dx \right| \geq \varepsilon \quad \Rightarrow \quad \left| \int_\Omega u_0^3(x)\, dx \right| \geq \varepsilon. \tag{13.61}$$

Thus, we conclude that $u_0(x) \in B_1^\varepsilon$.

2. Prove the weak continuity of F_2 on B_1^ε. Similarly to the proof of Lemma 13.1, we can prove that $t_2 = t_2(J_1, J_2, J_3)$ is a continuous function of the set $|J_3| \geq \varepsilon > 0$. Therefore, the functional $F_2 = F_2(v)$ is weakly continuous on B_1^ε.

3. By Theorem 11.3, the weakly continuous functional $F_2 = F_2(v)$ attains a minimum on the weakly compact set B_1^ε. We note that the functional $t_2 = t_2(v)$ has the following asymptotic behavior as $J_3 \to 0$:

$$t_2 = t_2(J_1, J_2, J_3) = \frac{1 - 2J_1}{J_3} - \frac{J_2}{1 - 2J_1} - \frac{J_2^2 J_3}{(1 - 2J_1)^3} + o(J_3); \tag{13.62}$$

therefore, the functional $F_2 = F_2(J_1, J_2, J_3)$ has the following asymptotic expansion:

$$F_2 = F_2(J_1, J_2, J_3) = \frac{(1 - 2J_1)^3}{3J_3^2} - 2\frac{J_2(1 - 2J_1)}{J_3}$$

$$+ \frac{J_2^2}{1 - 2J_1} - \frac{4}{3}\frac{J_2^3 J_3}{(1 - 2J_1)^3} + o(J_3); \tag{13.63}$$

here we use the inequalities (13.44) and (13.45), which imply that the functionals J_1 and J_2 are bounded as $J_3 \to 0$. Thus, we have obtained the following asymptotic behavior:

$$\lim_{J_3 \to 0} F_2 = +\infty. \tag{13.64}$$

Therefore, there exists small $\varepsilon > 0$ such that the minimum of the functional F_2 is attained on the weakly compact set $B_1^\varepsilon \subset B_1$. $\qquad\square$

Lemma 13.3. *The following inequalities hold:*

$$t_1(w_1) \neq 0, \quad t_2(w_2) \neq 0, \tag{13.65}$$

where w_1 and w_2 are minimum points of the functionals $F_1(v)$ and $F_2(v)$ respectively on the closed unit ball B_1.

Proof. 1. Prove that $t_2(w_2) \neq 0$. From the explicit form (13.41) and the conditions (13.51) and (13.53) we obtain that everywhere in the closed unit ball B_1, the numerator of the expression for $t_2(v)$ is always strictly positive and hence $t_2(w_2) \neq 0$.

2. To prove the inequality $t_1(w_1) \neq 0$, we assume the contrary, $t_1(w_1) = 0$. Then Eq. (13.50) implies $F_1(w_1) = 0$. We examine the behavior of the functional $F_1(v)$ as $\|v\| \to +0$, i.e., at the center of the closed ball B_1. Note that in this case, the following limit properties hold:

$$J_1(v), J_2(v), J_3(v) \to 0 \quad \text{as } \|v\| \to +0. \tag{13.66}$$

This follows from the formulas (13.55)–(13.56) in which we set $v_1 = v$ and $v_2 = 0$. Let $v_0(x) \in B_1$ be a function such that

$$J_{30} := \int_\Omega v_0^3(x)\, dx \neq 0.$$

Consider the function $v(x) = \lambda v_0(x)$. Then we have

$$J_1 = \lambda J_{10}, \quad J_2 = \lambda^2 J_{20}, \quad J_3 = \lambda^3 J_{30}, \tag{13.67}$$

$$J_{10} := \int\limits_{\Omega} H^2(x)v_0(x)\,dx, \quad J_{20} := \int\limits_{\Omega} H(x)v_0^2(x)\,dx. \tag{13.68}$$

Then from the explicit form (13.43) of the functional $t_1 = t_1(v)$ and the condition (13.51) we obtain the asymptotic behavior

$$t_1 = t_1(v) \sim \frac{J_2}{1 - 2J_1}, \quad F_1(v) \sim -\frac{J_2^2}{(1 - 2J_1)^2} < 0 \quad \text{as } \lambda \to +0. \tag{13.69}$$

Therefore, at the minimum point $w_1 \in B_1$, the inequality $F_1(w_1) < 0$ holds; a contradiction. Therefore, $t_1(w_1) \neq 0$. $\qquad\square$

Theorem 13.2. *Minimums $w_j \in B_1$ of the functionals $F_j = F_j(v)$ are attained on the sphere $S = \partial B_1$.*

Proof. 1. Prove the differentiability of $t_j = t_j(v)$ at the points $v = w_j$. Clearly, the functional $t_1 = t_1(v)$ is differentiable on the whole closed ball B_1 and, in particular, at the minimum point $v = w_1$ of the functional F_1. The minimum point w_2 of the functional F_2 is located on the part B_1^ε of the closed unit ball B_1, where $\varepsilon > 0$ is small, and the functional $t_2 = t_2(v)$ is also differentiable on this part.

2. Assume that the minimum w_j lie in $O(0,1) := \{w(x) \in H_0^1(\Omega) : \|w\| < 1\}$. The functionals $F_j = F_j(w)$ have the following explicit form:

$$F_j(w) = E_1(t_j(w)w) = t_j^2 + f_1(t_j(w)w), \tag{13.70}$$

$$f_1 = f_1(z) = \frac{2}{3}\int\limits_{\Omega} \left(-(z(x) + H(x))^3 + H^3(x)\right) dx. \tag{13.71}$$

The functionals F_j attain minimums at the points $w = w_j \in B_1$. Earlier, we have seen that $t_j = t_j(w)$ are solutions of the equation

$$2t_j(w_j) + \langle f_{1f}'(t_j(w_j)w_j), w_j \rangle = 0. \tag{13.72}$$

Now we consider the real-valued function of a real argument

$$\varphi(s) := F_j(sw_j), \quad s \in (1 - \delta, 1 + \delta), \tag{13.73}$$

where $\delta \in (0, 1)$ is sufficiently small. Since the functionals $F_j = F_j(w)$ attain their minimums at the points $w_j \in O(0, 1)$, we conclude that the function $\varphi(s)$ also attains its minimum at the point $s = 1$. Therefore, on the one hand,

$$\left.\frac{d\varphi(s)}{ds}\right|_{s=1} = 0. \tag{13.74}$$

On the other hand, the following equalities hold:

$$\left.\frac{d\varphi(s)}{ds}\right|_{s=1} = \left.\frac{d}{ds}F_j(sw_j)\right|_{s=1} = \left[2t_j(sw_j)\frac{dt_j(sw_j)}{ds}\right.$$
$$+ \left.\left\langle f_{1f}'(t_j(sw_j)sw_j), \frac{d(t_j(sw_j)sw_j)}{ds}\right\rangle\right]\bigg|_{s=1}$$

$$= \left[2t_j(w_j) + \langle f'_{1f}(t_j(w_j)w_j), w_j\rangle\right] \frac{dt_j(sw_j)}{ds}\Bigg|_{s=1}$$

$$+ \langle f'_{1f}(t_j(w_j)w_j), w_j\rangle t_j(w_j) = -2t_j^2(w_j) < 0, \qquad (13.75)$$

where we have used Eq. (13.72) and Lemma 13.3. The contradiction obtained means that the minimum points w_j of the functionals F_j lie on the unit sphere $S = \partial B_1$. □

Thus, we have proved that for $\|w_j\| = 1$, $j = 1, 2$, the points $u_j(x) = t_j(w_j)w_j(x)$ are critical points of the functional $E_1(u)$ and, therefore, of the functional $E(u)$. Note that the points $u_1(x)$ and $u_2(x)$ are distinct since in the opposite case

$$\|u_1\| = \|u_2\| \quad \Rightarrow \quad |t_1(w_1)| = |t_2(w_2)|,$$

but the last equality is invalid. Thus, the following theorem holds.

Theorem 13.3. *Let the conditions* (13.51) *and* (13.53) *be fulfilled. Then there exist two distinct nontrivial weak solutions of the problem* (13.11), (13.12).

13.4 Bibliographical Notes

The contents of this lecture is taken from [58, 59].

Lecture 14

M. A. Krasnosel'sky's Theory
of the Genus of a Set

M. A. Krasnosel'sky proposed a proof of L. A. Lyusternik's theorem on the existence of a countable set of critical points of a weakly continuous functional on the unit sphere of a Hilbert space, which does not require passing to projective space. This proof is based on a new topological invariant — the *genus* of a set. In this proof, the evenness of the functional is substantially used.

14.1 Genus of a Set

Let $\mathbb{H}$ be a real, separable Hilbert space with respect to the scalar product $(\cdot, \cdot)$. We denote the unit sphere of this space by

$$S := \left\{ u \in \mathbb{H} : \|u\| = 1 \right\}, \quad \|u\| = (u, u)^{1/2}.$$

For each set $F \subset S$, we introduce the set

$$F^* = \left\{ -u \in S : u \in F \subset S \right\},$$

which is symmetric to F with respect to $\theta \in \mathbb{H}$.

Definition 14.1. A compact set $F \subset S$ is said to be a *set of first kind* if each connected component of the set $F \cup F^*$ contains no symmetric points u and $-u$; notation $r(F) = 1$.

Definition 14.2. A compact set $F \subset S$ is said to be a *set of genus* $n \in \mathbb{N}$ if it can be represented as the union of n compact sets $\{H_j\}_{j=1}^n$ of first kind,

$$F = \bigcup_{j=1}^{n} H_j,$$

but cannot be represented as the union of $n - 1$ sets of first kind; notation $r(F) = n$.

Theorem 14.1 (see [35]). *The genus of the n-dimensional sphere is equal to $n + 1$.*

Corollary 14.1. *The sphere of an infinite-dimensional Hilbert space contains compact subsets of any finite genus.*

Remark 14.1. Under odd transformations, point symmetric with respect to θ pass into points that are also symmetric with respect to θ. Indeed, let $\{u, -u\}$ be a pair of points symmetric with respect to θ and φ be an odd transformation. Then $\{\varphi(u), \varphi(-u)\} = \{\varphi(u), -\varphi(u)\}$ is also a pair of symmetric points.

Lemma 14.1. *Let φ be an odd and continuous transformation of a compact set $F \subset S$ and $\varphi(F) \subset S$. Then $r(F) \leq r(\varphi(F))$.*

Proof. 1. Assume that $r(\varphi(F)) = 1$. We prove that $r(F) = 1$. Indeed, if $r(F) \neq 1$, then S contains a pair of points $\{u, -u\}$, which lies in a certain connected component $F_1 \subset F$. Then, due to the continuity of φ, $\varphi(F_1)$ is also connected and is contained in the set $\varphi(F)$; moreover, due to the oddness of φ, it contains the points $\varphi(u)$ and $-\varphi(u)$, which are symmetric with respect to $\theta \in \mathbb{H}$. Therefore, $r(\varphi(F)) > 1$; a contradiction. Therefore, $r(F) = 1$.

　　2. Now let $r(\varphi(F)) = n$. This means that the set $\varphi(F)$ can be represented as the union of first-kind compact sets $D_1, \ldots, D_n$:

$$\varphi(F) = \bigcup_{j=1}^{n} D_j$$

with preimages

$$H_j := \{u \in S : \varphi(u) \in D_j\}, \quad j = \overline{1, n}.$$

Due to the continuity of φ, the sets H_i are compact and cover the set F. Indeed, by the condition,

$$\varphi(F) = \bigcup_{j=1}^{n} D_j \quad \Rightarrow \quad F = \varphi^{-1}\left(\bigcup_{j=1}^{n} D_j\right) = \bigcup_{j=1}^{n} \varphi^{-1}(D_j) = \bigcup_{j=1}^{n} H_j.$$

Therefore, the set F is covered by the sets H_j. It remains to prove that the sets $\{H_j\}_{j=1}^{n}$ are compact. Since φ is a continuous mapping, the total preimage of a closed set is also closed. Therefore, the sets $\varphi^{-1}(D_j) = H_j$ are closed and lie in the compact set F. Hence the sets H_j are compact.

　　Due to the first step, we have $r(H_j) = 1$. Therefore, $r(F) \leq n$. □

Lemma 14.2. *The genus of the closed spherical layer $B_{\rho,R} := \{x \in \mathbb{R}^N : \rho \leq |x| \leq R\}$, $0 < \rho < R$, is equal to N.*

Proof. 1. *Case $N = 1$.* The set $B_{\rho,R} = [-R, -\rho] \cup [\rho, R]$ consists of two disconnected parts; moreover, both sets $[-R, -\rho]$ and $[\rho, R]$ do not contain numbers symmetric with respect to 0. Therefore, the genus of the set $B_{\rho,R}$ is equal to $N = 1$.

　　2. *Case $N = 2$.* In this case, the closed spherical layer $B_{\rho,R} := \{x \in \mathbb{R}^2 : \rho \leq |x| \leq R\}$ can be conveniently described in the polar coordinate system centered at the origin as follows:

$$B_{\rho,R} := \{(r, \varphi) : \rho \leq r \leq R, \ \varphi \in [0, 2\pi]\}. \tag{14.1}$$

This set can be partitioned into two parts

$$B^1_{\rho,R} := \{(r,\varphi): \ \rho \le r \le R, \ \varphi \in [0,\varphi_0] \cup [\pi, \pi+\varphi_0]\}, \tag{14.2}$$

$$B^2_{\rho,R} := \{(r,\varphi): \ \rho \le r \le R, \ \varphi \in [\varphi_0, \pi] \cup [\pi+\varphi_0, 2\pi]\}, \tag{14.3}$$

where $\varphi_0 \in (0, \pi/4)$. Clearly, $B_{\rho,R} = B^1_{\rho,R} \cup B^2_{\rho,R}$, and none of the connected components of the sets $B^1_{\rho,R}$ and $B^2_{\rho,R}$ contain points that are symmetric with respect to the origin. Therefore, the genus of the sets $B^1_{\rho,R}$ and $B^2_{\rho,R}$ is equal to 1. Obviously, the set $B_{\rho,R}$ cannot be represented as a single first-kind set; therefore, the genus of the set $B_{\rho,R}$ is equal to $N = 2$.

3. *Case $N \ge 3$.* We apply induction in N. Assume that

$$r(B_{\rho,R}) = N-1 \quad \text{if} \quad B_{\rho,R} = \{x \in \mathbb{R}^{N-1} : \rho \le |x| \le R\}.$$

We prove that then

$$r(B_{\rho,R}) = N \quad \text{if} \quad B_{\rho,R} = \{x \in \mathbb{R}^N : \rho \le |x| \le R\}.$$

Consider the general spherical coordinate system centered at the origin. The spherical coordinates $\{r, \varphi_1, \ldots, \varphi_{n-2}, \varphi_{n-1}\}$ and the Cartesian coordinates $\{x_1, x_2, \ldots, x_N\}$ are related by the formulas

$$x_1 = r \cos\varphi_1,$$

$$x_2 = r \sin\varphi_1 \cos\varphi_2,$$

$$x_3 = r \sin\varphi_1 \sin\varphi_2 \cos\varphi_3,$$

$$\ldots\ldots\ldots\ldots\ldots \tag{14.4}$$

$$x_{N-1} = r \sin\varphi_1 \sin\varphi_2 \cdots \sin\varphi_{N-2} \cos\varphi_{N-1},$$

$$x_N = r \sin\varphi_1 \sin\varphi_2 \cdots \sin\varphi_{N-2} \sin\varphi_{N-1}.$$

The N-dimensional spherical layer $B_{\rho,R}$ is described as follows:

$$B_{\rho,R} = \Big\{(r, \varphi_1, \ldots, \varphi_{N-2}, \varphi_{N-1}): \ \rho \le r \le R,$$

$$\varphi_1 \in [0,\pi], \ldots, \varphi_{N-2} \in [0,\pi], \ \varphi_{N-1} \in [0, 2\pi]\Big\}. \tag{14.5}$$

First, we partition the N-dimensional spherical layer $B_{\rho,R}$ into two parts

$$B_{\rho,R} = B^1_{\rho,R} \cup B^2_{\rho,R}, \tag{14.6}$$

where

$$B^1_{\rho,R} = \Big\{(r, \varphi_1, \ldots, \varphi_{N-2}, \varphi_{N-1}): \ \rho \le r \le R,$$

$$\varphi_1 \in [0, \varphi_0] \cup [\pi-\varphi_0, \pi], \ldots, \varphi_{N-2} \in [0,\pi], \ \varphi_{N-1} \in [0, 2\pi]\Big\}, \tag{14.7}$$

$$B^2_{\rho,R} = \Big\{(r, \varphi_1, \ldots, \varphi_{N-2}, \varphi_{N-1}): \ \rho \le r \le R,$$

$$\varphi_1 \in [\varphi_0, \pi-\varphi_0], \ldots, \varphi_{N-2} \in [0,\pi], \ \varphi_{N-1} \in [0, 2\pi]\Big\}, \tag{14.8}$$

where $\varphi_0 \in (0, \pi/4)$. It is easy to see that the set $B^1_{\rho,R}$ has genus 1.

Now we consider the section of the spherical layer $B_{\rho,R}$ by the plane $x_1 = 0$ ($\varphi_1 = \pi/2$). Obviously, the section obtained is an $(N-1)$-dimensional spherical layer. Choosing appropriately the ranges for angles $\{\varphi_2, \ldots, \varphi_{N-2}, \varphi_{N-1}\}$, we can partition this $(N-1)$-dimensional spherical layer into $N-1$ closed sets of genus 1. Choosing the corresponding ranges of angles, we can partition the set $B^2_{\rho,R}$ into $N-1$ closed parts whose genus is also 1. Thus, $r(B_{\rho,R}) = 1 + (N-1) = N$. □

14.2 Classes M_k

Let φ be an odd continuous mapping of the $(k-1)$-dimensional unit sphere $S^{k-1} := \{x \in \mathbb{R}^k : |x| = 1\}$ into the infinite-dimensional unit sphere $S \subset \mathbb{H}$, i.e., a mapping that transforms points of the sphere S^{k-1} symmetric with respect to $\theta_k \in \mathbb{R}^k$ to points of the sphere S symmetric with respect to $\theta \in \mathbb{H}$.

Let $\{e_1, \ldots, e_k\}$ be orts of a certain Cartesian rectangular coordinate system and $(\xi_1, \ldots, \xi_k)$ be the coordinates of a point in this coordinate system. The following assertion holds.

Lemma 14.3. $r(\varphi(S^{k-1})) \geq k$.

Proof. By Lemma 14.1, we have $r(\varphi(S^{k-1})) \geq r(S^{k-1}) = k$. □

Definition 14.3. We say that a set $A \in S$ belongs to the class M_k, if there exists an odd continuous mapping $\varphi : S^{k-1} \to S$ such that $A = \varphi(S^{k-1})$.

Lemma 14.4. *Let Φ be an odd continuous transformation of the sphere S. If F is a set of class M_k, then $\Phi(F)$ is also a set of class M_k.*

Proof. Let $F = \varphi(S^{k-1})$, where φ is an odd continuous mapping of S^{k-1} into S. The mapping $\varphi_1 = \Phi \cdot \varphi$ is also an odd continuous mapping of S^{k-1} into S. Therefore, $\Phi(F) = (\Phi \cdot \varphi)(S^{k-1})$ is a set of class M_k. □

14.3 Conditions for Functionals

Definition 14.4. A functional $\psi(u) : \mathbb{H} \to \mathbb{R}^1$ defined on a real Hilbert space $\mathbb{H}$ is said to be uniformly Fréchet differentiable on $\mathbb{H}$ if the following equality holds:

$$\psi(u + h) = \psi(u) + \langle \psi'_f(u), h \rangle + \omega(u, h), \tag{14.9}$$

where

$$\lim_{\|h\| \to +0} \sup_{u \in B_R} \frac{|\omega(u, h)|}{\|h\|} = 0, \quad B_R := \{u \in \mathbb{H} : \|u\| \leq R\}, \tag{14.10}$$

for any $R > 0$.

Definition 14.5. A functional $\psi : \mathbb{H} \to \mathbb{R}^1$ is said to be *weakly continuous* on a weakly closed subset $M \subset \mathbb{H}$ if for any sequence $\{u_n\} \subset M$ such that

$$u_n \rightharpoonup u_0 \quad \text{weakly in } \mathbb{H} \text{ as } n \to +\infty, \tag{14.11}$$

the following limit relation holds:

$$\psi(u_n) \to \psi(u_0) \quad \text{as } n \to +\infty. \tag{14.12}$$

Definition 14.6 (see Definition 5.3). The *gradient* of a Fréchet-differentiable functional $\psi(u) : \mathbb{H} \to \mathbb{R}^1$ is the following composition of operators:

$$\operatorname{grad} \psi(u) := J \psi'_f(u) : \mathbb{H} \to \mathbb{H}, \tag{14.13}$$

where J is the Riesz–Fréchet isometry for the mutually dual Hilbert spaces $\mathbb{H}$ and $\mathbb{H}^*$.

Remark 14.2. Note that Eq. (14.9) can be rewritten in the form

$$\psi(u+h) = \psi(u) + (\operatorname{grad}\psi(u), h) + \omega(u, h), \tag{14.14}$$

where we have used the equality

$$(\operatorname{grad}\psi(u), h) = (J\psi'_f(u), h) = \langle \psi'_f(u), h \rangle,$$

which is an immediate consequence of the Riesz–Fréchet theorem.

Definition 14.7 (see Definition 10.3). An operator $A(u) : \mathbb{H} \to \mathbb{H}$ is said to be *boundedly Lipschitz continuous* if for any $R > 0$, the inequality

$$\|A(u_1) - A(u_2)\| \leq \mu(R)\|u_1 - u_2\| \tag{14.15}$$

holds for all $u_k \in \mathbb{H}$ such that $\|u_k\| \leq R$, $k = 1, 2$; here $\mu(R) > 0$ is a function, which is bounded on compact sets of $[0, +\infty)$.

Lemma 14.5. *Let a functional $\psi(u) : \mathbb{H} \to \mathbb{R}^1$ be Fréchet differentiable on $\mathbb{H}$ and its gradient $\operatorname{grad}\psi(u)$ is a boundedly Lipschitz continuous mapping. Then the following estimate is valid:*

$$\left|\psi(u+h) - \psi(u) - (\operatorname{grad}\psi(u), h)\right| \leq C(R)\|h\|^2 \tag{14.16}$$

for all $\|u\| \leq R$, $\|h\| \leq R$, and $R > 0$.

Proof. We have the equality

$$\psi(u+h) - \psi(u) = \int_0^1 \frac{d}{dt}\psi(u+th)\, dt = \int_0^1 \langle \psi'_f(u+th), h \rangle\, dt$$

$$= \int_0^1 (\operatorname{grad}\psi(u+th), h)\, dt \quad \text{for all } \|u\| \leq R \text{ and } \|h\| \leq R.$$

$$\tag{14.17}$$

Therefore, the following inequality holds:

$$\left|\psi(u+h) - \psi(u) - (\operatorname{grad}\psi(u), h)\right|$$

$$\leq \int_0^1 \left|(\operatorname{grad}\psi(u+th) - \operatorname{grad}\psi(u), h)\right| dt$$

$$\leq \mu(2R)\int_0^1 t\|h\|^2\, dt = \mu(2R)\frac{1}{2}\|h\|^2. \tag{14.18}$$

The lemma is proved. $\qquad\square$

Remark 14.3. Note that the conditions of Lemma 14.5 yield sufficient conditions of the uniform differentiability of the functional $\psi(u)$ on $\mathbb{H}$.

Lemma 14.6. *Let a weakly continuous and Fréchet-differentiable functional $\psi(u)$ on $\mathbb{H}$ have a boundedly Lipschitz continuous gradient $\operatorname{grad}\psi(u)$. Then the weak convergence*

$$u_n \rightharpoonup u_0 \quad \text{weakly in } \mathbb{H} \text{ as } n \to +\infty \tag{14.19}$$

implies that

$$\operatorname{grad}\psi(u_n) \to \operatorname{grad}\psi(u_0) \quad \text{strongly in } \mathbb{H} \text{ as } n \to +\infty, \tag{14.20}$$

i.e., the operator $\operatorname{grad}\psi(u)$ is totally continuous.

Proof. Assume the contrary; then there exists a sequence

$$u_n \rightharpoonup u_0 \quad \text{weakly in } \mathbb{H} \text{ as } n \to +\infty \tag{14.21}$$

and a number $\varepsilon > 0$ such that

$$\|\operatorname{grad}\psi(u_n) - \operatorname{grad}\psi(u_0)\| \geq \varepsilon > 0 \quad \text{for all } n \in \mathbb{N}. \tag{14.22}$$

Note that due to (14.21) and Theorem 2.3, there exists $R > 0$ such that the inequality

$$\sup_{n\in\mathbb{N}} \|u_n\| \leq R < +\infty \tag{14.23}$$

holds. Consider the sequence

$$v_n := \frac{\operatorname{grad}\psi(u_n) - \operatorname{grad}\psi(u_0)}{\|\operatorname{grad}\psi(u_n) - \operatorname{grad}\psi(u_0)\|}. \tag{14.24}$$

Then $\|v_n\| = 1$ and

$$\Big(\operatorname{grad}\psi(u_n) - \operatorname{grad}\psi(u_0), v_n\Big) \geq \varepsilon > 0 \quad \text{for all } n \in \mathbb{N}. \tag{14.25}$$

Since the Hilbert space $\mathbb{H}$ is reflexive, there exists s subsequence $\{v_{n_k}\} \subset \{v_n\}$ such that

$$v_{n_k} \rightharpoonup v_0 \quad \text{weakly in } \mathbb{H} \text{ as } n_k \to +\infty, \tag{14.26}$$

and $\|v_0\| \leq 1$ since

$$\{v_{n_k}\} \subset B_1 := \{v \in \mathbb{H} : \|v\| \leq 1\}$$

and B_1 is weakly closed in the Hilbert space. For any $t \in (0,1]$, the following chain of equalities holds:

$$\psi(u_n + tv_n) - \psi(u_0 + tv_n)$$

$$= \Big[\psi(u_n + tv_n) - \psi(u_n) - t(\operatorname{grad}\psi(u_n), v_n)\Big]$$

$$- \Big[\psi(u_0 + tv_n) - \psi(u_0) - t(\operatorname{grad}\psi(u_0), v_n)\Big]$$

$$+ \psi(u_n) - \psi(u_0) + t\Big(\operatorname{grad}\psi(u_n) - \operatorname{grad}\psi(u_0), v_n\Big). \tag{14.27}$$

Due to (14.23), we obtain from (14.16) the following lower estimate:

$$\psi(u_n + tv_n) - \psi(u_0 + tv_n) \geq -2C(R)t^2 + \psi(u_n) - \psi(u_0) + t\varepsilon. \tag{14.28}$$

Passing to the limit as $n \to +\infty$, due to the weak continuity of the functional $\psi(u)$, we obtain the inequality

$$0 \geq -2C(R)t^2 + t\varepsilon > 0 \quad \text{for small } t > 0. \tag{14.29}$$

This contradiction completes the proof. $\qquad\qquad\square$

14.4 Bibliographical Notes

The contents of this lecture is taken from [35, 56, 67].

Auxiliary Results

15.1 Class $\mathscr{F}$ of Real-Valued Functionals

Introduce the class $\mathscr{F}$ of functionals $\psi(u) : \mathbb{H} \to \mathbb{R}^1$. Let $B_1 = \{u \in \mathbb{H} : \|u\| \leq 1\}$ and $S := \{u \in \mathbb{H} : \|u\| = 1\}$.

Definition 15.1. We say that a functional ψ belongs in the class $\mathscr{F}$ if ψ is a weakly continuous and Fréchet differentiable on $\mathbb{H}$, its gradient $\operatorname{grad}\psi(u)$ is boundedly Lipschitz continuous on $\mathbb{H}$, and the following conditions hold:

(a) $\psi(-u) = \psi(u)$ for all $u \in S \subset \mathbb{H}$;
(b) $\psi(u) > 0$ for all $u \in B_1 \backslash \{\theta\}$ and $\psi(\theta) = \theta$;
(c) $\operatorname{grad}\psi(u) \neq \theta \in \mathbb{H}$ for $u \neq \theta$ and $\operatorname{grad}\psi(\theta) = \theta$.

Remark 15.1. Note that the condition $\psi \in \mathscr{F}$ and Lemma 14.5 imply the inequality (14.16), which, in turn, implies the existence of a constant $c_1 > 0$ such that the inequality

$$|\omega(u, h)| \leq c_1 \|h\|^2 \tag{15.1}$$

holds for all $u \in B_1$. Therefore, we arrive at the following obvious assertion.

Lemma 15.1. *For any $\varepsilon > 0$, there exists $\delta = \delta(\varepsilon) := \varepsilon/c_1$ such that for all $\|h\| \leq \delta$, the inequality*

$$|\omega(u, h)| \leq \varepsilon \|h\| \tag{15.2}$$

holds for all $u \in B_1$.

Introduce the notation

$$D(u) := \operatorname{grad}\psi(u) - (\operatorname{grad}\psi(u), u)u, \quad u \in S. \tag{15.3}$$

Note that

$$(D(u), u) = 0 \quad \text{for all} \quad u \in S. \tag{15.4}$$

Let

$$l := \sup_{u \in S} \|D(u)\|, \quad m := \sup_{\substack{u \in S, \\ v \in B_1}} \left| \big(\operatorname{grad}\psi(u), u \big) + \big(v, D(u) \big) \right|. \tag{15.5}$$

Lemma 15.2. *A functional $\psi(u) \in \mathscr{F}$ is bounded on the ball B_1 and, moreover, $l > 0$ and $m > 0$ are also bounded.*

Proof. 1. Prove the boundedness of ψ. Since $\psi(u)$ is weakly continuous on $\mathbb{H}$, in particular, it is weakly continuous on the weakly closed set $B_1 \subset \mathbb{H}$. Assume that it is unbounded on B_1. Therefore, there exists a sequence $\{u_n\} \subset B_1$ such that

$$\psi(u_n) \to +\infty \quad \text{as } n \to +\infty. \tag{15.6}$$

Since B_1 is a bounded set in $\mathbb{H}$, due to the reflexivity of $\mathbb{H}$ we conclude that there exists a subsequence $\{u_{n_k}\} \subset \{u_n\}$ such that

$$u_{n_k} \rightharpoonup u_0 \in B_1 \quad \text{weakly in } \mathbb{H} \text{ as } n_k \to +\infty. \tag{15.7}$$

Due to the weak continuity of the functional ψ, we obtain the limit relation

$$\psi(u_{n_k}) \to \psi(u_0) \quad \text{as } n_k \to +\infty. \tag{15.8}$$

The limit properties (15.6) and (15.8) lead us to a contradiction. Therefore, if $\psi \in \mathscr{F}$, then it is bounded on B_1.

2. Prove the boundedness of l and m. The condition $\psi \in \mathscr{F}$ implies the bounded Lipschitz continuity of $\operatorname{grad} \psi(u)$ on $\mathbb{H}$. Hence we immediately obtain the boundedness of $\operatorname{grad} \psi(u)$ on the ball B_1. Indeed, the chain of inequalities

$$\|\operatorname{grad} \psi(u)\| \leq \|\operatorname{grad} \psi(\theta)\| + \|\operatorname{grad} \psi(u) - \operatorname{grad} \psi(\theta)\|$$
$$\leq C(1)\|u\| \leq C(1) \quad \text{for all } u \in B_1 \tag{15.9}$$

implies the upper boundedness of l and m. Clearly, $l > 0$ and $m > 0$. □

Introduce the following operator family:

$$\chi(u) := \frac{u + a(u)D(u)}{\|u + a(u)D(u)\|}, \quad u \in S, \tag{15.10}$$

where

$$a(u) := \min\left\{\frac{1}{2l}\delta\left(\frac{\|D(u)\|}{4}\right), \frac{1}{2l}, \frac{1}{4m}\right\}, \quad \delta(\varepsilon) = \frac{\varepsilon}{c_1}. \tag{15.11}$$

Lemma 15.3. *Let $\psi(u) \in \mathscr{F}$. Then the inequality*

$$\psi(\chi(u)) \geq \psi(u) + \frac{a(u)}{4}\|D(u)\|^2 \tag{15.12}$$

holds for all $u \in S$.

Proof. 1. Prove some auxiliary estimates. First, we introduce the following notation:

$$g := a(u)D(u), \quad h := \chi(u) - u = \frac{u + g}{\|u + g\|} - u. \tag{15.13}$$

Let $u \in S$, i.e., $\|u\| = 1$. Then we have the inequality

$$\|u + g\| \geq 1. \tag{15.14}$$

Indeed, by the definition (15.13) of the element $g(u)$ and the orthogonality property (15.4), we have

$$\|u + g\|^2 = (u + g, u + g) = \|u\|^2 + \|g\|^2 \geq 1. \tag{15.15}$$

Prove the inequalities

$$0 \leq \|u + g\| - 1 \leq \|g\|^2, \quad u \in S. \tag{15.16}$$

Indeed, due to (15.14), we have the following chain of expressions:

$$(\|u + g\| - 1)(\|u + g\| + 1) = \|u + g\|^2 - 1 = \|g\|^2$$
$$\Rightarrow \quad 0 \leq \|u + g\| - 1 \leq \|g\|^2. \tag{15.17}$$

Next, we prove the inequality

$$\|g\| \leq \frac{1}{2}, \quad u \in S. \tag{15.18}$$

Indeed, by (15.5), (15.11), and (15.13), we have

$$\|g\| \leq a(u)\|D(u)\| \leq a(u)l \leq \min\left\{\frac{1}{2}\delta\left(\frac{\|D(u)\|}{4}\right), \frac{1}{2}, \frac{l}{4m}\right\} \leq \frac{1}{2}. \tag{15.19}$$

The following inequalities are valid:

$$\|h\| \leq 2\|g\| \leq \delta\left(\frac{\|D(u)\|}{4}\right). \tag{15.20}$$

Indeed, by (15.13), (15.11), (15.16), and (15.18) we have

$$\|h\| = \frac{\|g + (1 - \|u + g\|)u\|}{\|u + g\|} \leq \frac{\|g\| + \|u + g\| - 1}{\|u + g\|}$$
$$\leq \|g\| + \|g\|^2 \leq 2\|g\| \leq \delta\left(\frac{\|D(u)\|}{4}\right), \tag{15.21}$$

for all $u \in S$, where we used the inequalities (15.19).

From the definition (15.3) of the operator $D(u)$ we obtain the equality

$$\operatorname{grad}\psi(u) = D(u) + (\operatorname{grad}\psi(u), u)u, \quad u \in S. \tag{15.22}$$

The following inequality holds:

$$|(\operatorname{grad}\psi(u), h) - (D(u), g)| \leq \frac{1}{4}a(u)\|D(u)\|^2, \quad u \in S. \tag{15.23}$$

Indeed, the following chain of equalities holds:

$$(\operatorname{grad}\psi(u), h) = \left(D(u) + (\operatorname{grad}\psi(u), u)u, \frac{u + g}{\|u + g\|} - u\right)$$
$$= \frac{(D(u), g)}{\|u + g\|} + (\operatorname{grad}\psi(u), u)\left(\frac{1}{\|u + g\|} - 1\right)$$
$$= \frac{1 - \|u + g\|}{\|u + g\|}\left[(\operatorname{grad}\psi(u), u) + (D(u), g)\right] + (D(u), g), \tag{15.24}$$

where we used the orthogonality properties

$$(D(u), u) = 0 \quad \Rightarrow \quad (g(u), u) = 0 \quad \text{for all } u \in S.$$

This fact, the inequalities (15.16), and the definitions (15.5), (15.11), and (15.18) imply the following chain of inequalities:

$$\left| (\operatorname{grad} \psi(u), h) - (D(u), g) \right| \leq \frac{\|u + g\| - 1}{\|u + g\|} \left| (\operatorname{grad} \psi(u), u) + (D(u), g) \right|$$

$$\leq m\|g\|^2 = ma^2(u)\|D(u)\|^2 \leq \frac{1}{4} a(u)\|D(u)\|^2,$$

$$(15.25)$$

since, due to the definition (15.11), we have the inequality $ma(u) \leq 1/4$.

The following estimate is valid for all $u \in S$:

$$|\omega(u, h)| \leq \frac{1}{2} a(u)\|D(u)\|^2.$$ (15.26)

Indeed, due to (15.2), (15.20), (15.21), and Lemma 15.1, the inequality

$$|\omega(u, h)| \leq \frac{\|D(u)\|}{4}\|h\| \leq \frac{\|D(u)\|}{4} 2\|g\| = \frac{1}{2} a(u)\|D(u)\|^2$$ (15.27)

holds for

$$\varepsilon = \frac{\|D(u)\|}{4}, \quad \delta(\varepsilon) = \frac{\varepsilon}{c_1} = \frac{1}{c_1}\frac{\|D(u)\|}{4}.$$

2. Due to the estimates (15.23) and (15.26) and the expression (15.13), the chain of relations

$$\psi(\chi(u)) - \psi(u) = (\operatorname{grad} \psi(u), h) + \omega(u, h)$$

$$= (D(u), g) + \left[(\operatorname{grad} \psi(u), h) - (D(u), g) \right] + \omega(u, h)$$

$$= a(u)\|D(u)\|^2 + \left[(\operatorname{grad} \psi(u), h) - (D(u), g) \right] + \omega(u, h)$$

$$\geq a(u)\|D(u)\|^2 - \frac{1}{4} a(u)\|D(u)\|^2 - \frac{1}{2} a(u)\|D(u)\|^2$$

$$= \frac{1}{4} a(u)\|D(u)\|^2, \quad u \in S,$$ (15.28)

proves the lemma. $\square$

Lemma 15.4. *If a functional $\psi(u) : \mathbb{H} \to \mathbb{R}^1$ is even, then its Fréchet derivative (if it exists) is an odd operator.*

Proof. Assume that the Fréchet derivative $\psi'_f(u) : \mathbb{B} \to \mathbb{B}^*$ exist at a pair of points $\{u, -u\}$. Then the equalities

$$\psi(u + h) = \psi(u) + \langle \psi'_f(u), h \rangle + \omega(u, h),$$ (15.29)

$$\psi(-u - h) = \psi(-u) + \langle \psi'_f(-u), -h \rangle + \omega(-u, -h)$$ (15.30)

hold for any $h \in \mathbb{B}$. Since the functional $\psi(u)$ is even, Eqs. (15.29) and (15.30) imply the equality

$$\langle \psi'_f(u) + \psi'_f(-u), h \rangle = \omega(-u, -h) - \omega(u, h) \tag{15.31}$$

for any $h \in \mathbb{B}$. Hence we have

$$\langle \psi'_f(u) + \psi'_f(-u), h \rangle = 0 \quad \text{for all } h \in \mathbb{B}; \tag{15.32}$$

indeed, in the opposite case, the right-hand side of this equality tends to zero as $\|h\| \to +0$ faster that the left-hand side. Thus, $\psi'_f(-u) = -\psi'_f(u)$. □

Corollary 15.1. *If a functional $\psi(u) : \mathbb{H} \to \mathbb{R}^1$ is even, then $\operatorname{grad}\psi(u) : \mathbb{H} \to \mathbb{H}$ is an odd operator.*

This immediately follows from Lemma 15.4 and the equalities

$$\operatorname{grad}\psi(-u) = J\psi'_f(-u) = J(-\psi'_f(u)) = -J\psi'_f(u) = -\operatorname{grad}\psi(u).$$

Corollary 15.2. *If a functional $\psi(u) : \mathbb{H} \to \mathbb{R}^1$ is even, then the operator*

$$D(u) = \operatorname{grad}\psi(u) - (\operatorname{grad}\psi(u), u)u$$

defined on the unit sphere $S \in \mathbb{H}$ is odd.

Indeed, by Corollary 15.1 we have

$$\begin{aligned} D(-u) &= \operatorname{grad}\psi(-u) - \big(\operatorname{grad}\psi(-u),\ -u\big)(-u) \\ &= -\operatorname{grad}\psi(u) + \big(\operatorname{grad}\psi(u), u\big)u = -D(u) \end{aligned}$$

for all $u \in S$.

Lemma 15.5. *The operator $\chi(u) : S \to S$ defined by Eq. (15.10) is continuous and odd for $u \in S$.*

Proof. The continuity follows from the fact that, due to the inequality (15.14), the denominator $\|u + a(u)D(u)\|$ is greater than or equal to 1 on the sphere $u \in S$ and the operator $D(u)$ is continuous on the sphere since $\psi \in \mathscr{F}$. Therefore, $\operatorname{grad}\psi(u)$ is continuous on B_1.

The oddness of $\chi(u)$ with respect to $u \in S$ follows from the definition (15.10) and the oddness of $D(u)$, which was proved in Corollary 15.2. □

Lemma 15.6. *The operator $\chi = \chi(u) : S \to S$ maps each set $F \subset S$ of the class M_k onto the set $\chi(F) \subset S$ of the class M_k.*

Lemma 15.6 follows from Lemmas 14.4 and 15.5.

We define on M_k the following real-valued functional $\hat\psi(F)$:

$$\hat\psi(F) = \inf_{u \in F} \psi(u), \quad F \in M_k, \tag{15.33}$$

where the functional $\psi(u)$ belongs to the class $\mathscr{F}$. Therefore, in particular, the functional $\psi(u)$ is weakly continuous and hence strongly continuous. Due to Theorem 11.2, the functional $\psi(u)$ attains its infimum and supremum on each compact set $F \subset S$.

Therefore,

$$\hat{\psi}(F) = \min_{u \in F} \psi(u), \quad F \in M_k. \tag{15.34}$$

By Lemma 15.2, the functional $\hat{\psi}(F)$ is upper bounded. Consider the number

$$c_k := \sup_{F \in M_k} \min_{u \in F} \psi(u). \tag{15.35}$$

Lemma 15.7. *Let $\psi(u) \in \mathscr{F}$. Then there exists a sequence*

$$\{u_n\} \subset \bigcup_{F \in M_k} F \subset S$$

such that

$$\lim_{n \to +\infty} \psi(u_n) = c_k, \quad \lim_{n \to +\infty} \|D(u_n)\| = 0.$$

Proof. Assume the contrary. Then there exist positive numbers α and β such that for all $u \in \bigcup_{F \in M_k} F$ satisfying the condition

$$|\psi(u) - c_k| < \beta, \tag{15.36}$$

the following inequality holds:

$$\inf_{u \in \bigcup_{F \in M_k} F} \|D(u)\| > \alpha. \tag{15.37}$$

By the definition (15.35) of the number c_k, there exists a set $F_0 \in M_k$ such that

$$\hat{\psi}(F_0) = \min_{u \in F_0} \psi(u) > c_k - \frac{\gamma \alpha^2}{4}. \tag{15.38}$$

Due to the monotonicity of the function $\delta = \delta(\varepsilon) = \varepsilon/c_1$, we can choose a number $\gamma > 0$ such that the following inequalities hold:

$$0 < \gamma < \min \left\{ \frac{1}{2l} \delta \left(\frac{\alpha}{4} \right), \frac{1}{2l}, \frac{1}{4m} \right\} \le a(u). \tag{15.39}$$

Consider the set $\chi(F_0)$. By Lemma 15.6, we have $\chi(F_0) \in M_k$. Therefore, due to (15.35), the following inequality holds:

$$\hat{\psi}(\chi(F_0)) \le c_k. \tag{15.40}$$

Let

$$L := \{ u \in F_0 : \psi(u) < c_k + \beta \}. \tag{15.41}$$

Then, on the one hand, by the inequality (15.12) from Lemma 15.3 we have

$$\psi(\chi(u)) \ge \psi(u) \ge c_k + \beta \quad \text{for all } u \in F_0 \backslash L. \tag{15.42}$$

On the other hand, using the inequalities (15.12), (15.36), and (15.37) we have

$$\|D(u)\| > \alpha \quad \Rightarrow \quad \exists \varepsilon > 0: \ \|D(u)\|^2 \ge \alpha^2 + \varepsilon^2 \quad \text{for } u \in L;$$

therefore, taking into account (15.39), we obtain

$$\psi(\chi(u)) \ge \psi(u) + \frac{a(u)}{4}\|D(u)\|^2 \ge c_k - \frac{\gamma\alpha^2}{4} + \frac{1}{4}a(u)(\alpha^2 + \varepsilon^2)$$

$$\ge c_k - \frac{\gamma\alpha^2}{4} + \frac{\gamma\alpha^2}{4} + \frac{\gamma\varepsilon^2}{4} = c_k + \frac{\gamma\varepsilon^2}{4}. \tag{15.43}$$

Thus, from (15.42) and (15.43) we obtain the inequality

$$\hat{\psi}(\chi(F_0)) \ge c_k + \varepsilon_1, \quad \varepsilon_1 = \min\left\{\beta, \frac{\gamma\varepsilon^2}{4}\right\} \quad \Rightarrow \quad \hat{\psi}(\chi(F_0)) > c_k. \tag{15.44}$$

The inequalities (15.40) and (15.44) are contradictory. $\qquad\square$

Lemma 15.8. *Let $\psi(u) \in \mathscr{F}$ and $\{u_n\} \subset S$ be a weakly converging sequence of points of the sphere S such that*

$$\lim_{n \to +\infty} \|D(u_n)\| = 0 \tag{15.45}$$

and, moreover,

$$\operatorname{grad}\psi(u_n) \to \psi_0 \ne \theta \quad \text{strongly in } \mathbb{H} \text{ as } n \to +\infty. \tag{15.46}$$

Then

$$u_n \to u_0 \quad \text{strongly in } \mathbb{H} \text{ as } n \to +\infty, \tag{15.47}$$

where $u_0 \in S$, the following equality is valid:

$$\operatorname{grad}\psi(u_0) = \lambda u_0, \quad \lambda \ne 0. \tag{15.48}$$

Proof. By the condition of the lemma we have

$$(\operatorname{grad}\psi(u_n), u_n) \to \lambda \in \mathbb{R}^1 \quad \text{as } n \to +\infty. \tag{15.49}$$

Recall that for points $v \in S$, the equality

$$D(v) := \operatorname{grad}\psi(v) - \big(\operatorname{grad}\psi(v), v\big)v \tag{15.50}$$

holds; therefore,

$$\big(D(u_n), \operatorname{grad}\psi(u_n)\big) = \big(\operatorname{grad}\psi(u_n), \operatorname{grad}\psi(u_n)\big) - \big(\operatorname{grad}\psi(u_n), u_n\big)^2. \tag{15.51}$$

From (15.51), (15.45), and (15.46) we obtain the limit equality

$$\lim_{n \to +\infty} \big(\operatorname{grad}\psi(u_n), u_n\big)^2 = \|\psi_0\|^2. \tag{15.52}$$

Comparing (15.49) and (15.52), we arrive at the following equality:

$$|\lambda| = \|\psi_0\| \ne 0 \quad \Rightarrow \quad \lambda \ne 0. \tag{15.53}$$

From Eq. (15.50) we obtain

$$u_n = \frac{\operatorname{grad}\psi(u_n)}{(\operatorname{grad}\psi(u_n), u_n)} - \frac{D(u_n)}{(\operatorname{grad}\psi(u_n), u_n)}. \tag{15.54}$$

By (15.53), there exists $n_0 \in \mathbb{N}$ such that for all $n \geq n_0$, the denominators in (15.54) are nonzero.

From (15.49), (15.53), and (15.54) we conclude that

$$u_n \to u_0 = \frac{\psi_0}{\lambda} \quad \text{strongly in } \mathbb{H} \text{ as } n \to +\infty. \tag{15.55}$$

Taking into account (15.53), we obtain

$$\|u_0\| = 1 \quad \Leftrightarrow \quad u_0 \in S.$$

Due to the strong continuity of $\operatorname{grad} \psi(u)$ on $\mathbb{H}$, from (15.55) we conclude that

$$\operatorname{grad} \psi(u_n) \to \operatorname{grad} \psi(u_0) = \psi_0 \quad \text{strongly in } \mathbb{H} \text{ as } n \to +\infty. \tag{15.56}$$

Thus, (15.55) and (15.56) imply the equality

$$\operatorname{grad} \psi(u_0) = \lambda u_0, \quad u_0 \in S, \quad \lambda \neq 0. \qquad \square$$

15.2 Bibliographical Notes

The contents of this lecture is taken from [35, 56, 67].

Applications of M. A. Krasnosel'sky's Theory of the Genus of a Set

16.1 Auxiliary Lemma

Let $E_1 \subset E_2 \subset \cdots$ be a sequence of finite-dimensional subspaces of a Hilbert space $\mathbb{H}$, $\dim E_n = n$, possessing the property

$$\mathbb{H} = \overline{\bigcup_{n=1}^{+\infty} E_n}. \tag{16.1}$$

Consider the linear operators of orthogonal projection of the Hilbert space $\mathbb{H}$ onto the subspaces E_n:

$$P_n : \mathbb{H} \to E_n. \tag{16.2}$$

By the Beppo Levi theorem, the following equalities hold:

$$u = u_n + u_n^{\perp}, \quad u_n = P_n u, \quad u_n^{\perp} = Q_n u := (I - P_n)u; \tag{16.3}$$

moreover, for each $u \subset \mathbb{H}$ and each $\varepsilon > 0$, there exists $n_\varepsilon \in \mathbb{N}$ such that, due to the Beppo Levi theorem, the following relations hold:

$$\|u_n^{\perp}\| = \|Q_n u\| = \|u - P_n u\| = \text{distance}(u, E_n) < \varepsilon \tag{16.4}$$

for $n \geq n_\varepsilon$.

Lemma 16.1. *If ψ is a weakly continuous functional on the closed ball $B_R := \{u \in \mathbb{H} : \|u\| \leq R\}$, then for each $\varepsilon > 0$, there exists a natural number $n_0 = n_0(\varepsilon)$ such that for all $n \geq n_0$, the inequality*

$$|\psi(P_n u) - \psi(u)| < \varepsilon \tag{16.5}$$

holds for all $u \in B_R$.

Proof. Assume the contrary. Then there exists a number $\alpha > 0$ and a sequence $\{u_n\} \subset B_R$ such that

$$|\psi(P_n u_n) - \psi(u_n)| \geq \alpha \quad \text{for all } n \in \mathbb{N}. \tag{16.6}$$

Since the closed ball $B_R := \{u \in \mathbb{H} : \|u\| \leq R\}$ is a bounded set, the sequence $\{u_n\} \subset B_R$ is also bounded. Since $\mathbb{H}$ is reflexive (it is a Hilbert space), there exists a subsequence $\{u_{n_k}\} \subset \{u_n\}$ such that

$$u_{n_k} \rightharpoonup u_0 \in B_R \quad \text{weakly in } \mathbb{H} \text{ as } n_k \to +\infty; \tag{16.7}$$

moreover, since $P_n : B_R \to B_R$, the following limit property also holds:

$$P_{n_k} u_{n_k} \rightharpoonup w_0 \in B_R \quad \text{weakly in } \mathbb{H} \text{ as } n_k \to +\infty. \tag{16.8}$$

We write the inequality (16.6) for subsequences found:

$$\left| \psi(P_{n_k} u_{n_k}) - \psi(u_{n_k}) \right| \geq \alpha \quad \text{for all } n_k \in \mathbb{N}. \tag{16.9}$$

Using the weak continuity of the functional ψ and passing to the limit as $n_k \to +\infty$ in the inequality (16.9), we arrive at the inequality

$$\left| \psi(w_0) - \psi(u_0) \right| \geq \alpha > 0 \quad \Rightarrow \quad w_0 \neq u_0. \tag{16.10}$$

Due to (16.4), we have the following limit property:

$$\lim_{n_k \to +\infty} \| Q_{n_k}(u_0 - w_0) \| = 0. \tag{16.11}$$

Therefore, we obtain the chain of relations

$$\begin{aligned}
\| u_0 - w_0 \|^2 = (u_0 - w_0, u_0 - w_0) &= \lim_{n_k \to +\infty} (u_{n_k} - P_{n_k} u_{n_k}, u_0 - w_0) \\
&= \lim_{n_k \to +\infty} (Q_{n_k} u_{n_k}, u_0 - w_0) = \lim_{n_k \to +\infty} (Q_{n_k} u_{n_k}, Q_{n_k}(u_0 - w_0)) \\
&= \limsup_{n_k \to +\infty} (Q_{n_k} u_{n_k}, Q_{n_k}(u_0 - w_0)) \\
&\leq \sup_{n_k \in \mathbb{N}} \| Q_{n_k} u_{n_k} \| \lim_{n_k \to +\infty} \| Q_{n_k}(u_0 - w_0) \| = 0, \tag{16.12}
\end{aligned}$$

since

$$\| Q_{n_k} u_{n_k} \| \leq \| u_{n_k} \| \leq R \quad \text{for all } n_k \in \mathbb{N}.$$

Thus, $u_0 = w_0$, which contradicts the formula (16.10). The lemma is proved. □

16.2   Critical Numbers of Functionals

Let $\psi \in \mathscr{F}$. Consider the functional

$$\hat{\psi}(F) := \min_{u \in F} \psi(F), \quad F \in M_k. \tag{16.13}$$

Here we write min instead of inf since the set $F \in M_k$ is compact and the functional $\psi(u)$ is weakly continuous and hence strongly continuous; therefore, it attains its minimum on F (see Theorem 11.2).

Definition 16.1. The numbers

$$c_k := \sup_{F \in M_k} \min_{u \in F} \psi(u), \quad k \in \mathbb{N}, \tag{16.14}$$

are called the *critical numbers* of the functional ψ.

We fix $\varepsilon > 0$ arbitrarily. By the definition of the numbers c_{k+1}, there exists a set $F_1 \in M_{k+1}$ such that

$$\hat{\psi}(F_1) > c_{k+1} - \varepsilon. \tag{16.15}$$

The set $F_1 \in M_{k+1}$ contains the subset $F_0 \in M_k$. Indeed, since $F_1 \in M_{k+1}$, there exists an odd continuous mapping $\varphi : S^k \to S$ such that $F_1 = \varphi(S^k)$. Let $S^{k-1} \subset S^k$ (for example, S^{k-1} is the equator of the sphere S^k). Then we have

$$F_0 := \varphi(S^{k-1}) \subset \varphi(S^k) = F_1.$$

For this subset $F_0 \in F_1$, we have the following chain of relations:

$$\hat{\psi}(F_0) = \min_{u \in F_0} \psi(u) \geq \min_{u \in F_1} \psi(u) = \hat{\psi}(F_1) > c_{k+1} - \varepsilon. \tag{16.16}$$

From (16.14) and (16.16) we obtain the inequality

$$c_k = \sup_{F \in M_k} \hat{\psi}(F) \geq \hat{\psi}(F_0) > c_{k+1} - \varepsilon. \tag{16.17}$$

Due to the arbitrariness of $\varepsilon > 0$, we arrive at the inequality

$$c_k \geq c_{k+1} \quad \text{for all } k \in \mathbb{N}. \tag{16.18}$$

Moreover, since $\psi(u) > 0$ on S, the minimum of the weakly continuous functional $\psi(u)$ on each compact set $F \subset S$ is positive. Therefore,

$$c_k > 0 \quad \text{for all } k \in \mathbb{N}. \tag{16.19}$$

Thus, we have proved the following assertion.

Lemma 16.2.

$$c_k \geq c_{k+1} > 0 \quad \text{for all } k \in \mathbb{N}. \tag{16.20}$$

Lemma 16.3.

$$\lim_{k \to +\infty} c_k = 0. \tag{16.21}$$

Proof. 1. Let $\varepsilon > 0$ be an arbitrary given positive number. The separable Hilbert space $\mathbb{H}$ can be represented as follows:

$$\mathbb{H} = \overline{\bigcup_{n=1}^{+\infty} E_n}, \quad E_n \subset E_{n+1}, \quad \dim E_n = n \in \mathbb{N}. \tag{16.22}$$

Denote by P_n the linear orthogonal projector of $\mathbb{H}$ onto E_n:

$$P_n : \mathbb{H} \to E_n, \quad u = u_n + u_n^\perp, \quad P_n u = u_n \in E_n, \quad (u_n, u_n^\perp) = 0. \tag{16.23}$$

Then due to Lemma 16.1, for given $\varepsilon > 0$, there exists $n_0 = n_0(\varepsilon) \in \mathbb{N}$ such that

$$\left| \psi(u) - \psi(P_{n_0} u) \right| < \frac{\varepsilon}{2} \quad \text{for all } u \in B_1. \tag{16.24}$$

Since the functional ψ is weakly continuous on B_1, it is also strongly continuous on B_1. Note that, since $\psi \in \mathscr{F}$, we have, in particular, $\psi(\theta) = 0$. Therefore, for $\varepsilon > 0$ fixed above, there exists $\rho \in (0, 1)$ such that

$$\psi(u) < \frac{\varepsilon}{2} \quad \text{if } \|u\| \leq \rho. \tag{16.25}$$

Therefore, for all $u \in B_1$ such that $P_{n_0} u \in B_\rho$, the following inequality holds:

$$\psi(P_{n_0} u) < \frac{\varepsilon}{2} \quad \text{if } \|P_{n_0} u\| \leq \rho. \tag{16.26}$$

Thus, from (16.24) and (16.26) we obtain the following inequalities:

$$\psi(u) \leq |\psi(u) - \psi(P_{n_0} u)| + \psi(P_{n_0} u) < \frac{\varepsilon}{2} + \frac{\varepsilon}{2} = \varepsilon \tag{16.27}$$

if $P_{n_0} u \in B_\rho$, $\rho \in (0, 1)$. The formula (16.27) implies that the image under the projector $P_{n_0} : F \subset \mathbb{H} \to E_{n_0}$ of each compact set F consisting of $u \in F$ such that

$$\sup_{u \in F} \psi(u) \geq \varepsilon \tag{16.28}$$

lies in the spherical layer

$$0 < \rho \leq \|P_{n_0} u\| \leq 1. \tag{16.29}$$

2. Note that, due to Lemma 14.2, each spherical layer in an n_0-dimensional space can be represented as the sum of n_0 closed sets of the first kind. Therefore, Lemma 14.1 implies that the compact set F defined by the inequality (16.28) has the genus

$$r(F) \leq r(P_{n_0} F) = n_0. \tag{16.30}$$

Let $k \geq n_0 + 1$. Then, by Lemma 14.3, each set $F \in M_k$ has genus $\geq n_0 + 1$:

$$r(F) \geq n_0 + 1 \quad \text{for all } F \in M_k, \ k \geq n_0 + 1. \tag{16.31}$$

Therefore, due to (16.28), (16.29), and (16.30) we see that on each set $F \in M_k$, the inequality $\psi(u) < \varepsilon$ holds at a certain point $u \in F$. Hence for any set $F \in M_k$, the following inequality holds:

$$\hat{\psi}(F) = \min_{u \in F} \psi(u) \leq \varepsilon \quad \text{for } k \geq n_0 + 1. \tag{16.32}$$

Consequently,

$$c_k = \sup_{F \in M_k} \min_{u \in F} \psi(u) \leq \varepsilon \quad \text{for all } k \geq n_0 + 1. \tag{16.33}$$

Thus, the limit relation (16.21) holds. $\qquad\qquad\qquad\qquad\qquad\qquad\square$

Lemmas 16.2 and 16.3 imply the following theorem.

Theorem 16.1. *Each functional $\psi(u)$ of the class $\mathscr{F}$ possesses a countable set of distinct, positive critical numbers $c_{n_k} > 0$, $n_k \in \mathbb{N}$.*

Remark 16.1. In fact, these solutions are linearly independent on the unit sphere.

16.3 Critical Points of Functionals

Definition 16.2. A point $u_0 \in S$ is called a *critical point* of a functional $\psi(u)$ on the sphere S if

$$\operatorname{grad} \psi(u_0) - \big(\operatorname{grad} \psi(u_0), u_0 \big) u_0 = 0. \tag{16.34}$$

Definition 16.3. The value of a functional $\psi(u_0)$ at its critical point $u_0 \in S$ is called a critical value of the functional ψ on the sphere S.

Theorem 16.2. *Let $\psi \in \mathscr{F}$. Then each critical number c_k defined by Eq. (16.14) is a critical value of the functional $\psi(u)$ on S.*

Proof. Let $\chi : S \to S$ be the operator defined by Eq. (15.10). Since $\psi \in \mathscr{F}$, due to Lemma 15.6, the operator χ maps each set F of the class M_k onto the set $\chi(F)$ of class M_k.

By Lemma 15.7, for each $k \in \mathbb{N}$, there exists a sequence

$$\{u_n\} \subset \bigcup_{F \in M_k} F \subset S \tag{16.35}$$

such that

$$\psi(u_n) \to c_k \quad \text{as } n \to +\infty, \tag{16.36}$$

$$D(u_n) \to \theta \quad \text{strongly in } \mathbb{H} \text{ as } n \to +\infty, \tag{16.37}$$

where $D(v) := \operatorname{grad} \psi(v) - \big(\operatorname{grad} \psi(v), v \big) v$, $v \in S$.

Since $\psi \in \mathscr{F}$, the operator $\operatorname{grad} \psi(u) : \mathbb{H} \to \mathbb{H}$ is totally continuous by Lemma 14.6. Recall that the ball B_1 is weakly closed in $\mathbb{H}$. Note that $\{u_n\} \subset S$ and hence the sequence $\{u_n\}$ is bounded. Therefore, there exists a subsequence $\{u_{n_k}\} \subset \{u_n\}$ such that the following limit properties hold:

$$u_{n_k} \rightharpoonup u_0 \in B_1 \qquad \text{weakly in } \mathbb{H} \text{ as } n_k \to +\infty, \tag{16.38}$$

$$\psi(u_{n_k}) \to \psi(u_0) \qquad \text{in } \mathbb{H} \text{ as } n_k \to +\infty, \tag{16.39}$$

$$\operatorname{grad} \psi(u_{n_k}) \to \operatorname{grad} \psi(u_0) \quad \text{as } n_k \to +\infty. \tag{16.40}$$

Due to (16.36) and Lemma 16.2, we obtain from (16.39) the following equality:

$$\psi(u_0) = c_k > 0 \quad \Rightarrow \quad u_0 \neq \theta. \tag{16.41}$$

Therefore, by Definition 15.1(c), we conclude that $\operatorname{grad} \psi(u_0) \neq \theta$. Therefore, by Lemma 15.8 implies that

$$\operatorname{grad} \psi(u_0) = \lambda_0 u_0, \quad \lambda_0 := (\operatorname{grad} \psi(u_0), u_0), \quad u_0 \in S. \tag{16.42}$$

Finally, using (16.41) and (16.42) we conclude that u_0 is a critical point of the functional $\psi(u)$ with the corresponding critical value c_k. The theorem is proved. $\square$

Theorems 16.1 and 16.2 imply the following well-known L. A. Lyusternik's theorem.

Theorem 16.3. *On the sphere $S \in \mathbb{H}$, a functional $\psi(u)$ of the class $\mathscr{F}$ has no less than a countable set of distinct critical points corresponding to distinct critical values of the functional $\psi(u)$.*

16.4   Example of a Countable Set of Solutions

Consider the following nonlinear eigenvalue problem:

$$\Delta u + \lambda |u|^{p-2}u = 0, \quad u|_{\partial\Omega} = 0, \quad 2 < p < \frac{2N}{N-2}, \tag{16.43}$$

where $\Omega \subset \mathbb{R}^N$ is a bounded smooth domain, $N \geq 3$. Consider the following functionals:

$$\varphi(u) := \frac{1}{2} \int_{\Omega} |D_x u|^2 \, dx, \quad \psi(u) := \frac{1}{p} \int_{\Omega} |u|^p \, dx. \tag{16.44}$$

Consider the unit sphere S of the Hilbert space $\mathbb{H} = H_0^1(\Omega)$:

$$S := \left\{ u \in H_0^1(\Omega) : \|\|D_x u\|\|_2 = 1 \right\} = \left\{ u \in H_0^1(\Omega) : \varphi(u) = \frac{1}{2} \right\}. \tag{16.45}$$

The following chain of dense and completely continuous embeddings holds:

$$H_0^1(\Omega) \overset{ds}{\hookrightarrow} L^p(\Omega) \overset{ds}{\subset} L^{p'}(\Omega) \overset{ds}{\hookrightarrow} H^{-1}(\Omega). \tag{16.46}$$

The Fréchet derivative of the functional

$$\psi(u) : L^p(\Omega) \to \mathbb{R}^1 \tag{16.47}$$

has the form

$$\psi_f'(u) : L^p(\Omega) \to L^{p'}(\Omega), \quad \langle \psi_f'(u), h \rangle_p = \int_{\Omega} |u(x)|^{p-2}u(x)h(x) \, dx \tag{16.48}$$

for any $h(x) \in L^p(\Omega)$, where $\langle \cdot, \cdot \rangle_p$ is the duality bracket between the Banach spaces $L^p(\Omega)$ and $L^{p'}(\Omega)$. By Theorem 1.4 and (16.46), we have

$$\langle f, u \rangle = \langle f, u \rangle_p \quad \text{for all } f(x) \in L^{p'}(\Omega) \text{ and } u(x) \in H_0^1(\Omega). \tag{16.49}$$

From this relation and (16.48) we conclude that the Fréchet derivative of the functional $\psi(u)$ (considered as a functional on $H_0^1(\Omega)$) is

$$\langle \psi_f'(u), h \rangle = \int_{\Omega} |u(x)|^{p-2}u(x)h(x) \, dx \tag{16.50}$$

for any $h(x) \in H_0^1(\Omega)$. Let

$$J = (-\Delta)^{-1} : H^{-1}(\Omega) \to H_0^1(\Omega) \tag{16.51}$$

be the linear Riesz–Fréchet isometry; then grad $\psi(u)$ has the following explicit form:

$$\text{grad}\,\psi(u) = J|u|^{p-2}u : H_0^1(\Omega) \to H_0^1(\Omega). \tag{16.52}$$

We prove that $\psi \in \mathscr{F}$.

Lemma 16.4. *The functional ψ is weakly continuous on $H_0^1(\Omega)$.*

Proof. Let $\{u_n\} \subset H_0^1(\Omega)$ be a weakly converging sequence

$$u_n \rightharpoonup u_0 \quad \text{weakly in } H_0^1(\Omega) \text{ as } n \to +\infty. \tag{16.53}$$

Since the embedding of $H_0^1(\Omega)$ into $L^p(\Omega)$ is completely continuous for $2 < p < 2N/(N-2)$, $N \geq 3$, the embedding operator is totally continuous. Therefore, from (16.53) we conclude that

$$u_n \to u_0 \quad \text{strongly in } L^p(\Omega) \text{ as } n \to +\infty. \tag{16.54}$$

Hence, in particular,

$$\|u_n\|_p \to \|u_0\|_p \quad \text{as } n \to +\infty \tag{16.55}$$

and

$$\psi(u_n) \to \psi(u_0) \quad \text{as } n \to +\infty. \tag{16.56}$$

The lemma is proved. $\qquad\square$

Lemma 16.5. *The gradient* $\operatorname{grad}\psi(u)$ *of the functional* $\psi(u)$ *is a boundedly Lipschitz continuous operator on* $H_0^1(\Omega)$.

Proof. First, we note that the following inequality holds:

$$\left| |u_1|^{p-2}u_1 - |u_2|^{p-2}u_2 \right| \leq (p-1)\max\left\{|u_1|^{p-2}, |u_2|^{p-2}\right\}|u_1-u_2|, \quad p > 2. \tag{16.57}$$

Indeed, raising to the power $p' = p/(p-1)$, we obtain the inequality

$$\left| |u_1|^{p-2}u_1 - |u_2|^{p-2}u_2 \right|^{p'}$$
$$\leq (p-1)^{p'}\max\left\{|u_1|^{(p-2)p'}, |u_2|^{(p-2)p'}\right\}|u_1-u_2|^{p'}. \tag{16.58}$$

Now let $u_1(x), u_2(x) \in L^p(\Omega)$. Then, integrating both sides of the inequality (16.58) by $x \in \Omega$ and applying the Hölder inequality with the parameters $q_1 = (p-1)/(p-2)$ and $q_2 = p-1$, we obtain the following inequality:

$$\left\| |u_1|^{p-2}u_1 - |u_2|^{p-2}u_2 \right\|_{p'} \leq (p-1)\max\left\{\|u_1\|_p^{p-2}, \|u_2\|_p^{p-2}\right\}\|u_1-u_2\|_p. \tag{16.59}$$

Applying (16.46), (16.52), and (16.59), we arrive at the inequality

$$\|\operatorname{grad}\psi(u_1) - \operatorname{grad}\psi(u_2)\| \leq \mu(R)\|u_1 - u_2\|, \tag{16.60}$$

where $\mu(R) = (p-1)c_2 c_1^{p-1} R^{p-2}$ for $\|u_1\| \leq R$, $\|u_2\| \leq R$, $\|v\| = \||D_x v|\|_2$, and $c_1 > 0$ and $c_2 > 0$ are the constants of continuous embeddings $H_0^1(\Omega) \subset L^p(\Omega)$ and $L^{p'}(\Omega) \subset H^{-1}(\Omega)$, respectively. The lemma is proved. $\qquad\square$

Verifying the conditions (a), (b), and (c), from Lemmas 16.4 and 16.5 and Theorem 16.3 we immediately obtain the following theorem.

Theorem 16.4. *The functional* ψ *belongs to the class* $\mathscr{F}$ *and hence it possesses a countable set of distinct critical points on the unit sphere* $S \subset H_0^1(\Omega)$.

Remark 16.2. Theorem 16.4 implies that there exists a countable set $\{u_n\} \subset S \subset H_0^1(\Omega)$ of distinct points such that

$$\operatorname{grad}\psi(u_n) = \mu_n u_n \quad \Leftrightarrow \quad J|u_n|^{p-2}u_n = \mu_n u_n$$

$$\Leftrightarrow \quad \Delta u_n + \lambda_n |u_n|^{p-2} u_n = 0, \quad \lambda_n = \frac{1}{\mu_n}, \quad \mu_n \neq 0 \quad (16.61)$$

(we used here the formula (16.51) for the Riesz–Fréchet operator $J = (-\Delta)^{-1}$). Thus, the nonlinear eigenvalue problem (16.43) has a countable number of linearly independent solutions on the unit sphere $S \subset H_0^1(\Omega)$.

16.5   Bibliographical Notes

The contents of this lecture is taken from [14, 23, 35, 56, 65, 67].

Mountain Pass Theorem

In this lecture, we consider the variational method proposed by A. Ambrosetti and P. Rabinowitz based on the so-called Mountain pass theorem, which has important applications in the theory of unbounded functionals.

17.1 Theorem on Deformations

Let $\mathbb{H}$ be a real Hilbert space and $\psi(u) \in C^{(1)}(\mathbb{H}; \mathbb{R}^1)$ be a functional satisfying the following condition: its gradient[1]

$$F(u) = \operatorname{grad} \psi(u) : \mathbb{H} \to \mathbb{H}$$

is boundedly Lipschitz-continuous. (The notion of bounded Lipschitz continuity was introduced in Definition 10.3 of Lecture 3).

Introduce the following notation:

$$A_c \overset{\text{def}}{=} \left\{ u \in \mathbb{H} : \psi(u) \leq c \right\},$$

$$K_c \overset{\text{def}}{=} \left\{ u \in \mathbb{H} : \psi(u) = c, \ F(u) = \operatorname{grad} \psi(u) = \theta \right\}.$$

We denote by $\mathscr{F}$ the family of functionals $\psi(u) \in C^{(1)}(\mathbb{H}; \mathbb{R}^1)$ whose gradients are boundedly Lipschitz-continuous.

Definition 17.1. An element $u \in \mathbb{H}$ is called a critical point of a functional $\psi(u)$ if $\operatorname{grad} \psi(u) = \theta$. A real number c is called a critical value of the functional $\psi(u)$ if $K_c \neq \varnothing$.

We prove that if a number c is not a critical value of a functional $\psi(u)$, then the set $A_{c+\varepsilon}$ "is deforming"[2] in $A_{c-\varepsilon}$ for some $\varepsilon > 0$.

Since, in general, the space $\mathbb{H}$ is infinite-dimensional, we need the following compactness condition.

Definition 17.2 (Palais–Smale condition). We say that a functional $\psi \in C^{(1)}(\mathbb{H}; \mathbb{R}^1)$ satisfies the Palais–Smale compactness condition (PS) if any sequence $\{u_k\}_{k=1}^{+\infty} \subset \mathbb{H}$ satisfying the conditions

---

[1]Recall that $\operatorname{grad} \psi(u) \overset{\text{def}}{=} J\psi_f'(u)$, where $J : \mathbb{H}^* \to \mathbb{H}$ is a Riesz isometry.

[2]The sense of this term will be clear in the following theorem.

  (i) $\{\psi(u_k)\}_{k=1}^{+\infty}$ is bounded;

  (ii) $\operatorname{grad}\psi(u_k) \to \theta$ strongly in $\mathbb{H}$

contains a strongly converging (in $\mathbb{H}$) subsequence.

Theorem 17.1 (Theorem on deformation). *Assume that a functional $\psi \in \mathscr{F}$ satisfies the Palais–Smale condition and*

$$K_c = \varnothing. \tag{17.1}$$

Then for any sufficiently small $\varepsilon > 0$, there exists a constant $0 < \delta < \varepsilon$ and a function $\eta(t,u) \in \mathbb{C}([0,1] \times \mathbb{H}; \mathbb{H})$ such that the mapping

$$\eta_t(u) \overset{\text{def}}{=} \eta(t,u), \quad 0 \le t \le 1, \ u \in \mathbb{H},$$

satisfies the following conditions:

  (i) $\eta_0(u) = u$ *for all* $u \in \mathbb{H}$;

  (ii) $\eta_1(u) = u$ *for all* $u \notin \psi^{-1}([c - \varepsilon, c + \varepsilon])$;

  (iii) $\psi(\eta_t(u)) \le \psi(u), \ u \in \mathbb{H}, \ 0 \le t \le 1$;

  (iv) $\eta_1(A_{c+\delta}) \subset A_{c-\delta}$.

Proof. First, we prove that there exist constants $0 < \sigma, \delta_1 < 1$ such that

$$\|\operatorname{grad}\psi(u)\|_{\mathbb{H}} \ge \sigma \quad \text{for all} \quad u \in A_{c+\delta_1} \backslash A_{c-\delta_1} \tag{17.2}$$

(in other words, $c - \delta_1 < \psi(u) \le c + \delta_1$). Assume the contrary. If (17.2) does not hold for all constants $\sigma, \delta_1 \in (0,1)$, then there exist sequences $\sigma_k \to 0$ and $\delta_k \to 0$ and elements

$$u_k \in A_{c+\delta_k} \backslash A_{c-\delta_k} \tag{17.3}$$

such that

$$\|\operatorname{grad}\psi(u_k)\|_{\mathbb{H}} < \sigma_k \quad \Rightarrow \quad \operatorname{grad}\psi(u_k) \to \theta \quad \text{strongly in } \mathbb{H} \text{ as } k \to +\infty. \tag{17.4}$$

Due to (17.3) we have

$$c - \delta_k < \psi(u_k) \le c + \delta_k, \tag{17.5}$$

i.e., the numerical sequence $\{\psi(u_k)\}$ is bounded. By the Palais–Smale condition, there exist a subsequence $\{u_{k_j}\}_{j=1}^{+\infty} \subset \{u_k\}_{k=1}^{+\infty}$ and an element $u \in \mathbb{H}$ such that

$$u_{k_j} \to u \quad \text{strongly in } \mathbb{H} \text{ as } k_j \to +\infty.$$

In particular, there exists a constant $M > 0$ such that

$$\|u_{k_j}\| \le M < +\infty, \tag{17.6}$$

where the constant $M > 0$ is independent of k_j. Since $\psi \in \mathbb{C}^{(1)}(\mathbb{H}; \mathbb{R}^1)$, we conclude that (17.5) and (17.4) imply

$$\psi(u) = c, \quad \operatorname{grad}\psi(u) = \theta.$$

Indeed, due to the bounded Lipschitz continuity of $\operatorname{grad} \psi(\cdot)$ and (17.6), the following limit property holds:

$$\operatorname{grad} \psi(u_{k_j}) \to \operatorname{grad} \psi(u) \quad \text{as } k_j \to +\infty;$$

moreover, as was stated above,

$$\operatorname{grad} \psi(u_{k_j}) \to \theta \quad \text{strongly in } \mathbb{H} \text{ as } k_j \to +\infty.$$

Therefore,

$$\operatorname{grad} \psi(u) = \theta. \tag{17.7}$$

On the other hand, due to the inequalities (17.5) and the continuity of the functional $\psi(u) \in \mathbb{C}^{(1)}(\mathbb{H}; \mathbb{R}^1)$, we have

$$\psi(u_{k_j}) \to c, \quad \psi(u_{k_j}) \to \psi(u) \quad \text{as } k_j \to +\infty.$$

Thus,

$$\psi(u) = c. \tag{17.8}$$

Therefore, from the formulas (17.7) and (17.8) imply that $K_c \neq \varnothing$. This contradicts the assumption $K_c = \varnothing$.

Now we fix constants $\delta_1 \in (0,1)$ and $\sigma \in (0,1)$ such that the following inequality holds:

$$\|\operatorname{grad} \psi(u)\|_{\mathbb{H}} \geq \sigma \quad \text{for all } u \in A_{c+\delta_1} \backslash A_{c-\delta_1}. \tag{17.9}$$

Clearly, for arbitrary $\delta \in (0, \delta_1)$ we have the embedding

$$B \subset A_{c+\delta_1} \backslash A_{c-\delta_1}, \quad B \overset{\text{def}}{=} \left\{ u \in \mathbb{H} \mid c - \delta \leq \psi(u) \leq c + \delta \right\}.$$

Therefore, the inequality (17.9) remains valid for all $u \in B$.

Now we fix arbitrary $\varepsilon > 0$ and choose $\delta > 0$ as follows:

$$0 < \delta < \varepsilon, \quad 0 < \delta < \sigma^2/2, \quad 0 < \delta < \delta_1. \tag{17.10}$$

We set

$$A \overset{\text{def}}{=} \left\{ u \in \mathbb{H} \mid \psi(u) \leq c - \varepsilon \text{ or } \psi(u) \geq c + \varepsilon \right\},$$

$$B \overset{\text{def}}{=} \left\{ u \in \mathbb{H} \mid c - \delta \leq \psi(u) \leq c + \delta \right\}, \quad A \cap B = \varnothing,$$

$$C \overset{\text{def}}{=} \mathbb{H} \backslash (A \cup B) \neq \varnothing.$$

Note that the operator $F(u) = \operatorname{grad} \psi(u)$ is bounded on bounded sets since $\psi \in \mathscr{F}$ and, therefore, $\operatorname{grad} \psi(u)$ is boundedly Lipschitz-continuous. This allows one to prove the bounded Lipschitz continuity of the functional ψ:

$$|\psi(u_1) - \psi(u_2)| \leq c_2(R)\|u_1 - u_2\| \quad \text{for all} \quad \|u_k\| \leq R \quad \text{for } k = 1, 2. \tag{17.11}$$

Indeed, the following equalities hold:

$$\psi(u_1) - \psi(u_2) = \int\limits_0^1 \frac{\partial \psi(su_1 + (1-s)u_2)}{\partial s} \, ds = \int\limits_0^1 \langle \psi_f'(su_1 + (1-s)u_2), u_1 - u_2 \rangle \, ds.$$

This implies

$$|\psi(u_1) - \psi(u_2)| \leq \int\limits_0^1 \left\|\psi'_f(su_1 + (1-s)u_2)\right\|_* ds \|u_1 - u_2\| \leq c_2(R)\|u_1 - u_2\|,$$

where we used the following inequalities:

$$\int\limits_0^1 \left\|\psi'_f(su_1 + (1-s)u_2)\right\|_* ds$$

$$\leq \int\limits_0^1 \left\|\psi'_f(su_1 + (1-s)u_2) - \psi'_f(\theta)\right\|_* ds + \int\limits_0^1 \left\|\psi'_f(\theta)\right\|_* ds$$

$$\leq c_1(R) \int\limits_0^1 \|su_1 + (1-s)u_2\| ds + \left\|\psi'_f(\theta)\right\|_*$$

$$\leq c_1(R)R + \left\|\psi'_f(\theta)\right\|_* := c_2(R)$$

for all $\|u_k\| \leq R$, $k = 1,2$; here we used the bounded Lipschitz continuity of the Fréchet derivative $\psi'_f(u)$ in the following explicit form:

$$\left\|\psi'_f(su_1 + (1-s)u_2) - \psi'_f(\theta)\right\|_* \leq c_1(R)\|su_1 + (1-s)u_2\|$$

for all $u_1, u_2 \in \mathbb{H}$ such that $\|u_k\| \leq R$, $k = 1,2$.

The inequality (17.11) implies the bounded Lipschitz continuity of the functional $\psi(u)$. In particular, we conclude that the sets A and B are strongly closed.

We prove that the mapping

$$u \mapsto \text{distance}(u, A) + \text{distance}(u, B)$$

is bounded from below by a constant $\delta > 0$ for all u from a bounded subset in $\mathbb{H}$, where

$$\text{distance}(u, K) \overset{\text{def}}{=} \inf_{v \in K \subset \mathbb{H}} \|u - v\| \geq 0.$$

Indeed, note that

$$\text{distance}(u, A) + \text{distance}(u, B) \geq 0$$

and assume that there exists a bounded sequence $\{u_n\} \subset \mathbb{H}$ (satisfying the condition $\|u_n\| \leq R$) such that

$$\text{distance}(u_n, A) \to +0, \quad \text{distance}(u_n, B) \to +0 \quad \text{as } n \to +\infty.$$

Then there exist sequences $\{a_n\} \subset A$ and $\{b_n\} \subset B$ such that

$$\|a_n - u_n\| \to +0, \quad \|b_n - u_n\| \to +0 \quad \text{as } n \to +\infty.$$

By the triangle inequality, we have

$$\|a_n - b_n\| \leq \|a_n - u_n\| + \|b_n - u_n\| \to +0 \quad \text{as } n \to +\infty.$$

Without loss of generality, we assume that

$$\|a_n - u_n\| \le R, \quad \|b_n - u_n\| \le R \quad \text{for all } n \in \mathbb{N}.$$

Therefore,

$$\|a_n\| \le \|a_n - u_n\| + \|u_n\| \le 2R,$$
$$\|b_n\| \le \|b_n - u_n\| + \|b_n\| \le 2R.$$

Then, due to (17.11) we obtain the inequality

$$|\psi(a_n) - \psi(b_n)| \le c_2(2R)\|a_n - b_n\| \to +0 \quad \text{as } n \to +\infty, \tag{17.12}$$

where $c_2 > 0$ is a constant independent of $n \in \mathbb{N}$.

Now we note that the following inequalities hold:

$$c - \delta \le \psi(b_n) \le c + \delta \quad \text{for all } n \in \mathbb{N}, \tag{17.13}$$

$$\psi(a_n) \ge c - \varepsilon \quad \text{or} \quad \psi(a_n) \ge c + \varepsilon \quad \text{for all } n \in \mathbb{N}. \tag{17.14}$$

Due to (17.13), the numerical sequence $\{\psi(b_n)\}$ is bounded and hence there exists a subsequence $\{b_{n_n}\} \subset \{b_n\}$ such that

$$\psi(b_{n_n}) \to b \quad \Rightarrow \quad c - \delta \le b \le c + \delta. \tag{17.15}$$

Now, due to (17.12), for the corresponding subsequence $\{a_{n_n}\} \subset \{a_n\}$ we have

$$|\psi(a_{n_n}) - b| \le |\psi(a_{n_n}) - \psi(b_{n_n})| + |\psi(b_{n_n}) - b| \to +0 \quad \text{as } n \to +\infty.$$

This and (17.14) imply that

$$b \le c - \varepsilon \quad \text{or} \quad b \ge c + \varepsilon. \tag{17.16}$$

Since $0 < \delta < \varepsilon$, we arrive at a contradiction between the inequalities (17.15) and (17.16). Therefore,

$$\text{distance}(u, A) + \text{distance}(u, B) \ge \delta(R) > 0$$

for all $u \in \{u \in \mathbb{H} : \|u\| \le R\}$. (Note that possibly $\delta(R) \to +0$ as $R \to +\infty$.)

Therefore, the function

$$g(u) \stackrel{\text{def}}{=} \frac{\text{distance}(u, A)}{\text{distance}(u, A) + \text{distance}(u, B)}, \quad u \in \mathbb{H},$$

satisfies the conditions

$$0 \le g \le 1, \quad g = 0 \quad \text{on } A, \quad g = 1 \quad \text{on } B. \tag{17.17}$$

Moreover, the function g is boundedly Lipschitz continuous on bounded sets. Indeed, consider the difference

$$g(u_1) - g(u_2)$$

$$= \frac{\text{distance}(u_1, A)}{\text{distance}(u_1, A) + \text{distance}(u_1, B)} - \frac{\text{distance}(u_2, A)}{\text{distance}(u_2, A) + \text{distance}(u_2, B)}$$

$$= \frac{\text{distance}(u_1, A)\,\text{distance}(u_2, B) - \text{distance}(u_2, A)\,\text{distance}(u_1, B)}{(\text{distance}(u_1, A) + \text{distance}(u_1, B))\,(\text{distance}(u_2, A) + \text{distance}(u_2, B))}.$$

This immediately implies that

$$|g(u_1) - g(u_2)| \leq \frac{1}{\delta} |\text{distance}(u_1, B) - \text{distance}(u_2, B)|$$
$$+ \frac{1}{\delta} |\text{distance}(u_1, A) - \text{distance}(u_2, A)|.$$

Due to the triangle inequality we have

$$\|u_1 - v\| \leq \|u_1 - u_2\| + \|u_2 - v\|,$$
$$\|u_2 - v\| \leq \|u_1 - u_2\| + \|u_1 - v\|.$$

Taking the infimum over $v \in A$, we obtain

$$\text{distance}(u_1, A) \leq \text{distance}(u_2, A) + \|u_1 - u_2\|,$$
$$\text{distance}(u_2, A) \leq \text{distance}(u_1, A) + \|u_1 - u_2\|.$$

This implies the inequality

$$|\text{distance}(u_1, A) - \text{distance}(u_2, A)| \leq \|u_1 - u_2\|.$$

Similarly, we prove the inequality

$$|\text{distance}(u_1, B) - \text{distance}(u_2, B)| \leq \|u_1 - u_2\|.$$

Hence we conclude that

$$|g(u_1) - g(u_2)| \leq \frac{2}{\delta(R)} \|u_1 - u_2\| \tag{17.18}$$

for all $u_1, u_2 \in \mathbb{H}$ such that $\|u_k\| \leq R$, $k = \overline{1, 2}$, and $R > 0$.

We set

$$h(t) \overset{\text{def}}{=} \begin{cases} 1, & 0 \leq t \leq 1, \\ 1/t, & t \geq 1, \end{cases} \tag{17.19}$$

and introduce the mapping

$$V : \mathbb{H} \to \mathbb{H}$$

by the formula

$$V(u) \overset{\text{def}}{=} -g(u)h\left(\|\text{grad}\,\psi(u)\|_{\mathbb{H}}\right) \text{grad}\,\psi(u) \quad (u \in \mathbb{H}). \tag{17.20}$$

Note that V is bounded. Indeed, $0 \leq g(u) \leq 1$,

$$\|\text{grad}\,\psi(u)\|_{\mathbb{H}} \leq 1 \quad \Rightarrow \quad \|V(u)\|_{\mathbb{H}} \leq 1,$$
$$\|\text{grad}\,\psi(u)\|_{\mathbb{H}} \geq 1 \quad \Rightarrow \quad \|V(u)\|_{\mathbb{H}} \leq 1.$$

For arbitrary $u \in \mathbb{H}$, we consider the Cauchy problem

$$\frac{d\eta}{dt}(t) = V(\eta(t)), \quad t > 0, \quad \eta(0) = u. \tag{17.21}$$

The mapping V is bounded and boundedly Lipschitz-continuous (see (17.18)) since it is a composition of boundedly Lipschitz-continuous mappings. Therefore, there exists a unique classical solution for all $t \in [0, +\infty)$. To emphasize the dependence of a solution on the time t and on the initial state $u \in \mathbb{H}$, we will write

$$\eta(t, u) = \eta_t(u), \quad u \in \mathbb{H}.$$

We restrict ourselves to the case where $0 \le t \le 1$. In this case, the mapping $\eta(t, u) \in \mathbb{C}([0, 1] \times \mathbb{H}; \mathbb{H})$ defined above satisfies the conditions (i) and (ii). Indeed, on the one hand, we have

$$\eta_0(u) = u \quad \text{for all } u \in \mathbb{H};$$

this follows from the initial condition in the Cauchy problem (17.21). On the other hand, let

$$\eta(0) = u \in D \stackrel{\text{def}}{=} \left\{ u \in \mathbb{H} : \psi(u) < c - \varepsilon \text{ or } \psi(u) > c + \varepsilon \right\} \subset A.$$

Since $g(u) = 0$ for all $u \in A$ and the solution $\eta(t, u)$ is continuous, we conclude that

$$\eta(t, u) \in D \subset A \text{ for all } t \in [0, t_1]$$
$$\Rightarrow \quad g(\eta(t, u)) = 0 \quad \Rightarrow \quad V(\eta(t, u)) = 0$$
$$\Rightarrow \quad \frac{d\eta}{dt}(t) = 0 \quad \Rightarrow \quad \eta(t, u) = u \text{ for all } t \in [0, t_1],$$

where $t_1 > 0$ is sufficiently small. Further, using the algorithm of extension of a solution in time, we obtain that $\eta(t, u) = u$ for all $t \ge 0$.

Now we have

$$\frac{d}{dt} \psi(\eta_t(u)) = \left(\operatorname{grad} \psi(\eta_t(u)), \frac{d}{dt}\eta_t(u) \right)_{\mathbb{H}} = (\operatorname{grad} \psi(\eta_t(u)), V(\eta_t(u)))_{\mathbb{H}}$$
$$= -g(\eta_t(u))h\left(\|\operatorname{grad} \psi(\eta_t(u))\|_{\mathbb{H}}\right) \|\operatorname{grad} \psi(\eta_t(u))\|_{\mathbb{H}}^2. \tag{17.22}$$

In particular,

$$\frac{d}{dt}\psi(\eta_t(u)) \le 0 \quad (u \in \mathbb{H}, \ 0 \le t \le 1) \quad \Rightarrow \quad \psi(\eta_t(u)) \le \psi(\eta_0(u)) = \psi(u).$$

Therefore, the assertion (iii) is proved.

Now we fix a point $u \in A_{c+\delta}$ and prove the relation

$$\eta_1(u) \in A_{c-\delta}, \tag{17.23}$$

which implies the assertion (iv). If $\eta_t(u) \notin B$ for some $t \in [0, 1]$, we immediately obtain the required assertion. Indeed, assume that there exists $t^* \in [0, 1]$ such that

$$\eta_{t^*}(u) \notin B \quad \Leftrightarrow q \quad \psi(\eta_{t^*}(u)) < c - \delta \quad \text{or} \quad \psi(\eta_{t^*}(u)) > c + \delta.$$

Due to (iii) we have $\psi(\eta_{t^*}(u)) \le \psi(u) \le c + \delta$ and hence

$$\psi(\eta_{t^*}(u)) < c - \delta, \quad \psi(\eta_t(u)) \le \psi(\eta_{t^*}(u)) \quad \text{for all } t \in [t^*, 1].$$

Therefore, in this case, we have the inequality $\psi(\eta_1(u)) < c - \delta$.

Now we assume that $\eta_t(u) \in B$ for all $0 \le t \le 1$. Then $g(\eta_t(u)) = 1$, $0 \le t \le 1$. Therefore, from (17.22) we have

$$\frac{d}{dt}\psi(\eta_t(u)) = -h\left(\|\operatorname{grad} \psi(\eta_t(u))\|_{\mathbb{H}}\right) \|\operatorname{grad} \psi(\eta_t(u))\|_{\mathbb{H}}^2. \tag{17.24}$$

If $\|\mathrm{grad}\,\psi(\eta_t(u))\|_{\mathbb{H}} \leq 1$, then from (17.2), (17.17), and (17.19), we obtain

$$\frac{d}{dt}\psi(\eta_t(u)) = -\|\mathrm{grad}\,\psi(\eta_t(u))\|_{\mathbb{H}}^2 \leq -\sigma^2.$$

Indeed, since $\delta \in (0, \delta_1)$, we have

$$\eta_t(u) \in B = \overline{A_{c+\delta} \backslash A_{c-\delta}} \subset A_{c+\delta_1} \backslash A_{c-\delta_1}$$

due to the choice of $\delta \in (0, \delta_1)$ for all $t \in [0, 1]$.

On the other hand, if $\|\mathrm{grad}\,\psi(\eta_t(u))\|_{\mathbb{H}} \geq 1$, then from (17.2) and (17.19) we obtain

$$\frac{d}{dt}\psi(\eta_t(u)) \leq -1 \leq -\sigma^2$$

(recall that $\sigma \in (0, 1)$). Due to these inequalities, from (17.10) and (17.24) we deduce the estimate

$$\psi(\eta_1(u)) \leq \psi(u) - \sigma^2 \leq c + \delta - \sigma^2 \leq c - \delta,$$

which implies (17.23). The assertion (iv) is proved. $\square$

17.2   Mountain Pass Theorem

Using the "minimax" technique and the deformation η constructed above, we prove the existence of a critical point. The proof will be based on the assertion called the "mountain pass theorem." First, we introduce the notion of the *set of admissible paths*.

Definition 17.3. The family

$$\Gamma \stackrel{\text{def}}{=} \left\{ g \in \mathbb{C}([0,1]; \mathbb{H}) \,\middle|\, g(0) = \theta, \; g(1) = v \right\}$$

for some $v \in \mathbb{H}$ is called the set of admissible paths.

Theorem 17.2. *Let $\psi \in \mathscr{F}$ satisfy the Palais–Smale condition. Also assume that the following conditions are fulfilled:*

(i) *$\psi(\theta) = 0$;*
(ii) *there exist constants $r, a > 0$ such that $\psi(u) \geq a$ if $\|u\| = r$;*
(iii) *there exists an element $v \in \mathbb{H}$ such that $\|v\| > r$ and $\psi(v) \leq 0$.*

Then

$$c \stackrel{\text{def}}{=} \inf_{g \in \Gamma} \max_{0 \leq t \leq 1} \psi(g(t))$$

is a critical value of the functional ψ.

Proof. First, since $\max\limits_{t\in[0,1]} \psi(g(t)) \geq a$, we have $c \geq a$. Indeed, $g(0) = \theta$ and $g(1) = v$ by Definition 17.3. Since $\|g(0)\| = \|\theta\| = 0$, $\|g(1)\| = \|v\| > r$, and $g(t) \in \mathbb{C}([0,1]; \mathbb{H})$, there exists $t_0 \in (0,1)$ such that

$$\|g(t_0)\| = r \quad \Rightarrow \quad \psi(g(t_0)) \geq a.$$

Assume that c is not a critical value of the functional $\psi(u)$, so that $K_c = \varnothing$. Choose a sufficiently small number

$$0 < \varepsilon < \frac{a}{2} \quad \Rightarrow \quad c - \varepsilon > 0.$$

Due to Theorem 17.1, there exist a constant $0 < \delta < \varepsilon$ and a homeomorphism $\eta_t(u) : \mathbb{H} \to \mathbb{H}$ such that

$$\eta_1(A_{c+\delta}) \subset A_{c-\delta}, \tag{17.25}$$

$$\eta_1(u) = u \quad \text{if} \quad u \notin \psi^{-1}([c-\varepsilon, c+\varepsilon]). \tag{17.26}$$

We choose $g \in \Gamma$ such that

$$\max\limits_{0 \leq t \leq 1} \psi(g(t)) \leq c + \delta. \tag{17.27}$$

Then

$$\hat{g}(t) \stackrel{\text{def}}{=} \eta_1(g(t)) \in \mathbb{C}([0,1]; \mathbb{H})$$

also belongs to Γ since

$$\eta_1(g(0)) = \eta_1(\theta) = \theta, \quad \psi(\theta) = 0 < c - \varepsilon,$$

and

$$\eta_1(g(1)) = \eta_1(v) = v, \quad \psi(v) \leq 0 < c - \varepsilon$$

due to (17.26). Indeed, we note that

$$\psi(\theta) = 0 < c - \varepsilon, \quad \psi(v) \leq 0 < c - \varepsilon \quad \Rightarrow \quad \theta, v \notin \psi^{-1}([c-\varepsilon, c+\varepsilon]).$$

Then from (17.27) we conclude that

$$\max\limits_{0 \leq t \leq 1} \psi(\hat{g}(t)) \leq c - \delta$$

and hence

$$c = \inf\limits_{g \in \Gamma} \max\limits_{0 \leq t \leq 1} \psi(g(t)) \leq c - \delta.$$

This leads to a contradiction since $\delta > 0$. The theorem is proved. $\qquad\square$

Note that the minimax point $u_0 \in K_c$ of the functional ψ is a nonzero element since $0 < c = \psi(u_0)$ and $\psi(\theta) = 0$.

Lecture 18

An Application
of the Mountain Pass Theorem

In this lecture, we apply the mountain pass theorem proved in the previous lecture to the proof of the existence (or nonexistence) of a solution to a certain boundary-value problem.

18.1 Theorem on the Existence of a Solution

Consider the following nonlinear boundary-value problem:

$$-\Delta u = |u|^{q-1} u \quad \text{in } \Omega, \qquad u = 0 \quad \text{on } \partial\Omega, \tag{18.1}$$

where $\Omega \subset \mathbb{R}^N$ is a bounded domain with sufficiently smooth boundary $\partial\Omega$. Assume that

$$1 < q < \frac{N+2}{N-2}, \quad N \geq 3. \tag{18.2}$$

Obviously, $u \equiv 0$ is a (trivial) solution of (18.1). We are interested in nontrivial solutions.

Definition 18.1. A *weak solution* of the problem (18.1) is a function $u(x) \in H_0^1(\Omega)$ satisfying the equality

$$\langle \Delta u + |u|^{q-1} u, \phi \rangle = 0 \tag{18.3}$$

for all $\phi(x) \in H_0^1(\Omega)$, where $\langle \cdot, \cdot \rangle$ is the duality bracket between the spaces $H_0^1(\Omega)$ and $H^{-1}(\Omega)$.

Theorem 18.1. *The boundary-value problem (18.1) has at least one nontrivial weak solution $u(x) \in H_0^1(\Omega)$.*

Proof. We introduce the Euler functional

$$\psi(u) \stackrel{\text{def}}{=} \int_\Omega \left[\frac{1}{2} |D_x u|^2 - \frac{1}{q+1} |u|^{q+1} \right] dx, \quad u(x) \in H_0^1(\Omega). \tag{18.4}$$

We apply the mountain pass theorem to the functional $\psi(u)$. Consider the Hilbert space $\mathbb{H} := H_0^1(\Omega)$ with the norm

$$\|u\| \stackrel{\text{def}}{=} \left(\int_\Omega |D_x u|^2 \, dx \right)^{1/2}.$$

Then

$$\psi(u) := \psi_1(u) - \psi_2(u) \stackrel{\text{def}}{=} \frac{1}{2}\|u\|^2 - \frac{1}{q+1}\int_\Omega |u|^{q+1}\, dx. \tag{18.5}$$

We verify that ψ belongs to the class $\mathscr{F}$.

First, we prove that $\psi_1 \in \mathscr{F}$. Note that for any $u, h \in \mathbb{H}$ we have

$$\psi_1(u+h) = \frac{1}{2}\|u+h\|^2 = \frac{1}{2}(u+h, u+h)$$

$$= \frac{1}{2}\|u\|^2 + (u,h) + \frac{1}{2}\|h\|^2 = \psi_1(u) + (u,h) + \omega_1(u,h),$$

$$\omega_1(u,h) = \frac{1}{2}\|h\|^2.$$

Therefore, ψ_1 is Fréchet differentiable at the point u and $\operatorname{grad}\psi_1(u) = u$. In particular, $\operatorname{grad}\psi_1(u)$ is Lipschitz-continuous and hence boundedly Lipschitz-continuous. Thus, $\psi_1 \in \mathscr{F}$.

Now we consider ψ_2. By the Browder–Minty theorem (see Theorem 20.1), due to the uniform monotonicity of the Laplace operator

$$-\Delta : H_0^1(\Omega) \to H^{-1}(\Omega),$$

the problem

$$\langle \Delta v + v^*, \phi \rangle = 0, \quad \phi \in H_0^1(\Omega),$$

has a unique weak solution $v \in H_0^1(\Omega)$ for any $v^* \in H^{-1}(\Omega)$. We set $v = Jv^*$, so that

$$J = (-\Delta)^{-1} : H^{-1}(\Omega) \to H_0^1(\Omega) \tag{18.6}$$

is a Riesz isometry. Note that by the Krasnosel'sky theorem the operator

$$F(u) := |u|^{q-1}u : L^{q+1}(\Omega) \to L^{(q+1)/q}(\Omega)$$

is bounded and strongly continuous as an operator acting from the Banach space $\mathbb{B} = L^{q+1}(\Omega)$ into the corresponding dual space $\mathbb{B}^* = L^{(q+1)/q}(\Omega)$. Moreover, if

$$2 < q+1 < 2^* = \frac{2N}{N-2} \quad \Rightarrow \quad 1 < q < \frac{N+2}{N-2}, \quad N \ge 3,$$

then the following chain of dense and continuous embeddings holds:

$$H_0^1(\Omega) \stackrel{ds}{\subset} L^{q+1}(\Omega) \stackrel{ds}{\subset} L^{(q+1)/q}(\Omega) \stackrel{ds}{\subset} H^{-1}(\Omega).$$

Now we prove that

$$\operatorname{grad}\psi_2(u) = J|u|^{q-1}u \tag{18.7}$$

for $u \in H_0^1(\Omega)$. On the one hand, for $q > 1$ we have

$$\psi'_{2f}(u) = |u|^{q-1}u, \quad \psi'_{2f}(\cdot) : \mathbb{B} = L^{q+1}(\Omega) \to \mathbb{B}^* = L^{(q+1)/q}(\Omega). \tag{18.8}$$

On the other hand, by the definition of grad,

$$\operatorname{grad}\psi_2(u) \stackrel{\text{def}}{=} J\psi'_{2f}(u).$$

Verify that the mapping

$$\operatorname{grad} \psi_2 : H_0^1(\Omega) \to H_0^1(\Omega)$$

is boundedly Lipschitz-continuous. Indeed, since $q > 1$, we have the inequality

$$\left| |u_1|^{q-1} u_1 - |u_2|^{q-1} u_2 \right| \le q \max \left\{ |u_1|^{q-1}, |u_2|^{q-1} \right\} |u_1 - u_2|$$

for all $u_1, u_2 \in \mathbb{R}^1$. Therefore, for the functions $u_1(x), u_2(x) \in L^{q+1}(\Omega)$ we obtain the inequalities

$$\int_\Omega \left| |u_1|^{q-1} u_1 - |u_2|^{q-1} u_2 \right|^{(q+1)/q} dx$$

$$\le q^{(q+1)/q} \int_\Omega \max \left\{ |u_1|^{(q-1)(q+1)/q}, |u_2|^{(q-1)(q+1)/q} \right\} |u_1 - u_2|^{(q+1)/q} dx$$

$$\le q^{(q+1)/q} \max \left\{ \left(\int_\Omega |u_1|^{q+1} dx \right)^{(q-1)/q}, \left(\int_\Omega |u_2|^{q+1} dx \right)^{(q-1)/q} \right\}$$

$$\times \left(\int_\Omega |u_1 - u_2|^{q+1} dx \right)^{1/q};$$

here we have applied the Hölder inequality with the parameters

$$q_1 = q, \quad q_2 = \frac{q}{q-1}, \quad q > 1.$$

Therefore, we obtain the inequality

$$\left\| |u_1|^{q-1} u_1 - |u_2|^{q-1} u_2 \right\|_{(q+1)/q} \le q \max \left\{ \|u_1\|_{q+1}^{q-1}, \|u_2\|_{q+1}^{q-1} \right\} \|u_1 - u_2\|_{q+1}. \quad (18.9)$$

Due to (18.8), the following inequalities are valid:

$$\|\psi_{2f}'(u_1) - \psi_{2f}'(u_2)\|_* \le k_1 \|\psi_{2f}'(u_1) - \psi_{2f}'(u_2)\|_{(q+1)/q}$$

$$\le q k_1 \max \left\{ \|u_1\|_{q+1}^{q-1}, \|u_2\|_{q+1}^{q-1} \right\} \|u_1 - u_2\|_{q+1}$$

$$\le q k_1 k_2^q \max \left\{ \|u_1\|^{q-1}, \|u_2\|^{q-1} \right\} \|u_1 - u_2\|.$$

Now we arrive at the required estimate:

$$\|\operatorname{grad} \psi_2(u_1) - \operatorname{grad} \psi_2(u)\| = \|\psi_{2f}'(u_1) - \psi_{2f}'(u_2)\|_*$$

$$\le k_3 \max \left\{ \|u_1\|^{q-1}, \|u_2\|^{q-1} \right\} \|u_1 - u_2\|, \quad (18.10)$$

where $k_3 := q k_1 k_2^q$. Therefore, $\psi_2 \in \mathscr{F}$.

Now we verify the Palais–Smale condition for the functional ψ. Assume that a sequence $\{u_k\}_{k=1}^{+\infty} \subset H_0^1(\Omega)$ is such that

the numeric sequence $\{\psi(u_k)\}_{k=1}^{+\infty}$ is bounded $\qquad (18.11)$

and the sequence $\{\operatorname{grad} \psi_2(u_k)\} \subset H_0^1(\Omega)$ satisfies the limit relation

$$\operatorname{grad} \psi(u_k) \to \theta \quad \text{strongly in } H_0^1(\Omega) \text{ as } k \to +\infty. \tag{18.12}$$

The limit relation (18.12) immediately implies that

$$u_k - J|u_k|^{q-1}u_k \to \theta \quad \text{strongly in } H_0^1(\Omega) \text{ as } k \to +\infty. \tag{18.13}$$

From the equality

$$\|\psi_f'(u_k)\|_* \overset{\text{def}}{=} \sup_{\|v\| \le 1} \left| \langle \psi_f'(u_k), v \rangle \right|.$$

we obtain

$$\|\operatorname{grad} \psi(u_k)\| = \|\psi_f'(u_k)\|_* = \sup_{\|v\| \le 1} \left| \langle \psi_f'(u_k), v \rangle \right| = \sup_{\|v\| \le 1} \left| \left(\operatorname{grad} \psi(u_k), v \right)_{H_0^1(\Omega)} \right|.$$

Due to (18.12) we conclude that for any $\varepsilon > 0$, there exists $n_0 \in \mathbb{N}$ such that the following chain of inequalities holds for all $k \ge n_0$ for all $v(x) \in H_0^1(\Omega)$, $\|v\| \le 1$:

$$\varepsilon \ge \|\operatorname{grad} \psi(u_k)\| \ge \left| \left(\operatorname{grad} \psi(u_k), v \right)_{H_0^1(\Omega)} \right|.$$

Without loss of generality we may assume that $\|u_k\| \ne 0$. Therefore, in the last inequality we may set

$$v := \frac{u_k}{\|u_k\|}$$

and obtain the inequality[1]

$$\left| \left(\operatorname{grad} \psi(u_k), u_k \right)_{H_0^1(\Omega)} \right| \le \varepsilon \|u_k\|. \tag{18.14}$$

Note that $J := (-\Delta)^{-1}$; integrating by parts, we obtain the following chain of equalities:

$$\left| \left(\operatorname{grad} \psi(u_k), u_k \right)_{H_0^1(\Omega)} \right|$$

$$= \left| \int_\Omega \left[(D_x u_k, D_x u_k) - (D_x J|u_k|^{q-1}u_k, D_x u_k) \right] dx \right|$$

$$= \left| \int_\Omega \left(|D_x u_k|^2 - |u_k|^{q+1} \right) dx \right|.$$

Indeed, the following chain of equalities holds:

$$\int_\Omega \left(D_x J|u_k|^{q-1}u_k, D_x u_k \right) dx = \left\langle (-\Delta)J|u_k|^{q-1}u_k, u_k \right\rangle$$

$$= \left\langle (-\Delta)(-\Delta)^{-1}|u_k|^{q-1}u_k, u_k \right\rangle$$

$$= \left\langle |u_k|^{q-1}u_k, u_k \right\rangle = \int_\Omega |u_k|^{q+1} \, dx.$$

---

[1] In the case where $\|u_k\| = 0$, this inequality also holds.

We have

$$\left| \int_\Omega \left[|D_x u_k|^2 - |u_k|^{q+1} \right] dx \right| \le \varepsilon \|u_k\|$$

for $\varepsilon > 0$ and $k \ge n_0$. In particular, if $\varepsilon = 1$, then

$$\int_\Omega |u_k|^{q+1} \, dx \le \|u_k\|^2 + \|u_k\| \tag{18.15}$$

for all sufficiently large $k \in \mathbb{N}$. The condition (18.11) implies

$$\left(\frac{1}{2} \|u_k\|^2 - \frac{1}{q+1} \int_\Omega |u_k|^{q+1} \, dx \right) \le c_1 < +\infty$$

for all $k \in \mathbb{N}$ and a certain constant c_1; therefore,

$$\frac{1}{2}\|u_k\|^2 \le \frac{1}{q+1} \int_\Omega |u_k|^{q+1} \, dx + c_1 \le \frac{1}{q+1}\|u_k\|^2 + \frac{1}{q+1}\|u_k\| + c_1.$$

Using the arithmetic Hölder inequality with parameters, we obtain the inequality

$$\frac{1}{q+1}\|u_k\| \le \varepsilon \|u_k\|^2 + c_2(\varepsilon), \quad c_2(\varepsilon) := \frac{1}{4\varepsilon(q+1)^2}, \quad \varepsilon > 0,$$

which implies the following estimate:

$$\|u_k\|^2 \le c_3(\varepsilon) := (c_2(\varepsilon) + c_1) \left(\frac{1}{2}\frac{q-1}{q+1} - \varepsilon \right)^{-1},$$

where

$$0 < \varepsilon < \frac{1}{2}\frac{q-1}{q+1}.$$

Therefore, the sequence $\{u_k\}_{k=1}^\infty$ is uniformly bounded in $H_0^1(\Omega)$ with respect to $k \in \mathbb{N}$. Hence, on the one hand, there exist a subsequence $\{u_{k_j}\}_{j=1}^{+\infty}$ and a function $u \in H_0^1(\Omega)$ such that

$$u_{k_j} \rightharpoonup u \quad \text{weakly in } H_0^1(\Omega) \text{ as } k_j \to +\infty.$$

On the other hand, due to the completely continuous embedding

$$H_0^1(\Omega) \hookrightarrow\hookrightarrow L^{q+1}(\Omega), \quad 1 < q < \frac{N+2}{N-2}, \quad N \ge 3,$$

which is totally continuous in fact,[2] we obtain the following limit property:

$$u_{k_j} \to u \quad \text{strongly in } L^{q+1}(\Omega) \text{ as } k_j \to +\infty.$$

Therefore, due to (18.9), we conclude that

$$|u_{k_j}|^{q-1}u_{k_j} \to |u|^{q-1}u \quad \text{strongly in } L^{(q+1)/q}(\Omega) \text{ as } k_j \to +\infty.$$

---

[2] By the Riesz–Fréchet, theorem, the Hilbert space is reflexive.

Then, due to the continuous embedding $L^{(q+1)/q}(\Omega) \subset H^{-1}(\Omega)$, we see that

$$|u_{k_j}|^{q-1} u_{k_j} \to |u|^{q-1} u \quad \text{strongly in } H^{-1}(\Omega) \text{ as } k_j \to +\infty$$

and hence

$$J|u_{k_j}|^{q-1} u_{k_j} \to J|u|^{q-1} u \quad \text{strongly in } H^1_0(\Omega) \text{ as } k_j \to +\infty.$$

Therefore, from (18.13) we obtain that

$$u_{k_j} \to u \quad \text{strongly in } H^1_0(\Omega) \text{ as } k_j \to +\infty. \tag{18.16}$$

Indeed, we first note that the limit function $u(x) \in H^1_0(\Omega)$ satisfies the equality

$$\left(u - J|u|^{q-1}u, \phi\right) = 0 \quad \text{for all } \phi \in H^1_0(\Omega)$$

due to the following facts:

$$\left(u_{k_j} - J|u_{n_k}|^{q-1}u_{k_j}, \phi\right) \to +0 \qquad \text{for all } \phi \in H^1_0(\Omega);$$
$$\left(u_{k_j}, \phi\right) \to \left(u, \phi\right) \qquad \text{for all } \phi \in H^1_0(\Omega);$$
$$\left(J|u_{k_j}|^{q-1}u_{k_j}, \phi\right) \to \left(J|u|^{q-1}u, \phi\right) \quad \text{for all } \phi \in H^1_0(\Omega).$$

This immediately implies the equality

$$u - J|u|^{q-1}u = \theta \in H^1_0(\Omega).$$

Moreover, the following chain of inequalities holds:

$$\left\|u_{k_j} - u\right\|$$
$$= \left\|u_{k_j} - J|u_{k_j}|^{q-1}u_{k_j} + J|u_{k_j}|^{q-1}u_{k_j} - J|u|^{q-1}u + J|u|^{q-1}u - u\right\|$$
$$= \left\|u_{k_j} - J|u_{k_j}|^{q-1}u_{k_j} + J|u_{k_j}|^{q-1}u_{k_j} - J|u|^{q-1}u\right\|$$
$$\leq \left\|u_{k_j} - J|u_{k_j}|^{q-1}u_{k_j}\right\| + \left\|J|u_{k_j}|^{q-1}u_{k_j} - J|u|^{q-1}u\right\| \to +0$$

as $k_j \to +\infty$. Thus, the functional $\psi(u)$ satisfies the Palais–Smale condition.

Now we verify the other conditions of the mountain pass theorem.

1. Obviously, $\psi(\theta) = 0$.
2. Let $u \in H^1_0(\Omega)$, $\|u\| = r$, where $r > 0$ will be specified below. Then

$$\psi(u) = \psi_1(u) - \psi_2(u) = \frac{r^2}{2} - \psi_2(u). \tag{18.17}$$

By (18.2),

$$\psi_2(u) = \frac{1}{q+1} \int_\Omega |u|^{q+1} \, dx \leq k_2^{q+1} \|u\|^{q+1} \leq c_4 r^{q+1}.$$

By (18.17),

$$\psi(u) \geq \frac{r^2}{2} - c_4 r^{q+1} \geq \frac{r^2}{4} = a > 0,$$

for sufficiently small $r > 0$, since $q + 1 > 2$.

3. Now we choose $u \in H_0^1(\Omega)$, $u \not\equiv 0$. We set $v := tu$, where $t > 0$ will be chosen below. Then

$$\psi(v) := \psi_1(tu) - \psi_2(tu) = t^2 \psi_1(u) - \frac{t^{q+1}}{q+1} \int_\Omega |u|^{q+1}\, dx < 0$$

for sufficiently large $t > 0$ since $q > 1$.

All the conditions of the mountain pass theorem are fulfilled, so we may conclude that there exists a function $u \in H_0^1(\Omega)$, $u \not\equiv 0$, such that

$$\operatorname{grad} \psi(u) = u - J|u|^{q-1}u = \theta \in H_0^1(\Omega).$$

In particular, for any $\phi(x) \in H_0^1(\Omega)$, we have the equality

$$\int_\Omega (D_x u, D_x \phi)\, dx = \int_\Omega |u|^{q-1} u \phi\, dx,$$

which implies that $u(x) \in H_0^1(\Omega)$ is a weak solution of the problem (18.1). The theorem is proved. □

18.2   Bibliographical Notes

The contents of this lecture is taken from $[2, 4, 5, 8, 13, 19, 28, 32, 38\text{–}41, 45, 55, 60]$.

PART 4

Methods of Monotonicity and Compactness

Galerkin Method
and Method of Monotonicity.
Elliptic Equations

In this lecture, we consider combinations of the Galerkin method with the method of monotonic operators, which can be successfully applied to problems with nonlinear and monotonic principal operators.

Consider the boundary-value problem

$$-\Delta u = f(x), \quad u|_{\partial\Omega} = 0, \tag{19.1}$$

where $\Omega \subset \mathbb{R}^3$ is a bounded domain with sufficiently smooth boundary $\partial\Omega \in C^{2,\delta}$, $\delta \in (0, 1]$. First, we consider the classical (i.e., pointwise) statement of the problem.

It is well known (see, e.g., [25]) that if $f(x) \in C^\alpha(\overline{\Omega})$, where $\alpha \in (0, 1]$, then there exists a unique classical solution of this problem in the Hölder space

$$u(x) \in C^{2+\alpha}(\overline{\Omega}).$$

However, in many physical problems, the function $f(x)$ is not continuous on a subset of the domain Ω, and hence it becomes necessary to generalize the notion of a solution.

First, we note that many physical boundary-value problems can be treated as integral equalities. Multiplying both sides of Eq. (19.1) by an arbitrary test function $\varphi(x) \in C_0^\infty(\Omega)$ and integrating over the domain Ω (in the Lebesgue sense) by parts, we obtain the following equality:

$$\int_\Omega (D_x u(x), D_x \varphi(x))\, dx = \int_\Omega f(x)\varphi(x)\, dx, \quad \varphi(x) \in C_0^\infty(\Omega). \tag{19.2}$$

Integral equalities of this type usually underlie the definition of a weak solution of a boundary-value problem.

In the sequel, we consider various boundary-value problems for nonlinear elliptic equations and discuss some methods for studying these problems. We focus on the weak formulation of problems considered and on the method of weak convergence.

19.1 Galerkin Method and Monotonicity Method

Consider the following boundary-value problem for one of the most famous nonlinear elliptic operators, which has the following classical statement:

$$-\operatorname{div}(|D_x u|^{p-2} D_x u) = f(x), \quad u|_{\partial\Omega} = 0, \quad p \geq 2. \tag{19.3}$$

Obviously, for $p = 2$, we have the problem (19.1). Since this monograph is mainly addressed to physicists, we present, for completeness, the derivation of the boundary-value problem (19.3).

Let a domain $\Omega \subset \mathbb{R}^3$ be surface simply connected and have a smooth boundary $\partial\Omega \in C^{2,\delta}$, $\delta \in (0, 1]$.

Consider the electric part of the Maxwell system in the quasi-stationary approximation (see, e.g., [42]):

$$\operatorname{div} \boldsymbol{D} = 4\pi n(x), \quad \operatorname{rot} \boldsymbol{E} = 0, \tag{19.4}$$

where the distribution of the density of free charges described by the function $n = n(x)$ is given. Now we assume that the nonlinear dependence $\boldsymbol{D} = \boldsymbol{D}(\boldsymbol{E})$ is described by the so-called Kerr nonlinearity:

$$\boldsymbol{D} = |\boldsymbol{E}|^{p-2}\boldsymbol{E}, \quad p \geq 2. \tag{19.5}$$

Since the domain Ω is surface simply connected, we may introduce the potential of the electric field by the formula

$$\boldsymbol{E} = -D_x \varphi. \tag{19.6}$$

Moreover, assume that the boundary of the domain Ω is "grounded," i.e., its potential is equal to zero; then we obtain the boundary condition

$$\varphi\big|_{\partial\Omega} = 0. \tag{19.7}$$

Equations (19.4)–(19.7) imply the problem (19.3), where $f(x) = 4\pi n(x)$.

Now we examine properties of the operator $\operatorname{div}(|D_x u|^{p-2} D_x u)$ called the p-*Laplacian*. First, we show that it acts as follows:

$$\operatorname{div}\left(|D_x u|^{p-2} D_x u\right) : W_0^{1,p}(\Omega) \to W^{-1,p'}(\Omega), \quad p' = \frac{p}{p-1}. \tag{19.8}$$

Recall that $W^{-1,p'}(\Omega)$ is the space of linear continuous functionals on the Sobolev space $W_0^{1,p}(\Omega)$. The p-Laplacian can be represented as the composition of the following three mappings:

$$\operatorname{div}(\xi) := \operatorname{div}\left(|D_x u|^{p-2} D_x u\right), \quad \xi := |\eta|^{p-2}\eta, \quad \eta := D_x u. \tag{19.9}$$

Let $u(x) \in W_0^{1,p}(\Omega)$. Then, first, by the definition of the space $W_0^{1,p}(\Omega)$, we have

$$\eta := D_x u : W_0^{1,p}(\Omega) \to L^p(\Omega) \otimes L^p(\Omega) \otimes L^p(\Omega). \tag{19.10}$$

Second, the following expression for the nonlinear operator $\xi := |\eta|^{p-2}\eta$ is valid:

$$\xi := |\eta|^{p-2}\eta : L^p(\Omega) \otimes L^p(\Omega) \otimes L^p(\Omega) \to L^{p'}(\Omega) \otimes L^{p'}(\Omega) \otimes L^{p'}(\Omega). \tag{19.11}$$

Indeed, consider the expressions for the coordinates of the vector ξ: if $\xi = (\xi_1, \xi_2, \xi_3)$ and $\eta = (\eta_1, \eta_2, \eta_3)$, then

$$\xi_1 = |\eta|^{p-2}\eta_1, \quad \xi_2 = |\eta|^{p-2}\eta_2, \quad \xi_3 = |\eta|^{p-2}\eta_3.$$

The following inequality holds:

$$|\xi_i|^{p'} = \left||\eta|^{p-2}\eta_i\right|^{p'} \leq \left||\eta|^{p-1}\right|^{p'} = |\eta|^p, \quad i = \overline{1,3}.$$

This implies that if $\eta \in L^p(\Omega) \otimes L^p(\Omega) \otimes L^p(\Omega)$, then $\xi_i \in L^{p'}(\Omega)$. Let

$$\langle f, u \rangle : W^{-1,p'}(\Omega) \otimes W_0^{1,p}(\Omega) \to \mathbb{R}^1 \tag{19.12}$$

be the duality bracket between the Banach spaces $W_0^{1,p}(\Omega)$ and $W^{-1,p'}(\Omega)$. The third operator div(ξ) acts as follows:

$$\langle \mathrm{div}(\xi(x)), \varphi(x) \rangle := - \int_\Omega (\xi(x), D_x\varphi(x))\, dx \tag{19.13}$$

for all $\varphi(x) \in W_0^{1,p}(\Omega)$ and any fixed vector-valued function

$$\xi(x) \in L^{p'}(\Omega) \otimes L^{p'}(\Omega) \otimes L^{p'}(\Omega).$$

It is easy to prove the linearity of the functional div($\xi(x)$) and its continuity in the strong topology of the Banach space $W_0^{1,p}(\Omega)$. We have

$$\mathrm{div} : L^{p'}(\Omega) \otimes L^{p'}(\Omega) \otimes L^{p'}(\Omega) \to W^{-1,p'}(\Omega). \tag{19.14}$$

Thus, the operator (19.9) as the composition of the operators (19.10)–(19.14) acts by the formula (19.8).

Let

$$\langle f, u \rangle : W^{-1,p'}(\Omega) \otimes W_0^{1,p}(\Omega) \to \mathbb{R}^1 \tag{19.15}$$

be the duality bracket between the Banach spaces $W_0^{1,p}(\Omega)$ and $W^{-1,p'}(\Omega)$. We prove the following important property of the p-Laplacian called the *strict monotonicity*. Introduce the brief notation

$$\Delta_p u \stackrel{\mathrm{def}}{=} \mathrm{div}\left(|D_x u|^{p-2} D_x u\right) \tag{19.16}$$

and state the following definition.

Definition 19.1. A mapping $F : \mathbb{B} \to \mathbb{B}^*$ is said to be *monotonic* with respect to the duality bracket $\langle \cdot, \cdot \rangle$ between Banach spaces $\mathbb{B}$ and $\mathbb{B}^*$ if for all $u, v \in \mathbb{B}$, the following inequality holds:

$$\langle F(u) - F(v), u - v \rangle \geq 0. \tag{19.17}$$

If the equality in the formula (19.17) holds only for $u = v$, then a mapping F is said to be *strictly monotonic*.

Note that, by the construction of the p-Laplacian (see the definition (19.13), the following formula of integration by parts for the functional div($\xi(x)$)) holds:

$$\langle -\Delta_p u, v \rangle = \int_\Omega |D_x u|^{p-2}\left(D_x u, D_x v\right) dx \tag{19.18}$$

for all $u, v \in W_0^{1,p}(\Omega)$.

Let $u_1(x), u_2(x) \in W_0^{1,p}(\Omega)$. Then, using (19.18), we obtain the formula

$$\langle -\Delta_p u_1 + \Delta_p u_2, u_1 - u_2 \rangle$$
$$= \int_\Omega \left(|D_x u_1|^{p-2} D_x u_1 - |D_x u_2|^{p-2} D_x u_2,\ D_x u_1 - D_x u_2\right) dx. \tag{19.19}$$

Using the inequality
$$(|\xi|^{p-2}\xi - |\eta|^{p-2}\eta, \ \xi - \eta) \geq C(p)|\xi - \eta|^p, \quad \xi, \eta \in \mathbb{R}^N \qquad (19.20)$$
(see Lemma 21.3), with some positive $C(p)$, we can prove the strict monotonicity of the operator $-\Delta_p$. We have

$$\langle -\Delta_p u_1 + \Delta_p u_2, u_1 - u_2 \rangle$$
$$= \int_\Omega \Big(|D_x u_1|^{p-2} D_x u_1 - |D_x u_2|^{p-2} D_x u_2, \ D_x u_1 - D_x u_2 \Big)$$
$$\geq C(p) \int_\Omega |D_x u_1 - D_x u_2|^p \, dx. \qquad (19.21)$$

Definition 19.2. A *weak solution* of the problem (19.3) under the condition $f \in W^{-1,p'}(\Omega)$ is a function $u(x) \in W_0^{1,p}(\Omega)$ satisfying the following equality:
$$\langle -\Delta_p u, v \rangle = \langle f, v \rangle, \quad v \in W_0^{1,p}(\Omega). \qquad (19.22)$$

We discuss the relationship between weak solutions and classical solutions of the problem (19.3). Let a solution of the problem (19.3) belong to the class $u(x) \in C^{2+\alpha}(\overline{\Omega})$ under the condition $f(x) \in C^\alpha(\overline{\Omega})$, $\alpha \in (0,1]$. Then this function $u(x)$ is a solution of the problem (19.22). Clearly, not all weak solutions are classical solutions.

We prove that the operator
$$A = -\Delta_p : \ W_0^{1,p}(\Omega) \to W^{-1,p'}(\Omega) \qquad (19.23)$$
is coercive. Indeed, integrating by parts (see (19.18)), we can prove the formula
$$\langle -\Delta_p u, u \rangle = \int_\Omega |D_x u|^p \, dx = \big\| |D_x u| \big\|_p^p, \quad p \geq 2, \qquad (19.24)$$
which implies the coerciveness.

Lemma 19.1 (the lemma on acute angles). *Let $T : \mathbb{R}^n \to \mathbb{R}^n$ be a continuous mapping satisfying the condition*
$$(Ta, a) \geq 0, \quad \text{where } |a| = R,$$
for some $R > 0$. Then there exists $a \in \mathbb{R}^n$ such that $|a| \leq R$ and $Ta = \theta$.

Proof. Assume that
$$Ta \neq \theta \quad \text{for all} \quad a \in K_R = \{a \mid a \in \mathbb{R}^n, \ |a| \leq R\}.$$
Then the mapping
$$a \mapsto -R \frac{Ta}{|Ta|}$$
is a continuous mapping from K_R to K_R. By the Brouwer fixed-point theorem, there exist $a \in K_R$ such that
$$a = -R \frac{Ta}{|Ta|}.$$
Obviously, $|a| = R$ and $(Ta, a) = -R|Ta| < 0$, which contradicts the assumption $(Ta, a) \geq 0$, where $|a| = R$. The lemma is proved. $\qquad\square$

Definition 19.3. We say that an operator $A : \mathbb{B} \to \mathbb{B}^*$ possesses the so-called S^+-*property* if the fact

$$u_m \rightharpoonup u \quad \text{weakly in } \mathbb{B} \text{ as } m \to +\infty$$

and the condition

$$\limsup_{m \to +\infty} \langle A(u_m), u_m - u \rangle \leq 0 \qquad (19.25)$$

imply that

$$u_m \to u \quad \text{strongly in } \mathbb{B} \text{ as } m \to +\infty.$$

Lemma 19.2. *The operator* (19.23) *possesses the S^+-property.*

Proof. Let

$$u_m \rightharpoonup u \quad \text{weakly in } W_0^{1,p}(\Omega) \text{ as } m \to +\infty.$$

Recall that $\Delta_p u(x) \in W^{-1,p'}(\Omega)$. In this case,

$$\langle \Delta_p u, u_m - u \rangle \to 0 \quad \text{as } m \to +\infty. \qquad (19.26)$$

Therefore, under the condition (19.25), we have

$$\limsup_{m \to +\infty} \langle -\Delta_p u_m + \Delta_p u, \ u_m - u \rangle \leq 0. \qquad (19.27)$$

On the other hand, similarly to (19.21), we obtain

$$\langle \Delta_p u - \Delta_p u_m, u_m - u \rangle$$

$$= \int_\Omega \left(|D_x u_m|^{p-2} D_x u_m - |D_x u|^{p-2} D_x u, \ D_x u_m - D_x u \right) dx$$

$$\geq C(p) \int_\Omega |D_x u_m - D_x u|^p \, dx \geq 0, \qquad (19.28)$$

where in the last inequality we have used the inequality (19.20). From (19.28) we obtain the limit relation

$$\liminf_{m \to +\infty} \langle -\Delta_p u_m + \Delta_p u, \ u_m - u \rangle \geq 0. \qquad (19.29)$$

Combining (19.27) and (19.29), we have

$$0 \leq \liminf_{m \to +\infty} \langle -\Delta_p u_m + \Delta_p u, \ u_m - u \rangle \leq \limsup_{m \to +\infty} \langle -\Delta_p u_m + \Delta_p u, \ u_m - u \rangle \leq 0$$

and hence

$$\lim_{m \to +\infty} \langle -\Delta_p u_m + \Delta_p u, \ u_m - u \rangle = 0.$$

The last relation together with (19.28) yields

$$\lim_{m \to +\infty} \int_\Omega |D_x u_m - D_x u|^p \, dx = 0.$$

Therefore,

$$u_m \to u \quad \text{strongly in } W_0^{1,p}(\Omega) \text{ as } m \to +\infty.$$

The lemma is proved. □

Now we prove the weak generalized solvability of the problem (19.3) in the sense of Definition 19.2 by using the Galerkin method.

1. Prove that the Banach space $W_0^{1,p}(\Omega)$ is separable, i.e., it contains a countable, everywhere dense set. Indeed, consider the linear isometric mapping (embedding)

$$L : v(x) \mapsto \left\{ v(x), \frac{\partial v(x)}{\partial x_1}, \ldots, \frac{\partial v(x)}{\partial x_N} \right\}. \tag{19.30}$$

Using this mapping, we can identify the space $W_0^{1,p}(\Omega)$ with the closed subspace in the uniformly convex ($p \geq 2$) separable space

$$\underbrace{L^p(\Omega) \otimes L^p(\Omega) \otimes \ldots \otimes L^p(\Omega)}_{N+1}. \tag{19.31}$$

Therefore, this countable, everywhere dense set in $L^p(\Omega) \otimes L^p(\Omega) \otimes \ldots \otimes L^p(\Omega)$ can be projected onto this subspace, which is isometric to $W_0^{1,p}(\Omega)$.

We denote this countable, everywhere dense set by $\{w_j\}$. Consider the following Galerkin approximation:

$$u_m(x) = \sum_{k=1}^{m} c_{mk} w_k(x), \quad c_{mk} \in \mathbb{R}^1, \tag{19.32}$$

where the functions $u_m(x)$ satisfy the following equality:

$$\langle -\Delta_p u_m, w_j \rangle = \langle f, w_j \rangle \quad \text{for all } j = \overline{1, m}. \tag{19.33}$$

2. Prove the solvability of this system of algebraic equations using Lemma 19.1. Consider the operator

$$T(c_m) : \mathbb{R}^m \to \mathbb{R}^m,$$

where

$$T(c_m) := \big(T_1(c_m), \ldots, T_m(c_m) \big), \quad c_m = (c_{m1}, \ldots, c_{mm}),$$
$$T_j(c_m) := -\langle \Delta_p u_m, w_j \rangle - \langle f, w_j \rangle, \quad j = \overline{1, m}.$$

We will use the standard scalar product $(\cdot, \cdot)$ in $\mathbb{R}^m$.

Below, we need the equality

$$\langle -\Delta_p v, v \rangle = \big\| |D_x v| \big\|_p^p, \quad v(x) \in W_0^{1,p}(\Omega).$$

The relations

$$
\begin{aligned}
\big(T(c_m), c_m \big) &= -\langle \Delta_p u_m, u_m \rangle - \langle f, w_m \rangle = \big\| |D_x u_m| \big\|_p^p - \langle f, w_m \rangle \\
&\geq \big\| |D_x u_m| \big\|_p^p - \| f \|_* \big\| |D_x u_m| \big\|_p \\
&= \big\| |D_x u_m| \big\|_p \Big(\big\| |D_x u_m| \big\|_p^{p-1} - \| f \|_* \Big) \geq 0 \tag{19.34}
\end{aligned}
$$

hold for sufficiently large r: $\left\||D_x u_m|\right\|_p = r > 0$, where $\|\cdot\|_*$ and $\left\||D_x u|\right\|_p$ are the norms of the Banach spaces $W^{-1,p'}(\Omega)$ and $W_0^{1,p}(\Omega)$:

$$\left\||D_x u|\right\|_p := \left(\int\limits_\Omega |D_x u|^p \, dx\right)^{1/p};$$

moreover, we have used the following general inequality:

$$\left|\langle f, u\rangle\right| \leq \|f\|_* \|u\| \quad \text{for all } f \in \mathbb{B}^* \text{ and } u \in \mathbb{B}.$$

3. It remains to note that on the finite-dimensional Euclidean space $\mathbb{R}^m$ all norms are equivalent. In particular, we can take the following norm on $\mathbb{R}^m$:

$$|c_m| := \left\||D_x u_m|\right\|_p.$$

Therefore, the estimate (19.34) implies that there exists a sufficiently large number $R > 0$ such that the following inequality holds:

$$\left(T(c_m), c_m\right) \geq 0 \quad \text{for all } |c_m| = R > 0.$$

Therefore, by Lemma 19.1, there exists $c_m \in \mathbb{R}^m$ such that

$$T(c_m) = 0 \quad \text{for } |c_m| \leq R,$$

i.e., the algebraic system (19.33) has a solution $u_m \in W_0^{1,p}(\Omega)$. Thus, the sequence $\{u_m\}$ of Galerkin approximations is well defined.

4. Now we prove that there exists a subsequence $\{u_{m_k}\} \subset \{u_m\}$ such that

$$u_{m_k} \rightharpoonup u \quad \text{weakly in } W_0^{1,p}(\Omega) \text{ as } m_k \to +\infty,$$

where $u(x)$ satisfies Eq. (19.22).

First, multiplying Eq. (19.33) by c_{mj} and summing by $j = \overline{1, m}$, we obtain the equality

$$\langle -\Delta_p u_m, u_m \rangle = \langle f, u_m \rangle. \tag{19.35}$$

Integrating by parts, we obtain the inequality

$$\left\||D_x u_m|\right\|_p^p = \langle f, u_m \rangle \leq \|f\|_* \left\||D_x u_m|\right\|_p,$$

which implies

$$\left\||D_x u_m|\right\|_p \leq \|f\|_*^{1/(p-1)} \quad \text{for all } m \in \mathbb{N}. \tag{19.36}$$

Therefore, the sequence $\{u_m\}$ is uniformly bounded in the Banach space $W_0^{1,p}(\Omega)$ and hence by Theorem 2.1, there exists a subsequence $\{u_{m_k}\} \subset \{u_m\}$ such that

$$u_{m_k} \rightharpoonup u \quad \text{weakly in } W_0^{1,p}(\Omega) \text{ as } m_k \to +\infty. \tag{19.37}$$

In what follows, we will denote this subsequence by $\{u_m\}$.

5. Now we prove that the sequence $\{u_m\}$ possesses the property (19.25) (the condition from the S^+-property of the p-Laplacian). We choose a sequence of the form

$$v_m := \sum_{j=1}^m k_{mj} w_j \tag{19.38}$$

such that

$$v_m \to u \quad \text{strongly in } W_0^{1,p}(\Omega) \text{ as } m \to +\infty. \tag{19.39}$$

Multiplying both sides of Eq. (19.33) by k_{mj} and summing by $j = \overline{1, m}$, we obtain the equality

$$\langle -\Delta_p u_m, v_m \rangle = \langle f, v_m \rangle. \tag{19.40}$$

Then due to (19.35) and (19.40), we have

$$
\begin{aligned}
\langle -\Delta_p u_m, u_m - u \rangle &= \langle f, u_m \rangle - \langle -\Delta_p u_m, u \rangle \\
&= \langle f, u_m \rangle - \langle -\Delta_p u_m, u - v_m \rangle - \langle -\Delta_p u_m, v_m \rangle \\
&= \langle f, u_m \rangle - \langle -\Delta_p u_m, u - v_m \rangle - \langle f, v_m \rangle \\
&= \langle f, u_m - v_m \rangle - \langle -\Delta_p u_m, u - v_m \rangle =: I_{1m} + I_{2m}. \tag{19.41}
\end{aligned}
$$

Consider the terms in the right-hand side of this equality separately.

The limit property

$$|I_{1m}| \le |\langle f, u_m - v_m \rangle| \to 0 \quad \text{as } m \to +\infty \tag{19.42}$$

is valid since

$$u_m - v_m = (u_m - u) - (v_m - u) \rightharpoonup 0 \quad \text{weakly in } W_0^{1,p}(\Omega).$$

Now we estimate the second term I_{2m}. We have the estimate

$$|I_{2m}| \le |\langle -\Delta_p u_m, u - v_m \rangle| \le \|\Delta_p u_m\|_* \|u - v_m\| \to 0 \quad \text{as } m \to +\infty \tag{19.43}$$

since the property (19.39) and the following chain of inequalities hold:

$$
\begin{aligned}
\|\Delta_p u_m\|_* &= \sup_{\||D_x\varphi|\|_p \le 1} \left| \langle -\Delta_p u_m, \varphi \rangle \right| \\
&= \sup_{\||D_x\varphi|\|_p \le 1} \left| \int_\Omega |D_x u_m|^{p-2} (D_x u_m, D_x\varphi) \, dx \right| \\
&\le \sup_{\||D_x\varphi|\|_p \le 1} \int_\Omega |D_x u_m|^{p-1} |D_x\varphi| \, dx \\
&\le \sup_{\||D_x\varphi|\|_p \le 1} \left(\int_\Omega |D_x u_m|^p \, dx \right)^{1/p'} \left(\int_\Omega |D_x\varphi|^p \, dx \right)^{1/p} \\
&\le \left(\int_\Omega |D_x u_m|^p \, dx \right)^{1/p'} \le \{(19.36)\} \le \|f\|_*.
\end{aligned}
$$

Therefore, due to (19.41)–(19.43), we see that the property (19.25) is fulfilled. Applying Lemma 19.2 on the S^+-property of the p-Laplacian, we arrive at the following important result:

$$u_m \to u \quad \text{strongly in } W_0^{1,p}(\Omega) \text{ as } m \to +\infty. \tag{19.44}$$

6. Applying Lemma 21.5, we see that, due to the property (19.36), the following inequality holds:

$$\mu(R_m) \le c_1 \max\left\{\left\|\|D_x u\|\right\|_p, \|f\|_*^{1/(p-1)}\right\}.$$

Therefore, due to (19.44) and (21.20), we conclude that

$$\Delta_p u_m \to \Delta_p u \quad \text{strongly in } W^{-1,p'}(\Omega) \text{ as } m \to +\infty. \tag{19.45}$$

Passing to the limit as $m \to +\infty$ in Eq. (19.33) and taking into account (19.45), we obtain the following result:

$$\langle -\Delta_p u, w_j \rangle = \langle f, w_j \rangle \quad \text{for all } j = 1, 2, 3, \dots \tag{19.46}$$

Due to the fact that the countable family $\{w_j\}$ is dense in $W_0^{1,p}(\Omega)$, this implies that the function $u(x) \in W_0^{1,p}(\Omega)$ constructed above satisfies Eq. (19.22) from the definition of the weak solution (Definition 19.2).

Indeed, let $\varphi(x) \in W_0^{1,p}(\Omega)$ be an arbitrary fixed function. Then, due to the completeness of $\{w_j\} \subset W_0^{1,p}(\Omega)$, there exists a sequence

$$\varphi_m = \sum_{j=1}^m d_{mj} w_j, \quad \left\|\|D_x \varphi_m - D_x \varphi\|\right\|_p \to +0 \quad \text{as } m \to +\infty. \tag{19.47}$$

Multiplying Eq. (19.46) by d_{mj} and summing by $j = \overline{1, m}$, we obtain the equality

$$\langle -\Delta_p u, \varphi_m \rangle = \langle f, \varphi_m \rangle, \tag{19.48}$$

which implies the equivalent equality

$$\langle -\Delta_p u, \varphi \rangle + \langle -\Delta_p u, \varphi_m - \varphi \rangle = \langle f, \varphi \rangle + \langle f, \varphi_m - \varphi \rangle. \tag{19.49}$$

Note that the following inequalities hold:

$$\left|\langle -\Delta_p u, \varphi_m - \varphi \rangle\right| \le \|\Delta_p u\|_* \left\|\|D_x \varphi_m - D_x \varphi\|\right\|_p, \tag{19.50}$$

$$\left|\langle f, \varphi_m - \varphi \rangle\right| \le \|f\|_* \left\|\|D_x \varphi_m - D_x \varphi\|\right\|_p. \tag{19.51}$$

Due to (19.47), passing to the limit as $m \to +\infty$ in (19.48), we obtain the required relation:

$$\langle -\Delta_p u, \varphi \rangle = \langle f, \varphi \rangle \quad \text{for all } \varphi(x) \in W_0^{1,p}(\Omega). \tag{19.52}$$

7. It remains to prove the uniqueness of the weak solution. For this, taking in the inequality (19.21) any two weak solutions in the sense of Definition 19.2 as $u_1, u_2 \in W_0^{1,p}(\Omega)$, we obtain

$$\int_\Omega \left|D_x u_1 - D_x u_2\right|^p dx = 0.$$

This implies the uniqueness of the weak solution of the problem (19.3). Thus, we have proved the following assertion.

Theorem 19.1. *For any function $f(x) \in W^{-1,p'}(\Omega)$, there exists a unique weak solution $u(x) \in W_0^{1,p}(\Omega)$ of the problem (19.3) (i.e., a solution in the sense of Definition 19.2).*

19.2 Bibliographical Notes

The contents of this lecture is taken from [13, 15, 16, 22, 37, 44–46, 64].

Lecture 20

Method of Monotonic Operators. General Results

20.1 Fundamentals of the Theory of Monotonic Operators

Before further reading, the readers are advised to refresh Definitions 1.1–1.3 in their memory.

Lemma 20.1. *Each monotonic operator $A : \mathbb{B} \to \mathbb{B}^*$ is locally bounded.*

Proof. 1. Assume that A is not locally bounded. Then there exist an element u and a sequence $\{u_n\}$ such that $u_n \to u$ strongly in $\mathbb{B}$ and $\|Au_n\|_* \to +\infty$ as $n \to +\infty$. Without loss of generality, we can assume that

$$\|Au_n\|_* > 1 \quad \text{for all } n \in \mathbb{N}.$$

2. For $n \in \mathbb{N}$, we set $\alpha_n := 1 + \|Au_n\|_* \|u_n - u\|$. Due to the monotonicity of A, for any $v \in \mathbb{B}$ we have

$$\langle Au_n - A(u + v), u_n - u - v \rangle \geq 0,$$

and hence the following chain of inequalities holds:

$$\frac{1}{\alpha_n} \langle Au_n, v \rangle \leq \frac{1}{\alpha_n} \Big(\langle Au_n, v \rangle + \langle Au_n - A(u + v), u_n - u - v \rangle \Big)$$

$$\leq \frac{1}{\alpha_n} \Big(\langle Au_n, u_n - u \rangle + \langle A(u + v), v + u - u_n \rangle \Big)$$

$$\leq \frac{\|Au_n\|_* \|u_n - u\|}{\alpha_n} + \frac{1}{\alpha_n} \|A(u + v)\|_* \big(\|v\| + \|u - u_n\| \big)$$

$$\leq 1 + \frac{1}{\alpha_n} \|A(u + v)\|_* \Big(\|v\| + \|Au_n\|_* \|u - u_n\| \Big) \leq M_1, \qquad (20.1)$$

where the constant M_1 depends on u and v but does not depend on n.

3. Substituting $-v$ instead of v, we obtain the estimate

$$-\frac{1}{\alpha_n} \langle Au_n, v \rangle \leq 1 + \frac{1}{\alpha_n} \|A(u - v)\|_* \Big(\|v\| + \|Au_n\|_* \|u - u_n\| \Big) \leq M_2.$$

The estimates obtained imply

$$\limsup_{n \to +\infty} \left| \frac{1}{\alpha_n} \langle Au_n, v \rangle \right| < +\infty \quad \forall v \in \mathbb{B}.$$

By the Banach–Steinhaus theorem, we obtain the inequality

$$\frac{1}{\alpha_n}\|Au_n\|_* \le M \equiv \max\{M_1, M_2\},$$

i.e.,

$$\|Au_n\|_* \le M_1\alpha_n = M_1\Big(1 + \|Au_n\|_*\|u - u_n\|\Big).$$

Let $n_0 \in \mathbb{N}$ be chosen so that the condition $M\|u - u_n\| \le 1/2$ for all $n \ge n_0$. Then the last inequality implies that $\|Au_n\|_* \le 2M$ for $n \ge n_0$. This contradicts the assumption $\|Au_n\|_* \to +\infty$. The lemma is proved. □

Lemma 20.2. *Each linear monotonic operator $A : \mathbb{B} \to \mathbb{B}^*$ is strongly continuous.*

Proof. Let $u_n \to u$ strongly in $\mathbb{B}$. We set

$$v_n = \begin{cases} \dfrac{u_n - u}{\|u_n - u\|^{1/2}} & \text{for } u_n \ne u, \\ \theta & \text{for } u_n = u. \end{cases}$$

Then $v_n \to \theta$ strongly in $\mathbb{B}$ and by Lemma 20.1 we have $\|Av_n\|_* \le M = \text{const}$. This implies

$$\|Au_n - Au\|_* = \|u_n - u\|^{1/2}\|Av_n\|_* \le M\|u_n - u\|^{1/2} \to +0.$$

The lemma is proved. □

Lemma 20.3. *Let $A : \mathbb{B} \to \mathbb{B}^*$ be a monotonic operator. Then the following assertions are equivalent:*

(1) *the operator A is radially continuous;*
(2) *the condition*

$$\langle f - Av, u - v\rangle \ge 0 \quad \text{for all } v \in \mathbb{B}$$

  implies $Au = f$;
(3) *the relations*

$$\text{(a)}\quad u_n \rightharpoonup u \quad \text{weakly in } \mathbb{B} \text{ as } n \to +\infty;$$

$$\text{(b)}\quad Au_n \overset{*}{\rightharpoonup} f \quad *\text{-weakly in } \mathbb{B}^* \text{ as } n \to +\infty,$$

$$\text{(c)}\quad \limsup_{n\to+\infty}\langle Au_n, u_n\rangle \le \langle f, u\rangle$$

  imply that $Au = f$;
(4) *the operator A is demicontinuous.*

Proof. 1. Prove the implication $(1) \Rightarrow (2)$. Let v be an arbitrary element of $\mathbb{B}$ and $v_t = u - tv$, $t > 0$. We have

$$0 \le \langle f - Av_t, u - v_t\rangle = \langle f - Av_t, tv\rangle = t\langle f - Av_t, v\rangle;$$

or, after dividing by t, $0 \leq \langle f - Av_t, v \rangle$. Letting t tend to 0, due to the radial continuity of the operator A we obtain the inequality

$$0 \leq \langle f - Au, v \rangle \quad \text{for all } v \in \mathbb{B}.$$

Since $v \in \mathbb{B}$ is arbitrary, this inequality implies that $Au = f$. Indeed, let, for example, $\langle f - Au, v \rangle > 0$ for some $v \in \mathbb{B}$. Then for $-v$ we obtain the inequality

$$0 \leq \langle f - Au, -v \rangle \quad \Rightarrow \quad 0 < \langle f - Au, v \rangle \leq 0.$$

This contradiction shows that

$$\langle f - Au, v \rangle = 0 \quad \text{for all } v \in \mathbb{B} \quad \Rightarrow \quad f - Au = \vartheta^* \in \mathbb{B}^*.$$

2. Prove the implication (2) $\Rightarrow$ (3). Let

$$u_n \rightharpoonup u \quad \text{weakly in } \mathbb{B} \text{ as } n \to +\infty,$$

$$Au_n \overset{*}{\rightharpoonup} f \quad \text{*-weakly in } \mathbb{B}^* \text{ as } n \to +\infty,$$

$$\limsup_{n \to +\infty} \langle Au_n, u_n \rangle \leq \langle f, u \rangle.$$

Then for arbitrary $v \in \mathbb{B}$ we have

$$\langle f - Av, u - v \rangle = \langle f, u \rangle - \langle f, v \rangle - \langle Av, u - v \rangle$$

$$\geq \limsup_{n \to +\infty} \Big(\langle Au_n, u_n \rangle - \langle f, v \rangle - \langle Av, u - v \rangle \Big)$$

$$= \limsup_{n \to +\infty} \Big(\langle Au_n, u_n \rangle - \langle Au_n, v \rangle - \langle Av, u_n - v \rangle \Big)$$

$$= \limsup_{n \to +\infty} \langle Au_n - Av, u_n - v \rangle \geq 0.$$

The last inequality follows from the monotonicity of the operator A. Thus, due to the property (2), we conclude that $Au = f$.

3. Prove the implication (3) $\Rightarrow$ (4).

Let $u_n \to u$ strongly in $\mathbb{B}$ as $n \to +\infty$. Due to the local boundedness of the operator A, the sequence $\{\|Au_n\|_*\}$ is bounded. Then there exists a subsequence $\{v_n\}$ of the sequence $\{u_n\}$ such that

$$Av_n \overset{*}{\rightharpoonup} f \quad \text{*-weakly in } \mathbb{B}^* \text{ as } n \to +\infty; \tag{20.2}$$

here we used the separability of the space $\mathbb{B}$. Clearly, we see that

$$v_n \to u \quad \text{strongly in } \mathbb{B} \text{ as } n \to +\infty; \tag{20.3}$$

therefore,

$$v_n \rightharpoonup u \quad \text{weakly in } \mathbb{B} \text{ as } n \to +\infty.$$

The following equality holds:

$$\langle Av_n, v_n \rangle = \langle Av_n, v_n - u \rangle + \langle Av_n, u \rangle;$$

moreover, due to (20.2) and (20.3) the boundedness of $\{\|Au_n\|_*\}$, we have

$$\big| \langle Av_n, v_n - u \rangle \big| \leq \|Av_n\|_* \|v_n - u\| \leq M_1 \|v_n - u\| \to +0,$$

$$\langle Av_n, u \rangle \to \langle f, u \rangle$$

as $n \to +\infty$. Therefore,

$$\lim_{n \to +\infty} \langle Av_n, v_n \rangle = \langle f, u \rangle$$

and further, due to the property (3), $Au = f$ and

$$Av_n \overset{*}{\rightharpoonup} Au \quad \text{*-weakly in } \mathbb{B}^* \text{ as } n \to +\infty.$$

Now we assume that there exists a subsequence $\{w_n\} \subset \{u_n\}$ such that the sequence $\{Aw_n\}$ does not converge *-weakly to Au in $\mathbb{B}^*$. Then there exists $\varepsilon > 0$ and an element $z \in \mathbb{B}$ such that for some subsequence $\{w_n\} \subset \{u_n\}$, the following inequality holds:

$$\left| \langle Aw_n, z \rangle - \langle Au, z \rangle \right| > \varepsilon \quad \text{for all } n \in \mathbb{N}. \tag{20.4}$$

Note that

$$\|Aw_n\|_* \leq M_1, \quad w_n \to u \quad \text{strongly in } \mathbb{B} \text{ as } n \to +\infty.$$

Then, repeating the reasoning, we obtain that there exists a subsequence $\{w_{n_n}\}$ such that

$$Aw_{n_n} \overset{*}{\rightharpoonup} Au \quad \text{*-weakly in } \mathbb{B}^* \text{ as } n \to +\infty$$

and $\{w_{n_n}\}$ satisfies the inequality (20.4). The contradiction obtained shows that

$$Au_n \overset{*}{\rightharpoonup} Au \quad \text{*-weakly in '} \mathbb{B}^* \text{ as } n \to +\infty.$$

4. The implication (4) $\Rightarrow$ (1) was proved above (see Remark 1.1). The lemma is proved. $\qquad \square$

Lemma 20.4. *Let $A : \mathbb{B} \to \mathbb{B}^*$ be a radially continuous monotonic operator. Then for any $f \in \mathbb{B}^*$, the set $K(f)$ of solutions of the equation $Au = f$ is convex and weakly closed.*

Proof. 1. Let $u_1, u_2 \in K(f)$ and $u_t = tu_1 + (1 - t)u_2$, $t \in [0, 1]$. Then for any $v \in \mathbb{B}$ we have

$$
\begin{aligned}
\langle f - Av, u_t - v \rangle &= \langle f - Av, u_t - tv - (1 - t)v \rangle \\
&= \langle f - Av, tu_1 - tv \rangle + \langle f - Av, (1 - t)u_2 - (1 - t)v \rangle \\
&= t\langle Au_1 - Av, u_1 - v \rangle + (1 - t)\langle Au_2 - Av, u_2 - v \rangle \geq 0.
\end{aligned}
$$

By Lemma 20.3(2), this implies $Au_t = f$, i.e., $K(f)$ is convex.

2. Let $\{u_n\}$ be a sequence of elements $u_n \in K$ such that

$$u_n \rightharpoonup u \quad \text{weakly in } \mathbb{B} \text{ as } n \to +\infty.$$

For any $v \in \mathbb{B}$ we have

$$\langle f - Av, u - v \rangle = \lim_{n \to +\infty} \langle f - Av, u_n - v \rangle = \lim_{n \to +\infty} \langle Au_n - Av, u_n - v \rangle \geq 0$$

since $Au_n = f$ for all $n \in \mathbb{N}$. Thus, Lemma 20.3(2) implies the equality $Au = f$, i.e., $K(f)$ is weakly closed. The lemma is proved. $\qquad \square$

20.2 Browder–Minty Existence Theorem

In this section, we present the theory of monotonic coercive operators, which has important applications in the theory of elliptic boundary-value problems.

Assume that $\mathbb{B}$ is a separable and reflexive real Banach space.

Theorem 20.1 (Browder–Minty theorem). *Let* $A : \mathbb{B} \to \mathbb{B}^*$ *be a radially continuous, monotonic, and coercive operator. Then for any* $f \in \mathbb{B}^*$, *the set of solutions of the equation*

$$Au = f \tag{20.5}$$

is nonempty, weakly closed, and convex.

Proof. Due to Lemma 20.4, we must prove only the existence of a solution $u \in \mathbb{B}$ of the problem

$$\langle Au, v \rangle = \langle f, v \rangle \tag{20.6}$$

for all $v \in \mathbb{B}$.

1. We prove that (20.6) has at least one solution. Let $\{w_n\} \subset \mathbb{B}$ be a countable and dense in $\mathbb{B}$ system of linearly independent elements and let $\mathbb{B}_n$ be a closed linear hull of the vectors $\{w_1, \dots, w_n\}$. Then the formula

$$L : \mathbb{R}^n \ni c_n = \{c_{n1}, \dots, c_{nn}\} \mapsto \sum_{i=1}^{n} c_{ni} w_i =: u_n$$

defines a bijective continuous mapping L of the space $\mathbb{R}^n$ on $\mathbb{B}_n$. Obviously, $|c|_1 := \|Lc\|$, $c \in \mathbb{R}^n$, is a norm on $\mathbb{R}^n$. Indeed, since $\{w_j\}_{j=1}^{+\infty}$ is a linearly independent system, we have

$$|c_n|_1 = \|Lc_n\| = 0 \quad \Leftrightarrow \quad \sum_{j=1}^{n} c_{nj} w_j = 0 \quad \Leftrightarrow \quad c_n = (c_{n1}, \dots, c_{nn}) = (0, \dots, 0).$$

Moreover, the following relations hold:

$$\begin{aligned}
|\lambda_1 c_n^1 + \lambda_2 c_n^2| &= \|L(\lambda_1 c_n^1 + \lambda_2 c_n^2)\| = \|L(\lambda_1 c_n^1) + L(\lambda_2 c_n^2)\| \\
&\le \|L(\lambda_1 c_n^1)\| + \|L(\lambda_2 c_n^2)\| = \lambda_1 \|Lc_n^1\| + \lambda_2 \|Lc_n^2\| \\
&= \lambda_1 |c_n^1|_1 + \lambda_2 |c_n^2|_1.
\end{aligned} \tag{20.7}$$

Due to the equivalence of all norms on a finite-dimensional space, we have

$$|c| \le a|c|_1 = a\|Lc\|, \quad a > 0.$$

2. To the problem (20.6), we assign the following finite-dimensional problem on the existence of Galerkin approximations

$$u_n := \sum_{i=1}^{n} c_{ni} w_i, \quad \langle Au_n, w_k \rangle = \langle f, w_k \rangle, \quad k = \overline{1, n}. \tag{20.8}$$

Introduce the operator $T : \mathbb{R}^n \to \mathbb{R}^n$ by the rule

$$Tc = \{b_1, \dots, b_n\}, \quad b_k = \langle ALc - f, w_k \rangle.$$

Since the operator A is radially continuous and monotonic, it is also demicontinuous (see Lemma 20.3(4)). Therefore, the operator T is continuous.

The operator $L : \mathbb{R}^n \to \mathbb{B}_n$ is continuous in the corresponding strong topologies of the spaces specified and the operator $A : \mathbb{B}_n \subset \mathbb{B} \to \mathbb{B}^*$ is demicontinuous. Therefore, their composition is a demicontinuous operator.

The coercivity of A implies that for sufficiently large $R_1 > 0$ we have

$$\left(\frac{\langle Au_n, u_n \rangle}{\|u_n\|} - \|f\|_* \right) \|u_n\| \geq 0 \quad \text{for } \|u_n\| \geq R_1.$$

Hence

$$(Tc, c) = \sum_{i=1}^{n} b_i c_i = \langle Au_n, u_n \rangle - \langle f, u_n \rangle \geq \left(\frac{\langle Au_n, u_n \rangle}{\|u_n\|} - \|f\|_* \right) \|u_n\| \geq 0$$

for $|c| = R = aR_1$, where we used the inequality

$$\langle f, u_n \rangle \leq \|f\|_* \|u_n\|.$$

Therefore, by Lemma 19.1, there exists $c \in \mathbb{R}^n$ such that $Tc = 0$. Consequently,

$$\langle Au_n, w_k \rangle = \langle f, w_k \rangle, \quad k = \overline{1, n}, \tag{20.9}$$

for $u_n = Lc$. Thus, there exists a solution of the problem (20.8).

3. Multiplying both sides of Eq. (20.9) by c_k and summing by $k = \overline{1, n}$, we obtain the equality

$$\langle Au_n, u_n \rangle = \langle f, u_n \rangle, \tag{20.10}$$

which implies the inequality

$$\langle Au_n, u_n \rangle \leq \|f\|_* \|u_n\| \quad \Rightarrow \quad \frac{\langle Au_n, u_n \rangle}{\|u_n\|} \leq \|f\|_*. \tag{20.11}$$

From the estimate (20.11) and the coercivity of A we conclude that there exists $M_1 \in \mathbb{R}$ such that

$$\|u_n\| \leq M_1. \tag{20.12}$$

Indeed, in the opposite case, due to the coercivity of the operator A we have

$$\lim_{\|u_n\| \to +\infty} \frac{\langle Au_n, u_n \rangle}{\|u_n\|} = +\infty.$$

From (20.11) and (20.12) we obtain the following inequality:

$$\langle Au_n, u_n \rangle \leq M_2, \quad n \in \mathbb{N}. \tag{20.13}$$

4. Now we prove that the inequalities (20.12) and (20.13) imply the estimate

$$\|Au_n\|_* \leq M_3, \quad n \in \mathbb{N}. \tag{20.14}$$

Indeed, since the operator A is monotonic, it is locally bounded at zero due to Lemma 20.1. Therefore, there exist constants $\varepsilon > 0$ and $M_4 > 0$ such that the following inequality is fulfilled:

$$\|Av\|_* \leq M_4 \quad \text{for all } \|v\| \leq \varepsilon.$$

Due to the monotonicity of the operator A we have

$$\langle Au_n - Av, u_n - v \rangle \geq 0.$$

Therefore,

$$\|Au_n\|_* = \sup_{\|w\| \leq 1} |\langle Au_n, w \rangle| = \left\{ w = \frac{v}{\varepsilon} \right\}$$

$$= \sup_{\|v\| \leq \varepsilon} \frac{1}{\varepsilon} |\langle Au_n, v \rangle| = \frac{1}{\varepsilon} \sup_{\|v\| \leq \varepsilon} \langle Au_n, v \rangle$$

$$\leq \frac{1}{\varepsilon} \sup_{\|v\| \leq \varepsilon} \left[\langle Au_n, v \rangle + \langle Au_n - Av, u_n - v \rangle \right]$$

$$= \frac{1}{\varepsilon} \sup_{\|v\| \leq \varepsilon} \left| \langle Au_n, u_n \rangle + \langle Av, v \rangle - \langle Av, u_n \rangle \right|$$

$$\leq \frac{1}{\varepsilon} \sup_{\|v\| \leq \varepsilon} \left[M_2 + \|Av\|_* \|v\| + \|Av\|_* \|u_n\| \right]$$

$$\leq \frac{1}{\varepsilon} \left(M_2 + M_4 \varepsilon + M_4 M_1 \right) =: M$$

(recall that $\mathbb{B}$ is a real Banach space).

5. Further, due to (20.9) and (20.14),

$$\lim_{n \to +\infty} \langle Au_n, w \rangle = \langle f, w \rangle \quad \forall w \in \bigcup_{n=1}^{+\infty} \mathbb{B}_n \equiv \mathbb{B}. \tag{20.15}$$

Indeed,

$$\langle Au_n - f, w \rangle = \langle Au_n - f, [w]_n \rangle + \langle Au_n - f, \{w\}_n \rangle \overset{(20.9)}{=\!=} 0 + \langle Au_n - f, \{w\}_n \rangle,$$

$$\left| \langle Au_n - f, \{w\}_n \rangle \right| \leq \{(20.14)\} \leq \left(M_3 + \|f\|_* \right) \|\{w\}_n\| \to 0,$$

where

$$[w]_n := \sum_{i=1}^{n} c_{ni} w_i, \quad \{w\}_n = w - [w]_n,$$

$$[w]_n \to w \quad \text{strongly as } n \to +\infty.$$

From (20.15) we conclude that

$$Au_n \overset{*}{\rightharpoonup} f \quad \text{*-weakly in } \mathbb{B}^* \text{ as } n \to +\infty. \tag{20.16}$$

By the *a priori* estimate (20.12), there exists a subsequence $\{u_{n_k}\}$ of the sequence $\{u_n\}$ such that

$$u_{n_k} \rightharpoonup u \quad \text{weakly in } \mathbb{B} \text{ as } k \to +\infty. \tag{20.17}$$

From Eq. (20.10) specified for the subsequence $\{u_{n_k}\}$ and (20.17) we have

$$\lim_{k \to +\infty} \langle Au_{n_k}, u_{n_k} \rangle = \lim_{k \to +\infty} \langle f, u_{n_k} \rangle = \langle f, u \rangle. \tag{20.18}$$

Then due to (20.16)–(20.18) and Lemma 20.3(3), the element $u \in \mathbb{B}$ is a solution of the equation $Au = f$. The theorem is proved. $\qquad\square$

Theorem 20.2. *Let $A : \mathbb{B} \to \mathbb{B}^*$ be a radially continuous, strictly monotonic, and coercive operator. Then there exists the inverse operator $A^{-1} : \mathbb{B}^* \to \mathbb{B}$, which is strictly monotonic and bounded.*

Proof. 1. First, we prove that the operator $A^{-1} : \mathbb{B}^* \to \mathbb{B}$ exists. Obviously, it suffices to show that the equation $Au = f$ has a unique solution for any $f \in \mathbb{B}^*$. Theorem 20.1 guarantees the existence of at least one solution u. Let v be another solution; then

$$\langle Au - Av, u - v \rangle = 0.$$

The strict monotonicity of A implies that $u = v$.

2. Now we prove that the operator A^{-1} is strictly monotonic in the following sense:

$$\langle f - g, A^{-1}f - A^{-1}g \rangle > 0.$$

for any $f, g \in \mathbb{B}^*$, $f \neq g$.

We set $u = A^{-1}f$ and $v = A^{-1}g$. Note that $u \neq v$. Indeed, in the opposite case, we have

$$u = v \quad \Rightarrow \quad A^{-1}f = A^{-1}g \quad \Rightarrow \quad f = g.$$

If $u \neq v$, then the strict monotonicity of A implies the inequality

$$\langle f - g, A^{-1}f - A^{-1}g \rangle = \langle Au - Av, u - v \rangle > 0 \quad \text{for all } f \neq g.$$

3. Finally, we prove that the operator A^{-1} is bounded. Let $Au = f$ (respectively, $u = A^{-1}f$) and $\|f\|_* \leq M$. Since

$$\langle Au, u \rangle \geq \gamma(\|u\|)\|u\|, \quad \langle Au, u \rangle = \langle f, u \rangle \leq \|f\|_*\|u\|,$$

we arrive at the inequality

$$\gamma(\|u\|) \leq \|f\|_*. \tag{20.19}$$

Since $\gamma(s) \to +\infty$ as $s \to +\infty$, we conclude that

$$\|u\| = \|A^{-1}f\| \leq K \quad \text{for all } \|f\|_* \leq M,$$

where the constant K depends only on M. Therefore, the operator A^{-1} is bounded. The theorem is proved. $\qquad\square$

20.3   Bibliographical Notes

The contents of this lecture is taken from [22, 46, 64, 67].

Lecture 21

Properties of the p-Laplacian

21.1 Important Auxiliary Inequalities

Lemma 21.1. *For any $x, y \in \mathbb{R}_+$ and $q > 0$, the following inequality holds:*

$$(x + y)^q \leq C_1(q)(x^q + y^q), \tag{21.1}$$

where

$$C_1(q) = \begin{cases} 1, & q \in (0, 1], \\ 2^{q-1}, & q \in (1, +\infty). \end{cases}$$

Proof. For $q = 1$, the assertion is trivial. Let $q > 1$. For $x = 0$, the assertion is trivial. In the opposite case, due to the homogeneity of the inequality (21.1), it suffices to prove the inequality

$$(1 + z)^q \leq 2^{q-1}(1 + z^q), \quad z > 0, \tag{21.2}$$

for $z = y/x$. Consider the function

$$f(z) := 2^{q-1}(1 + z^q) - (1 + z)^q.$$

Obviously, $f(1) = 0$. Clearly, $f(z) \in C(0, +\infty)$, $f'(z) < 0$ for $z \in (0, 1)$, and $f'(z) > 0$ for $z > 1$. Therefore, $z = 1$ is a minimum point of the function $f(z)$ and hence $f(z) \geq f(1) = 0$ on both intervals.

Now let $0 < q < 1$. Arguing similarly, we see that it suffices to prove the inequality

$$(1 + z)^q \leq 1 + z^q, \tag{21.3}$$

which is trivial for $z = 0$. Consider the function $g(z) := 1 + z^q - (1 + z)^q$; clearly, $g'(z) > 0$ for $z > 0$ and $\lim\limits_{z \to +0} g(z) = 0$, which implies (21.3). The lemma is proved. $\qquad \square$

Let $a, b \in \mathbb{R}^N$; we denote by $(\cdot, \cdot)$ the scalar product in $\mathbb{R}^N$.

Lemma 21.2. *The following identity is valid:*

$$\left(|b|^{p-2}b - |a|^{p-2}a, b - a\right)$$
$$= \frac{1}{2}\left(|b|^{p-2} + |a|^{p-2}\right)|b - a|^2 + \frac{1}{2}\left(|b|^{p-2} - |a|^{p-2}\right)\left(|b|^2 - |a|^2\right). \tag{21.4}$$

Proof. We denote the left and right hand sides of Eq. (21.4) by I_1 and I_2, respectively. On the one hand, we have

$$I_1 = (|b|^{p-2}b - |a|^{p-2}a, b - a) = |b|^p + |a|^p - (|b|^{p-2} + |a|^{p-2})(a,b).$$

On the other hand,

$$
\begin{aligned}
I_2 &= \frac{1}{2}(|b|^{p-2} + |a|^{p-2})|b - a|^2 + \frac{1}{2}(|b|^{p-2} - |a|^{p-2})(|b|^2 - |a|^2) \\
&= \frac{1}{2}(|b|^{p-2} + |a|^{p-2})(|b|^2 + |a|^2 - 2(a,b)) + \frac{1}{2}(|b|^{p-2} - |a|^{p-2})(|b|^2 - |a|^2) \\
&= |b|^p + |a|^p - (|b|^{p-2} + |a|^{p-2})(a,b). \qquad\qquad\qquad \square
\end{aligned}
$$

Since the function $t \mapsto t^{p-2}$ strictly increases on $[0, +\infty)$ for $p - 2 > 0$, the numbers $|b|^{p-2}$ and $|a|^{p-2}$ are related by the same inequalities that the numbers $|b|$ and $|a|$. Therefore,

$$(|b|^{p-2} - |a|^{p-2})(|b|^2 - |a|^2) \geq 0$$

and from (21.4) we immediately obtain

$$(|b|^{p-2}b - |a|^{p-2}a, b - a) \geq \frac{1}{2}(|b|^{p-2} + |a|^{p-2})|b - a|^2. \qquad (21.5)$$

Lemma 21.3. *For $p \geq 2$, the following inequalities hold:*

$$(|b|^{p-2}b - |a|^{p-2}a, b - a) \geq \frac{1}{2}(|b|^{p-2} + |a|^{p-2})|b - a|^2 \geq C_2(p)|b - a|^p, \qquad (21.6)$$

where

$$C_2(p) = \begin{cases} 2^{-p/2}, & p \in [2,4], \\ 2^{2-p}, & p \geq 4. \end{cases}$$

Proof. For $p = 2$, these inequalities are trivial. Let $p > 2$. Then we have

$$
\begin{aligned}
|b - a|^p &= |b - a|^2 |b - a|^{p-2} = |b - a|^2 (|b - a|^2)^{(p-2)/2} \\
&= |b - a|^2 (|b|^2 - 2(a,b) + |a|^2)^{(p-2)/2} \\
&\leq |b - a|^2 2^{(p-2)/2} (|b|^2 + |a|^2)^{(p-2)/2} \\
&\leq |b - a|^2 2^{(p-2)/2} C_1\left(\frac{p-2}{2}\right)(|b|^{p-2} + |a|^{p-2}).
\end{aligned}
$$

Applying the last inequality to (21.5), we arrive at the estimate

$$(|b|^{p-2}b - |a|^{p-2}a, b - a) \geq C_2(p)|b - a|^p. \qquad\qquad \square$$

21.2 Boundedness and Continuity of the p-Laplacian

Let Ω be a bounded domain, $p \geq 2$.

Lemma 21.4. *For any $u(x) \in W_0^{1,p}(\Omega)$, the following equality is valid:*

$$\|\Delta_p u\|_{W^{-1,p'}(\Omega)} = \left\| |D_x u| \right\|_p^{p-1}, \tag{21.7}$$

where

$$\left\| |D_x u| \right\|_p = \left(\int_\Omega \left(\sum_{i=1}^N (D_{x_i} u)^2 \right)^{p/2} dx \right)^{1/p}. \tag{21.8}$$

Proof. We have

$$\|\Delta_p u\|_{W^{-1,p'}(\Omega)} = \sup_{\left\| |D_x \varphi| \right\|_p \leq 1} |\langle \Delta_p u, \varphi \rangle|$$

$$= \sup_{\left\| |D_x \varphi| \right\|_p \leq 1} \left| \int_\Omega |D_x u|^{p-2} (D_x u, D_x \varphi)\, dx \right|$$

$$\leq \sup_{\left\| |D_x \varphi| \right\|_p \leq 1} \left| \int_\Omega |D_x u|^{p-1} |D_x \varphi|\, dx \right|$$

$$\leq \sup_{\left\| |D_x \varphi| \right\|_p \leq 1} \left\| |D_x u| \right\|_p^{p-1} \left\| |D_x \varphi| \right\|_p \leq \left\| |D_x u| \right\|_p^{p-1}. \tag{21.9}$$

The following lower estimates hold:

$$\|\Delta_p u\|_{W^{-1,p'}(\Omega)} = \sup_{\left\| |D_x \varphi| \right\|_p \leq 1} |\langle \Delta_p u, \varphi \rangle| \tag{21.10}$$

$$\geq \left| \left\langle \Delta_p u, \frac{u}{\left\| |D_x u| \right\|_p} \right\rangle \right| = \left\| |D_x u| \right\|_p^{p-1}. \tag{21.11}$$

The estimates (21.9) and (21.11) imply Eq. (21.7). The lemma is proved. $\square$

Lemma 21.5. *The operator $\Delta_p : W_0^{1,p}(\Omega) \to W^{-1,p'}(\Omega)$ is continuous for $p \geq 2$.*

Proof. We must prove that

$$\Delta_p u_m \to \Delta_p u \quad \text{strongly in } W^{-1,p'}(\Omega) \text{ as } m \to +\infty \tag{21.12}$$

if

$$u_m \to u \quad \text{strongly in } W_0^{1,p}(\Omega) \text{ as } m \to +\infty. \tag{21.13}$$

For this end, we prove the bounded Lipschitz continuity of the p-Laplacian. For the expression

$$\left| |\xi|^{p-2}\xi - |\eta|^{p-2}\eta \right|$$

where $\xi, \eta \in \mathbb{R}^n$ and $\min\{|\xi|, |\eta|\} > 0$, the following two "rough" estimates hold:

$$\left||\xi|^{p-2}\xi - |\eta|^{p-2}\eta\right| = \left||\xi|^{p-2}[\xi - \eta] + \eta\left[|\xi|^{p-2} - |\eta|^{p-2}\right]\right|$$

$$\leq |\xi|^{p-2}|\xi - \eta| + (p-2)|\eta| \max\left\{|\xi|^{p-3}, |\eta|^{p-3}\right\}|\xi - \eta|, \quad (21.14)$$

$$\left||\eta|^{p-2}\eta - |\xi|^{p-2}\xi\right| = \left||\eta|^{p-2}[\eta - \xi] + \xi\left[|\eta|^{p-2} - |\xi|^{p-2}\right]\right|$$

$$\leq |\eta|^{p-2}|\eta - \xi| + (p-2)|\xi| \max\left\{|\eta|^{p-3}, |\xi|^{p-3}\right\}|\eta - \xi|. \quad (21.15)$$

These estimates imply

$$\left||\xi|^{p-2}\xi - |\eta|^{p-2}\eta\right| \leq \max\left\{|\xi|^{p-2}, |\eta|^{p-2}\right\}|\xi - \eta|$$

$$+ (p-2) \min\left\{|\xi|, |\eta|\right\} \frac{\max\left\{|\xi|^{p-2}, |\eta|^{p-2}\right\}}{\min\left\{|\xi|, |\eta|\right\}}|\xi - \eta|$$

$$= (p-1) \max\left\{|\xi|^{p-2}, |\eta|^{p-2}\right\}|\xi - \eta| \quad (21.16)$$

for all $\xi, \eta \in \mathbb{R}^n \backslash \{O\}$ and $p \geq 2$. The case where $\min\{|\xi|, |\eta|\} = 0$ can be considered similarly.

Now, by the definition of the norm of the Banach space $W^{-1,p'}(\Omega)$, we have the following relations:

$$\left\|\Delta_p u - \Delta_p u_m\right\|_{W^{-1,p'}(\Omega)} = \sup_{\left\||D_x w|\right\|_p \leq 1} \left|\langle \Delta_p u - \Delta_p u_m, w \rangle\right|$$

$$\leq \sup_{\left\||D_x w|\right\|_p \leq 1} \left|\int_\Omega \left||D_x u|^{p-2}D_x u - |D_x u_m|^{p-2}D_x u_m\right||D_x w| \, dx\right|$$

$$\leq (p-1) \sup_{\left\||D_x w|\right\|_p \leq 1} \int_\Omega \left|D_x u_m - D_x u\right| \max\left\{|D_x u|^{p-2}, |D_x u_m|^{p-2}\right\}|D_x w| \, dx$$

$$= (p-1) \sup_{\left\||D_x w|\right\|_p \leq 1} I, \quad (21.17)$$

where

$$I := \int_\Omega \left|D_x u_m - D_x u\right| \max\left\{|D_x u|^{p-2}, |D_x u_m|^{p-2}\right\}|D_x w| \, dx; \quad (21.18)$$

here we used the inequality (21.16).

Now in the chain (21.17) we apply the generalized Hölder inequality (4.29) to the last integral. We set

$$p_1 = p, \quad p_2 = \frac{p}{p-2}, \quad p_3 = p, \quad r = 1, \quad \frac{1}{p_1} + \frac{1}{p_2} + \frac{1}{p_3} = 1, \quad p \geq 2$$

(if $p = 2$, then $p_1 = p_3 = 2$ and $p_2 = +\infty$). Then we obtain the following inequality for I:

$$I \leq \left(\int_\Omega |D_x u - D_x u_m|^p \, dx\right)^{1/p}$$

$$\times \left(\int_\Omega \max\left\{|D_x u|^p, |D_x u_m|^p\right\} dx\right)^{(p-2)/p} \times \left(\int_\Omega |D_x w|^p \, dx\right)^{1/p}. \quad (21.19)$$

Thus, the inequalities (21.17) and (21.19) imply the following estimate:

$$\left\|\Delta_p u - \Delta_p u_m\right\|_{W^{-1,p'}(\Omega)} \le \mu(R_m) \left\|\left|D_x u - D_x u_m\right|\right\|_p, \tag{21.20}$$

where

$$\mu(R_m) = c_1 R_m^{p-2}, \quad R_m = \max\left\{\left\|\left|D_x u\right|\right\|_p, \left\|\left|D_x u_m\right|\right\|_p\right\},$$

which shows that the p-Laplacian is boundedly Lipschitz continuous. Due to the limit property (21.13), there exists a constant $M > 0$ such that

$$\left\|\left|D_x u_m\right|\right\|_p \le M \quad \text{for all } m \in \mathbb{N}.$$

Thus the limit property (21.12) holds. The lemma is proved. $\qquad\square$

Lemma 21.6. *For the operator $\Delta_p : W_0^{1,p}(\Omega) \to W^{-1,p'}(\Omega)$, where $p \ge 2$, there exists a continuous and bounded inverse operator $(\Delta_p)^{-1} : W^{-1,p'}(\Omega) \to W_0^{1,p}(\Omega)$.*

Proof. We fix the following properties of the p-Laplacian.

1. *The operator $-\Delta_p$ is radially continuous.* Indeed, let $u(x), v(x) \in W_0^{1,p}(\Omega)$. Consider the function $\varphi(s) := \langle -\Delta_p(u + sv), v \rangle$ on the segment $s \in [0, 1]$. For any $s_0, s_1 \in [0, 1]$ we have

$$\left|\langle -\Delta_p(u + s_0 v), v \rangle - \langle -\Delta_p(u + s_1 v), v \rangle\right|$$

$$= \left|\langle -\Delta_p(u + s_0 v) + \Delta_p(u + s_1 v), v \rangle\right|$$

$$\le \left\| -\Delta_p(u + s_0 v) + \Delta_p(u + s_1 v)\right\|_{W^{-1,p'}(\Omega)} \left\|\left|D_x v\right|\right\|_p.$$

We have

$$\left\|(u + s_0 v) - (u + s_1 v)\right\|_{W_0^{1,p}(\Omega)} = \left\|(s_0 - s_1)v\right\|_{W_0^{1,p}(\Omega)}$$

$$\equiv \left\|\left|D_x((s_0 - s_1)v)\right|\right\|_p \to 0$$

as $s_1 \to s_0$; then by Lemma 21.5

$$\left\| -\Delta_p(u + s_0 v) + \Delta_p(u + s_1 v)\right\|_{W^{-1,p'}(\Omega)} \to 0.$$

Therefore, the function $\varphi(s)$ is continuous on the segment $[0, 1]$.

2. *The operator $-\Delta_p$ is coercive.* This follows from the identity

$$\langle -\Delta_p u, u \rangle = \int_\Omega |D_x u|^{p-2}(D_x u, D_x u)\, dx = \int_\Omega |D_x u|^p\, dx = \left\|\left|D_x u\right|\right\|_p^p. \tag{21.21}$$

3. *The operator $-\Delta_p$ is strictly monotonic.* Indeed, due to the inequality (21.6), we have

$$\langle -\Delta_p u_1 + \Delta_p u_2, u_1 - u_2 \rangle$$

$$= \int_\Omega \left(|D_x u_1|^{p-2} D_x u_1 - |D_x u_2|^{p-2} D_x u_2, \ D_x u_1 - D_x u_2\right) dx$$

$$\ge C_2(p) \left\|\left|D_x u_1 - D_x u_2\right|\right\|_p^p. \tag{21.22}$$

By the Browder–Minty theorem (see Theorem 20.2), the operator Δ_p has an inverse operator $(\Delta_p)^{-1}$. Due to Eq. (21.7), the equality

$$\left\| D_x (\Delta_p)^{-1} v \right\|_p = \|v\|_{W^{-1,p'}(\Omega)}^{1/(p-1)}$$

holds for all $v(x) \in W^{-1,p'}(\Omega)$. Therefore, the operator $(\Delta_p)^{-1}$ is bounded.

Let $v_1 = -\Delta_p u_1$ and $v_2 = -\Delta_p u_2$; then $u_1 = (-\Delta_p)^{-1} v_1$ and $u_2 = (-\Delta_p)^{-1} v_2$. Due to the inequality (21.22), we obtain the following inequality:

$$\left\| \left| D_x (-\Delta_p)^{-1} v_1 - D_x (-\Delta_p)^{-1} v_2 \right| \right\|_p^p$$

$$\leq \frac{1}{C_2(p)} \| v_1 - v_2 \|_{W^{-1,p'}(\Omega)} \left\| \left| D_x (-\Delta_p)^{-1} v_1 - D_x (-\Delta_p)^{-1} v_2 \right| \right\|_p, \quad (21.23)$$

which implies the inequality

$$\left\| \left| D_x (-\Delta_p)^{-1} v_1 - D_x (-\Delta_p)^{-1} v_2 \right| \right\|_p \leq \frac{1}{(C_2(p))^{1/(p-1)}} \| v_1 - v_2 \|_{W^{-1,p'}(\Omega)}^{1/(p-1)} \quad (21.24)$$

for all $v_1, v_2 \in W^{-1,p'}(\Omega)$. Therefore, the operator $(\Delta_p)^{-1}$ is continuous. Note that it is a Lipschitz continuous operator only in the trivial case where $p = 2$. The lemma is proved. $\qquad\qquad\square$

Theorem 21.1. *The operator Δ_p, $p \geq 2$, transforms a strongly measurable function $v(t) : [0, T] \to W_0^{1,p}(\Omega)$ into a strongly measurable function $\Delta_p v(t) : [0, T] \to W^{-1,p'}(\Omega)$.*

Proof. Note that the operator $-\Delta_p$ is radially continuous and monotonic and hence by Lemma 20.3 it is demicontinuous, i.e., for any sequence $\{u_n\} \subset W_0^{1,p}(\Omega)$ such that

$$u_n \to u \quad \text{strongly in } W_0^{1,p}(\Omega) \text{ as } n \to +\infty, \quad (21.25)$$

the following limit property holds:

$$\Delta_p u_n \rightharpoonup \Delta_p u \quad \text{weakly in } W^{-1,p'}(\Omega) \text{ as } n \to +\infty. \quad (21.26)$$

Let a function $v(t) : [0, T] \to W_0^{1,p}(\Omega)$ be strongly measurable; then there exists a sequence of simple functions $\{h_n(t)\}$ such that

$$h_n(t) \to v(t) \quad \text{strongly in } W_0^{1,p}(\Omega) \text{ as } n \to +\infty \quad (21.27)$$

for almost all $t \in [0, T]$. Due to the demicontinuity of the operator Δ_p, we conclude that

$$\Delta_p h_n(t) \rightharpoonup \Delta_p v(t) \quad \text{weakly in } W^{-1,p'}(\Omega) \text{ as } n \to +\infty \quad (21.28)$$

for almost all $t \in [0, T]$. Consider the real-valued functions $\varphi_n(t) := \langle \Delta_p h_n(t), w \rangle$ for arbitrary fixed $w \in W_0^{1,p}(\Omega)$. The functions $\varphi_n(t)$ are simple and hence Lebesgue measurable on $[0, T]$; moreover,

$$\varphi_n(t) \to \varphi(t) := \langle \Delta_p v(t), w \rangle \quad \text{as } n \to +\infty \quad (21.29)$$

for almost all $t \in [0, T]$ and for all $w \in W_0^{1,p}(\Omega)$. Therefore, the function $\varphi(t)$ is also Lebesgue measurable on $[0, T]$ and hence the function $\Delta_p v(t) : [0, T] \to W^{-1,p'}(\Omega)$ is weakly measurable (here we take into account the reflexivity of the space $W^{1,p}(\Omega)$). It remains to apply Theorem 3.2, which, together with the separability of the space $W^{-1,p'}(\Omega)$ implies the strong measurability of the function $\Delta_p v(t)$. The theorem is proved. $\square$

From Theorem 21.1 we obtain the following important assertion.

Theorem 21.2. *The operator Δ_p acts as follows:*

$$\Delta_p : L^p(0, T; W_0^{1,p}(\Omega)) \to L^{p'}(0, T; W^{-1,p'}(\Omega)), \quad p \geq 2. \tag{21.30}$$

Proof. We set

$$A(v) := - \operatorname{div}(|D_x v|^{p-2} D_x v). \tag{21.31}$$

We represent the operator $A(v)$ as the composition of three operators:

$$A(v) = - \operatorname{div} \eta, \quad \eta = |\xi|^{p-2}\xi, \quad \xi = D_x v. \tag{21.32}$$

Let $v(t) \in L^p(0, T; W_0^{1,p}(\Omega))$; then

$$\xi = D_x v : \ L^p(0, T; W_0^{1,p}(\Omega)) \to \underbrace{L^p(0, T; L^p(\Omega)) \otimes \cdots \otimes L^p(0, T; L^p(\Omega))}_{N},$$

$$\eta = |\xi|^{p-2}\xi : \ \underbrace{L^p(0, T; L^p(\Omega)) \otimes \cdots \otimes L^p(0, T; L^p(\Omega))}_{N}$$

$$\to \underbrace{L^{p'}(0, T; L^{p'}(\Omega)) \otimes \cdots \otimes L^{p'}(0, T; L^{p'}(\Omega))}_{N},$$

$$\operatorname{div} \eta : \ \underbrace{L^{p'}(0, T; L^{p'}(\Omega)) \otimes \cdots \otimes L^{p'}(0, T; L^{p'}(\Omega))}_{N} \to L^{p'}(0, T; W^{-1,p'}(\Omega)).$$

These facts, the relations (21.32), and Theorem 21.1 imply Theorem 21.2. The theorem is proved. $\square$

21.3 Bibliographical Notes

The contents of this lecture is taken from [13, 22, 44–46].

Lecture 22

Galerkin Method and Method of Compactness. Parabolic Equations, I

22.1 Parabolic Equation with the p-Laplacian

Consider the following initial-boundary-value problem:

$$\frac{\partial u}{\partial t} = \Delta_p u + f(x,t) \quad \text{in } D = \Omega \otimes (0,T], \quad p \geq 2, \tag{22.1}$$

$$u(x,t) = 0 \qquad \text{on } \partial\Omega \otimes [0,T], \tag{22.2}$$

$$u(x,0) = u_0(x) \qquad \text{on } x \in \Omega, \tag{22.3}$$

where $\Omega \subset \mathbb{R}^N$ is a bounded domain with sufficiently smooth boundary $\partial\Omega$. Recall that

$$\Delta_p u := \operatorname{div}(|D_x u|^{p-2} D_x u).$$

Definition 22.1. A function $u(x,t)$ of the class[1]

$$u(x,t) \in L^\infty(0,T; W_0^{1,p}(\Omega)), \quad u'(x,t) \in L^2(0,T; L^2(\Omega))$$

is called a *weak solution* of the problem (22.1)–(22.3) if for any $\varphi(x,t) \in L^p(0,T; W_0^{1,p}(\Omega))$, the following equality holds:

$$\int_0^T \int_\Omega \left[u'(x,t)\varphi(x,t) + |D_x u(x,t)|^{p-2}\big(D_x u(x,t), D_x\varphi(x,t)\big) \right.$$

$$\left. - f(x,t)\varphi(x,t) \right] dx\, dt = 0, \tag{22.4}$$

$$u(x,0) = u_0(x) \in W_0^{1,p}(\Omega), \quad f(x,t) \in C([0,T]; L^2(\Omega)). \tag{22.5}$$

Remark 22.1. Note that due to Theorem 4.4, the space

$$W_1(0,T) \stackrel{\text{def}}{=\!=} \Big\{ v(x,t) : \; v(x,t) \in L^\infty(0,T; W_0^{1,p}(\Omega)),$$

$$v'(x,t) \in L^2(0,T; L^2(\Omega)) \Big\}$$

---

[1] The time derivative is treated in the weak sense (see Definition 4.6).

is embedded into the space $C([0, T]; L^2(\Omega))$. Indeed, since $p \geq 2$, the following continuous embedding hold:

$$L^\infty(0, T; W_0^{1,p}(\Omega)) \subset L^p(0, T; W_0^{1,p}(\Omega)),$$
$$L^2(0, T; L^2(\Omega)) \subset L^{p'}(0, T; W^{-1,p'}(\Omega)).$$

Therefore, the continuous embedding $W_1(0, T) \subset W_{pp'}(0, T) \subset C([0, T]; L^2(\Omega))$ holds and the initial condition (22.5) makes sense.

We must prove that for some small $T > 0$, there exists a weak solution of the problem (22.1)–(22.3) in the sense of Definition 22.1. We use the Galerkin method and the method of compactness, taking into account the monotonicity of the p-Laplacian.

1. Galerkin approximations

As was proved above, the Banach space $W_0^{1,p}(\Omega)$ is separable (see p. 212). We denote a countable, everywhere dense set in $W_0^{1,p}(\Omega)$ by

$$\{w_j(x)\}_{j=1}^{+\infty} \overset{ds}{\subset} W_0^{1,p}(\Omega).$$

For convenience, we may assume that it is a family of linearly independent, orthonormal functions:

$$\int_\Omega w_{j_1}(x) w_{j_2}(x)\, dx = \delta_{j_1 j_2}, \quad j_1, j_2 \in \overline{1, m}, \tag{22.6}$$

where $\delta_{j_1 j_2}$ is the Kronecker delta. Consider the function

$$u_m(x, t) = \sum_{k=1}^m c_{mk}(t) w_k(x), \quad c_{mk}(t) \in C^{(1)}[0, T_m], \quad k = \overline{1, m}. \tag{22.7}$$

Definition 22.2. A collection of functions $c_{mk}(t) \in C^{(1)}([0, T_m])$, $k = \overline{1, m}$, is called a *solution* of the Galerkin system if these functions satisfy the following Cauchy problem for the system of first-order differential equations:

$$\int_\Omega \Big[u'_m(x, t) w_j(x) + |D_x u_m(x, t)|^{p-2} \big(D_x u_m(x, t), D_x w_j(x) \big)$$

$$- f(x, t) w_j(x) \Big]\, dx = 0, \quad j = \overline{1, m}, \tag{22.8}$$

where we used the notation (22.7), under the condition that the initial values $c_{mk}(0)$ satisfy the following limit property:

$$u_{m0} := u_m(0) = \sum_{k=1}^m c_{mk}(0) w_k(x) \to u_0(x) \quad \text{strongly in } W_0^{1,p}(\Omega) \text{ as } m \to +\infty.$$

$$\tag{22.9}$$

(Clearly, $u_0(x) \in W_0^{1,p}(\Omega)$.)

First, we prove that the system of Galerkin equations has a solution of the class $c_{mk}(t) \in C^{(1)}[0, T_m]$ for some $T_m > 0$, which, in general, depends on $m \in \mathbb{N}$. Note that, due to (22.6), the system (22.8) can be represented in the following form:

$$\frac{dc_{mj}(t)}{dt} = F_j(t, c_m), \quad a_{kj} := \int_{\Omega} w_k(x) w_j(x) \, dx, \tag{22.10}$$

$$F_j(t, c_m) \stackrel{\text{def}}{=} F_{1j}(c_m) + F_{2j}(t), \tag{22.11}$$

$$F_{1j}(c_m) \stackrel{\text{def}}{=} -\int_{\Omega} |D_x u_m(x, t)|^{p-2} \big(D_x u_m(x, t), D_x w_j(x)\big) \, dx, \tag{22.12}$$

$$F_{2j}(t) \stackrel{\text{def}}{=} \int_{\Omega} f(x, t) w_j(x) \, dx; \tag{22.13}$$

where the matrix $A = (a_{kj})_{1,1}^{m,m}$ of the coefficient of the time derivative is nondegenerate since the family $\{w_1(x), \ldots, w_m(x)\}$ is linearly independent for any $m \in \mathbb{N}$. We prove that the functions $F_{1j}(c_m)$ are boundedly Lipschitz continuous. Indeed, we have

$$\left| F_{1j}(c_m^1) - F_{1j}(c_m^2) \right|$$

$$\leq \int_{\Omega} \left| |D_x u_m^1(x,t)|^{p-2} D_x u_m^1(x,t) - |D_x u_m^2(x,t)|^{p-2} D_x u_m^2(x,t) \right| |D_x w_j| \, dx$$

$$\leq (p-1) \int_{\Omega} \max \left\{ |D_x u_m^1|^{p-2}, \ |D_x u_m^2|^{p-2} \right\} |D_x u_m^1 - D_x u_m^2| |D_x w_j| \, dx$$

$$\leq \mu_{1j}(R) \left\| |D_x u_m^1 - D_x u_m^2| \right\|_p \leq a_1(m) \mu_{1j}(R) |c_m^1 - c_m^2|, \tag{22.14}$$

where we applied the generalized Hölder inequality with the parameters

$$\frac{1}{q_1} + \frac{1}{q_2} + \frac{1}{q_3} = 1, \quad q_1 = \frac{p}{p-2}, \quad q_2 = p, \quad q_3 = p$$

and introduced the notation

$$\mu_{1j}(R) := (p-1) \left\| |D_x w_j| \right\|_p R^{p-2}, \quad R := \max \left\{ \left\| |D_x u_m^1| \right\|_p, \ \left\| |D_x u_m^2| \right\|_p \right\}.$$

Moreover, we used the inequality

$$\left\| |D_x u_m^1 - D_x u_m^2| \right\|_p \leq a_1(m) |c_m^1 - c_m^2|; \tag{22.15}$$

here

$$\left| c_m^1(t) - c_m^2(t) \right| := \left(\sum_{j=1}^{m} \left| c_{mj}^1(t) - c_{mj}^2(t) \right|^2 \right)^{1/2}$$

and

$$a_1(m) := \left(\int_{\Omega} \left(\sum_{k=1}^{N} \sum_{j=1}^{m} \left(\frac{\partial w_j(x)}{\partial x_k} \right)^2 \right)^{p/2} dx \right)^{1/p}. \tag{22.16}$$

Indeed, we have the following relations:

$$\left\| \left| D_x u_m^1 - D_x u_m^2 \right| \right\|_p = \left(\int\limits_\Omega \left(\left| D_x u_m^1 - D_x u_m^2 \right|^2 \right)^{p/2} dx \right)^{1/p}, \tag{22.17}$$

and

$$\left| D_x u_m^1 - D_x u_m^2 \right|^2 = \sum_{k=1}^N \left(\sum_{j=1}^m (c_{mj}^1 - c_{mj}^2) \frac{\partial w_j(x)}{\partial x_k} \right)^2$$

$$\leq \sum_{j=1}^m (c_{mj}^1 - c_{mj}^2)^2 \sum_{k=1}^N \sum_{j=1}^m \left(\frac{\partial w_j(x)}{\partial x_k} \right)^2$$

$$= |c_m^1 - c_m^2|^2 \sum_{k=1}^N \sum_{j=1}^m \left(\frac{\partial w_j(x)}{\partial x_k} \right)^2, \tag{22.18}$$

where we used the following inequality for nonnegative numbers:

$$\sum_{j=1}^m a_j b_j \leq \left(\sum_{j=1}^m a_j^2 \right)^{1/2} \left(\sum_{j=1}^m b_j^2 \right)^{1/2}.$$

From the inequalities (22.17) and (22.18) we obtain the estimate (22.15).

Note that the following inequality holds:

$$\left\| \left| D_x u_m^l \right| \right\|_p^{p-2} \leq a_1(m) |c_m^l|^{p-2}, \quad l = 1, 2, \tag{22.19}$$

where $a_1(m)$ is defined by Eq. (22.16).

The function $F_{2j}(t, c_m)$ is independent of c_m. We also recall that $f(x, t) \in C([0, T]; L^2(\Omega))$.

Thus, the right-hand side $F_j(t, c_m)$ is a functions continuous with respect to $t \in [0, T_m]$ and boundedly Lipschitz continuous with respect to $c_m \in \mathbb{R}^m$; therefore, the Cauchy problem for the system (22.10) has a unique solution $c_{km}(t) \in C^{(1)}([0, T_m])$, $k = \overline{1, m}$, for some small $T_m > 0$, which, in general, depends on $m \in \mathbb{N}$.

Remark 22.2. Note that we can use the algorithm of extension of solutions of nonlinear equations in time (in the abstract form, this algorithm is presented in Chaps. 29–33). Then we can prove that for any $\{c_{m1}(0), \ldots, c_{mm}(0)\} \in \mathbb{R}^m$, there exists the maximal number $T_0 = T_0(c_{m1}(0), \ldots, c_{mm}(0)) > 0$ such that for each $T \in (0, T_0)$, there exists a classical solution of the Cauchy problem for the system (22.10) under the additional limit condition for the initial functions (22.9) of the class $c_{mj}(t) \in C^{(1)}[0, T]$; moreover, either $T_0 = +\infty$ or $T_0 < +\infty$, and in the last case, the following limit relation holds:

$$\lim_{t \uparrow T_0} \sqrt{\sum_{j=1}^m c_{mj}^2(t)} = +\infty. \tag{22.20}$$

Below, we obtain a priori estimates, which will imply that for some small $T > 0$ independent of $m \in \mathbb{N}$, the inequality

$$\sup_{t \in [0,T]} \sqrt{\sum_{j=1}^{m} c_{mj}^2(t)} < +\infty \qquad (22.21)$$

is fulfilled. Therefore, $T < T_0$ for any $m \in \mathbb{N}$ and $c_{mj}(t) \in C^{(1)}[0, T]$, $j = \overline{1, m}$, for any $m \in \mathbb{N}$, i.e., for fixed small $T > 0$ independent of $m \in \mathbb{N}$, there exists a classical solution of the Cauchy problem (22.10) under the condition (22.9).

2. A priori estimates

To obtain a priori estimates, we multiply both sides of Eq. (22.8) by $c_{mj}(t) \in C^{(1)}[0, T_m]$, $T_m > 0$, and sum by $j = \overline{1, m}$. Then we arrive at the equality

$$\frac{1}{2}\frac{d}{dt}\|u_m\|_2^2 + \||D_x u_m|\|_p^p = \int_{\Omega} f(x,t) u_m(x,t)\, dx. \qquad (22.22)$$

Since $f(x,t) \in C([0,T]; L^2(\Omega))$ and $W_0^{1,p}(\Omega) \subset L^2(\Omega)$, $p \geq 2$, using the three-parameter Young inequality (11.18), we obtain the following estimate:

$$\left| \int_{\Omega} f(x,t) u_m(x,t)\, dx \right| \leq \|f\|_2 \|u_m\|_2 \leq \|f\|_2 K_1 \||D_x u_m|\|_p$$

$$\leq \varepsilon \||D_x u_m|\|_p^p + c_1(\varepsilon)\|f\|_2^{p'}, \qquad (22.23)$$

where

$$c_1(\varepsilon) := \frac{1}{p'(\varepsilon p)^{p'/p}} K_1^{p'}, \quad \varepsilon \in (0, 1);$$

here we used the notation

$$\|v(x,t)\|_q = \left(\int_{\Omega} |v(x,t)|^q\, dx \right)^{1/q}.$$

Thus, we arrive at the following inequality:

$$\frac{1}{2}\frac{d}{dt}\|u_m\|_2^2 + (1 - \varepsilon)\||D_x u_m|\|_p^p \leq c_1(\varepsilon)\|f\|_2^{p'}. \qquad (22.24)$$

Integrating by time, we obtain the inequality

$$\frac{1}{2}\|u_m\|_2^2(t) + (1 - \varepsilon)\int_0^t \||D_x u_m|\|_p^p(s)\, ds$$

$$\leq \frac{1}{2}\|u_{m0}\|_2^2 + c_1(\varepsilon)\int_0^t \|f\|_2^{p'}(s)\, ds, \quad t \in [0, T_m]. \qquad (22.25)$$

By the construction of Galerkin approximations,

$$u_m(0) = u_{m0} \to u_0 \quad \text{strongly in } W_0^{1,p}(\Omega) \text{ as } m \to +\infty. \tag{22.26}$$

Since $W_0^{1,p}(\Omega)$, $p \geq 2$, is continuously embedded in $L^2(\Omega)$, we conclude that

$$u_m(0) = u_{m0} \to u_0 \quad \text{strongly in } L^2(\Omega) \text{ as } m \to +\infty.$$

Therefore, the numerical sequence $\{\|u_{m0}\|_2\}$ is bounded. Moreover, the right-hand side of the inequality (22.25) is bounded by a constant $A_1(T) < +\infty$ independent of $m \in \mathbb{N}$, where $t \in [0,T]$ and $T > 0$ is a time moment from the condition $f(x,t) \in C([0,T]; L^2(\Omega))$. Thus, from (22.25) we obtain the following two a priori estimates:

$$\|u_m\|_2^2(t) \leq A_1(T), \tag{22.27}$$

$$\int_0^t \||D_x u_m|\|_p^p(t)\, dt \leq A_1(T), \tag{22.28}$$

where $t \in [0, T_m]$ and $T_m \leq T$.

Remark 22.3. Note that, taking into account (22.6), from the inequality (22.27) we obtain the following a priori estimate for the vector-valued function $c_m(t) = (c_{m1}(t), \dots, c_{mm}(t))$:

$$\sum_{j=1}^m c_{mj}^2(t) = \sum_{j_1,j_2=1,1}^{m,m} \int_\Omega c_{mj_1}(t) w_{j_1}(x) c_{mj_2}(t) w_{j_2}(x)\, dx$$

$$= \int_\Omega |u_m(x,t)|^2\, dx \leq A_1(T) < +\infty \tag{22.29}$$

for $t \in [0, T_m]$, $T_m \leq T$. Comparing the a priori estimate (22.29) and the limit relation (22.20), we obtain the lower estimate $T < T_0$ for the existence time T_0 of the classical solution of the system (22.10)–(22.13) of Galerkin approximations and hence $c_{mk}(t) \in C^{(1)}[0,T]$ for all $k = \overline{1,m}$ and $m \in \mathbb{N}$ (see Remark 22.1).

Now we obtain two *basic a priori estimates*. Multiplying both sides of Eq. (22.8) by $c'_{mj}(t) \in C[0,T]$ and summing by $j = \overline{1,m}$, we obtain the equality

$$\|u'_m\|_2^2 + \frac{1}{p}\frac{d}{dt}\||D_x u_m|\|_p^p = \int_\Omega f(x,t) u'_m(x,t)\, dx. \tag{22.30}$$

Since $f(x,t) \in C(0,T; L^2(\Omega))$, the following inequality holds:

$$\left| \int_\Omega f(x,t) u'_m\, dx \right| \leq \|f\|_2 \|u'_m\|_2 \leq \varepsilon \|u'_m\|_2^2 + c_2(\varepsilon)\|f\|_2^2, \tag{22.31}$$

where $c_2(\varepsilon) := 1/(4\varepsilon)$ and $\varepsilon \in (0,1)$. Thus, the inequalities (22.30) and (22.31) imply the following final result:

$$(1-\varepsilon)\|u'_m\|_2^2 + \frac{1}{p}\frac{d}{dt}\||D_x u_m|\|_p^p \leq c_2(\varepsilon)\|f\|_2^2. \tag{22.32}$$

Integrating by time, we arrive at the following inequality:

$$(1-\varepsilon)\int\limits_0^t \|u_m'\|_2^2(s)\,ds + \frac{1}{p}\||D_x u_m|\|_p^p \le \frac{1}{p}\||D_x u_{m0}|\|_p^p + c_2(\varepsilon)\int\limits_0^t \|f\|_2^2(s)\,ds. \quad (22.33)$$

Due to the limit relation (22.26), the numerical sequence $\{\||D_x u_{m0}|\|_p^p\}$ is bounded. Therefore, from (22.33) we obtain the following two a priori estimates:

$$\||D_x u_m|\|_p^p(t) \le A_2(T), \quad t \in [0,T], \quad (22.34)$$

$$\int\limits_0^T \|u_m'\|_2^2(t)\,dt \le A_2(T), \quad (22.35)$$

where the constant $A_2(T) > 0$ is independent of $m \in \mathbb{N}$.

3. Passing to the limit

The a priori estimates (22.34) and (22.35) imply that the functional sequence $\{u_m\}$ is uniformly bounded in $L^\infty(0,T;W_0^{1,p}(\Omega))$ and the functional sequence $\{u_m'\}$ is uniformly bounded in $L^2(0,T;L^2(\Omega))$. Therefore, there exist functions $u(t) \in L^\infty(0,T;W_0^{1,p}(\Omega))$ and $w(t) \in L^2(0,T;L^2(\Omega))$ and a subsequence of the sequence $\{u_m(t)\}$ (we denote it by the same symbol $\{u_m(t)\}$) such that

$$u_m \overset{*}{\rightharpoonup} u \quad *\text{-weakly in } L^\infty(0,T;W_0^{1,p}(\Omega)), \quad (22.36)$$

$$u_m' \rightharpoonup w \quad \text{weakly in } L^2(0,T;L^2(\Omega)). \quad (22.37)$$

Indeed, the Banach space $L^\infty(0,T;W_0^{1,p}(\Omega))$ is dual to the Banach space $L^1(0,T;W^{-1,p'}(\Omega))$, i.e., in the case considered,

$$\mathbb{B} = L^1(0,T;W^{-1,p'}(\Omega)), \quad \mathbb{B}^* = L^\infty(0,T;W_0^{1,p}(\Omega)).$$

Due to Theorem 2.2, we conclude that there exists a subsequence of the sequence $\{u_m\}$ such that the limit relation (22.36) is fulfilled. Finally, we note that the Banach space $L^2(0,T;L^2(\Omega))$ is reflexive and separable and hence, due to Theorem 2.1, there exists a subsequence of the sequence $\{u_m'\}$ satisfying the limit relation (22.37).

Lemma 22.1. *If*

$$u_m \overset{*}{\rightharpoonup} u \quad *\text{-weakly in } L^\infty(0,T;W_0^{1,p}(\Omega)) \text{ as } m \to +\infty,$$

then

$$u_m \rightharpoonup u \quad \text{weakly in } L^2(0,T;L^2(\Omega)) \text{ as } m \to +\infty.$$

Proof. Let $\langle \cdot, \cdot \rangle_1$ be the duality bracket between the spaces $L^1(0,T;W^{-1,p'}(\Omega))$ and $\left(L^1(0,T;W^{-1,p'}(\Omega))\right)^* = L^\infty(0,T;W_0^{1,p}(\Omega))$. Note that $(L^2(0,T;L^2(\Omega)))^* =$

$L^2(0, T; L^2(\Omega))$. Since $W_0^{1,p}(\Omega) \overset{ds}{\subset} L^2(\Omega)$ and the space $W_0^{1,p}(\Omega)$ is reflexive, Theorem 1.3 implies the dense embedding

$$L^2(\Omega) \overset{ds}{\subset} W^{-1,p'}(\Omega);$$

consequently,

$$L^2(0, T; L^2(\Omega)) \overset{ds}{\subset} L^1(0, T; W^{-1,p'}(\Omega)). \tag{22.38}$$

Moreover, the Banach space $L^2(0, T; L^2(\Omega))$ is reflexive and, due to Theorem 1.3, the following dense embedding holds:

$$L^\infty(0, T; W_0^{1,p}(\Omega)) = \left(L^1(0, T; W^{-1,p'}(\Omega))\right)^* \overset{ds}{\subset} L^2(0, T; L^2(\Omega)). \tag{22.39}$$

By the condition of the lemma,

$$\langle u_m - u, v \rangle_1 \to 0 \quad \text{as } m \to +\infty \text{ for all } v \in L^1(0, T; W^{-1,p'}(\Omega)). \tag{22.40}$$

Since $u_m - u \in L^2(0, T; L^2(\Omega))$, for any $v \in L^2(0, T; L^2(\Omega)) \subset L^1(0, T; W^{-1,p'}(\Omega))$ due to the embedding (22.38) and Theorem 1.4, we obtain the following equality of the duality brackets:

$$\langle u_m - u, v \rangle_1 = \int\limits_0^T \int\limits_\Omega \left(u_m(x, t) - u(x, t)\right) v(x, t) \, dx \, dt, \tag{22.41}$$

where we take into account the explicit form of the duality bracket between the spaces $L^2(0, T; L^2(\Omega))$ and $\left(L^2(0, T; L^2(\Omega))\right)^* = L^2(0, T; L^2(\Omega))$. Owing to the limit relation (22.40), we conclude that

$$u_m \rightharpoonup u \quad \text{weakly in } L^2(0, T; L^2(\Omega)) \text{ as } m \to +\infty. \tag{22.42}$$

The lemma is proved. $\qquad\qquad\qquad\qquad\qquad\qquad\qquad\qquad\qquad\qquad\qquad\square$

Lemma 22.2. *The functions* (22.36) *and* (22.37) *satisfy the equality* $u'(t) = w(t)$.

Proof. We use the notation from the theory of $\mathbb{B}$-valued distributions. Since $u_m(t) \in L^\infty(0, T; W_0^{1,p}(\Omega)) \subset L^\infty(0, T; L^2(\Omega))$, we consider the following equality, which is treated in the sense of $L^2(\Omega)$-valued distributions:

$$\langle\!\langle u_m'(t), \varphi(t) \rangle\!\rangle = -\langle\!\langle u_m(t), \varphi'(t) \rangle\!\rangle, \quad \varphi(t) \in \mathscr{D}(0, T). \tag{22.43}$$

Taking into account the limit properties (22.36) and (22.37) and Lemma 22.1, we have

$$\langle\!\langle u_m'(t), \varphi(t) \rangle\!\rangle = \int\limits_0^T u_m'(t) \varphi(t) \, dt \rightharpoonup \int\limits_0^T w(t) \varphi(t) \, dt, \tag{22.44}$$

$$\langle\!\langle u_m(t), \varphi'(t) \rangle\!\rangle = \int\limits_0^T u_m(t) \varphi'(t) \, dt \rightharpoonup \int\limits_0^T u(t) \varphi'(t) \, dt \tag{22.45}$$

weakly in $L^2(\Omega)$ as $m \to +\infty$ for any function $\varphi(t) \in \mathscr{D}(0,T)$.

Indeed, we prove, for example, (22.44). For any function $h(x) \in L^2(\Omega)$, we have

$$
\int\limits_\Omega h(x) \left[\int\limits_0^T u'_m(x,t)\varphi(t)\,dt - \int\limits_0^T w(x,t)\varphi(t)\,dt \right] dx
$$

$$
= \int\limits_0^T \int\limits_\Omega \left[u'_m(x,t) - w(x,t) \right] \varphi(t) h(x)\,dx\,dt \to +0 \quad \text{as } m \to +\infty.
$$

Owing to the obvious separability of the weak topology of the Banach space $L^2(\Omega)$, we obtain from (22.43)–(22.45) the following equality:

$$
\int\limits_0^T w(t)\varphi(t)\,dt = - \int\limits_0^T u(t)\varphi'(t)\,dt \quad \text{for any } \varphi(t) \in \mathscr{D}(0,T), \tag{22.46}
$$

where the integrals are $L^2(\Omega)$-valued Bochner integrals. Therefore, the function $u(t)$ has weak derivative $u'(t) = w(t)$. The lemma is proved. □

Now we note that, due to the a priori estimates (22.34) and (22.35), the functional sequence $\{u_m\}$ is uniformly (with respect to $m \in \mathbb{N}$) bounded in the Banach space

$$
W_{22}(0,T) \overset{\text{def}}{=} \left\{ v : v \in L^2(0,T;W_0^{1,p}(\Omega)), \ v' \in L^2(0,T;L^2(\Omega)) \right\}. \tag{22.47}
$$

We have the completely continuous embedding

$$
W_0^{1,p}(\Omega) \hookrightarrow\hookrightarrow L^2(\Omega), \quad p \geq 2.
$$

Therefore, by the Lions–Aubin compactness theorem (see Theorem 4.7), we also have the completely continuous embedding

$$
W_{22}(0,T) \hookrightarrow\hookrightarrow L^2(0,T;L^2(\Omega))
$$

and, therefore, the following limit relation holds:

$$
u_m \to u \quad \text{strongly in } L^2(0,T;L^2(\Omega)) \text{ as } m \to +\infty. \tag{22.48}
$$

Here we used Theorem 8.1 on the relationship between the complete continuity and the total continuity of linear operators.

Recall the equality

$$
\|\Delta_p u_m\|_* = \||D_x u_m|\|_p^{p-1} \quad \Rightarrow \quad \|\Delta_p u_m\|_*^{p'} = \||D_x u_m|\|_p^p, \tag{22.49}
$$

where $\|\cdot\|_*$ is the norm of the Banach space $W^{-1,p'}(\Omega)$ dual to $W_0^{1,p}(\Omega)$ (see Lemma 21.4). Due to the a priori estimate (22.28), the functional sequence $\{\Delta_p u_m\}$ is uniformly (with respect to $m \in \mathbb{N}$) bounded in the reflexive Banach space $L^{p'}(0,T;W^{-1,p'}(\Omega))$. (Recall that Theorem 21.1 asserts that the operator

Δ_p transforms strongly measurable functions into strongly measurable functions.) Therefore, by Theorem 2.1, there exists a subsequence[2] such that

$$-\Delta_p u_m \rightharpoonup \chi \quad \text{weakly in } L^{p'}(0, T; W^{-1, p'}(\Omega)), \; p' = \frac{p}{p-1}. \tag{22.50}$$

Remark 22.4. Recall that for the Banach space $L^p(0, T; W_0^{1,p}(\Omega))$, $p > 1$, its dual Banach space is $L^{p'}(0, T; W^{-1, p'}(\Omega))$ and the duality bracket between these spaces has the explicit form

$$\int_0^T \langle f(t), g(t) \rangle \, dt,$$

where $f(t) \in L^{p'}(0, T; W^{-1, p'}(\Omega))$, $g(t) \in L^p(0, T; W_0^{1,p}(\Omega))$, and $\langle \cdot, \cdot \rangle$ is the duality bracket between $W_0^{1,p}(\Omega)$ and $W^{-1, p'}(\Omega)$.

The limit relation (22.48) means that

$$\int_0^T \|u_m(t) - u(t)\|_2^2 \, dt \to +0 \quad \text{as } m \to +\infty.$$

Therefore, there exists a subsequence $\{u_m\}$ such that

$$u_m(t) \to u(t) \quad \text{strongly in } L^2(\Omega) \text{ for almost all } t \in [0, T]. \tag{22.51}$$

4. Method of monotonicity

Now we prove that the function $\chi(t) \in L^{p'}(0, T; W^{-1, p'}(\Omega))$ in the limit property (22.50) can be represented as

$$\chi(t) = -\Delta_p u \in L^{p'}(0, T; W^{-1, p'}(\Omega)). \tag{22.52}$$

First, we rewrite the system of Galerkin approximations (22.8) in the equivalent form

$$\langle u_m' - \Delta_p u_m - f(x, t), w_j \rangle = 0, \quad j = \overline{1, m}. \tag{22.53}$$

Multiplying both sides of this equality by a function $\varphi(t) \in C_0^\infty(0, T)$, integrating by time, passing to the limit as $m \to +\infty$, and taking into account the limit relations (22.37), (22.50), we obtain the equality

$$\int_0^T \langle u' + \chi - f(x, t), w_j \rangle \varphi(t) \, dt = 0, \quad j = \overline{1, +\infty}, \tag{22.54}$$

for all $\varphi(t) \in C_0^\infty(0, T)$.

Remark 22.5. Here we used the fact that the dense embedding

$$L^p(0, T; W_0^{1,p}(\Omega)) \overset{ds}{\subset} L^2(0, T; L^2(\Omega))$$

---

[2] We have integrated both sides of Eq. (22.49) by $t \in [0, T]$.

holds for $p \geq 2$ and hence, due to the reflexivity of $L^2(0,T;L^2(\Omega))$, the following dense embedding holds:

$$L^2(0,T;L^2(\Omega)) \overset{ds}{\subset} L^{p'}(0,T;W^{-1,p'}(\Omega)).$$

Therefore, by Theorem 1.4, we have the equality of the duality brackets

$$\int_0^T \langle u'_m - u', w_j \rangle \varphi(t)\, dt = \int_0^T \int_\Omega (u'_m(x,t) - u'(x,t))\, w_j(x)\varphi(t)\, dt, \tag{22.55}$$

and due to the limit relation (22.37), the right-hand side of Eq. (22.55) tends to zero as $m \to +\infty$.

Now, applying the fundamental lemma of calculus of variations, we obtain that

$$\langle u' + \chi - f(x,t), w_j \rangle = 0, \quad j = \overline{1,+\infty}, \tag{22.56}$$

for almost all $t \in [0,T]$. Since $\{w_j\}$ is a Galerkin basis in $W_0^{1,p}(\Omega)$, we conclude that Eq. (22.56) is equivalent to the equality

$$\langle u' + \chi - f(x,t), w \rangle = 0 \tag{22.57}$$

for all $w \in W_0^{1,p}(\Omega)$ and for almost all $t \in [0,T]$. In this equality, we set $w(t) = u(x)(t)$. Integrating by time $t \in [0,T]$, we obtain the equality

$$\int_0^T \langle \chi, u \rangle\, dt = \int_0^T \int_\Omega f(x,t)u\, dx\, dt + \frac{1}{2}\|u_0\|_2^2 - \frac{1}{2}\|u\|_2^2(T); \tag{22.58}$$

here we used Theorem 4.4(iii). Introduce the notation

$$A(w) \overset{\text{def}}{=} -\Delta_p w.$$

Let

$$X_m \overset{\text{def}}{=} \int_0^T \langle A(u_m) - A(v), u_m - v \rangle\, dt \tag{22.59}$$

for all $v(t) \in L^p(0,T;W_0^{1,p}(\Omega))$. Since the operator

$$-\Delta_p : L^p(0,T;W_0^{1,p}(\Omega)) \to L^{p'}(0,T;W^{-1,p'}(\Omega))$$

is monotonic, the inequality $X_m \geq 0$ holds (it is a consequence of the inequalities (21.22)).

The equality for X_m can be rewritten as follows:

$$X_m = \int_0^T \langle A(u_m), u_m \rangle\, dt - \int_0^T \langle A(u_m), v \rangle\, dt - \int_0^T \langle A(v), u_m - v \rangle\, dt. \tag{22.60}$$

Multiplying both sides of Eq. (22.8) by c_{mj}, summing by $j = \overline{1, m}$, and integrating by $t \in [0, T]$, we obtain the equality

$$\int_0^T \langle A(u_m), u_m \rangle \, dt = \int_0^T \int_\Omega f(x, t) u_m \, dx \, dt + \frac{1}{2} \|u_{m0}\|_2^2 - \frac{1}{2} \|u_m\|_2^2(T). \quad (22.61)$$

Note that, due to (22.51),

$$u_m(t) \to u(t) \quad \text{strongly in } L^2(\Omega) \quad (22.62)$$

for almost all $t \in [0, T]$. Moreover, by (22.36), (22.37), and Lemmas 4.4 and 22.2, we have

$$u(t) \in W_1(0, T) \stackrel{\text{def}}{=} \left\{ v(x, t) : \; v(x, t) \in L^\infty(0, T; W_0^{1,p}(\Omega)), \right.$$

$$\left. v'(x, t) \in L^2(0, T; L^2(\Omega)) \right\} \subset C([0, T]; L^2(\Omega)).$$

Therefore, (22.62) implies that

$$\|u_m\|_2(T) \to \|u\|_2(T) \quad \text{as } m \to +\infty. \quad (22.63)$$

Indeed, we first prove that the sequence $\{u_m(t)\}$ is uniformly continuous with respect to $t \in [0, T]$ in the norm of $L^2(\Omega)$. By (22.35), we have

$$u_m(t_2) - u_m(t_1) = \int_{t_1}^{t_2} u_m'(\tau) \, d\tau \quad \Rightarrow \quad \|u_m(t_2) - u_m(t_1)\|_2 \leq \int_{t_1}^{t_2} \|u_m'(\tau)\|_2 \, d\tau,$$

which implies the inequalities

$$\|u_m(t_2) - u_m(t_1)\|_2 \leq (t_2 - t_1)^{1/2} \left(\int_{t_1}^{t_2} \|u_m'(\tau)\|^2 \, d\tau \right)^{1/2}$$

$$\leq (t_2 - t_1)^{1/2} \left(\int_0^T \|u_m'(\tau)\|^2 \, d\tau \right)^{1/2}$$

$$\leq A_2^{1/2}(T)(t_2 - t_1)^{1/2}, \quad (22.64)$$

where the constant $A_2(T) > 0$ is independent of $m \in \mathbb{N}$. For any $\varepsilon > 0$, there exists $\delta = \delta(\varepsilon) > 0$ such that, on the one hand, the limit equality (22.62) holds for $t = T - \delta$ and hence

$$\|u_m(T - \delta) - u(T - \delta)\|_2 < \frac{\varepsilon}{3}, \quad m \geq m_0 \in \mathbb{N}. \quad (22.65)$$

On the other hand, due to the fact $u(t) \in C([0, T]; L^2(\Omega))$ we have the inequality

$$\|u(T) - u(T - \delta)\|_2 < \frac{\varepsilon}{3} \quad (22.66)$$

for sufficiently small $\delta > 0$. Finally, by the inequality (22.64), we have

$$\|u_m(T) - u_m(T - \delta)\|_2 \leq A_2(T)\delta^{1/2} < \frac{\varepsilon}{3}. \quad (22.67)$$

Thus, the following chain of inequalities holds:

$$\|u(T) - u_m(T)\|_2 \leq \|u(T) - u(T-\delta)\|_2$$
$$+ \|u(T-\delta) - u_m(T-\delta)\|_2 + \|u_m(T-\delta) - u_m(T)\|_2 < \varepsilon \quad (22.68)$$

for all $m \geq m_0$ and some sufficiently large $m_0 = m_0(\varepsilon) \in \mathbb{N}$.

From the limit relation (22.9) we conclude that

$$\|u_{m0}\|_2 \to \|u_0\|_2 \quad \text{as } m \to +\infty. \quad (22.69)$$

Then, due to (22.51), (22.63), and (22.69), we obtain from (22.61) the limit relation

$$\lim_{m \to +\infty} \int_0^T \langle A(u_m), u_m \rangle \, dt = \int_0^T \int_\Omega f(x,t)u(x,t) \, dx \, dt + \frac{1}{2}\|u_0\|_2^2 - \frac{1}{2}\|u\|_2^2(T). \quad (22.70)$$

Due to (22.58) we have

$$\lim_{m \to +\infty} \int_0^T \langle A(u_m), u_m \rangle \, dt = \int_0^T \langle \chi, u \rangle \, dt. \quad (22.71)$$

Passing to the limit as $m \to +\infty$ in (22.60), we obtain the inequality

$$0 \leq \int_0^T \langle \chi, u \rangle \, dt - \int_0^T \langle \chi, v \rangle \, dt - \int_0^T \langle A(v), u - v \rangle \, dt = \int_0^T \langle \chi - A(v), u - v \rangle \, dt \quad (22.72)$$

for all $v(t) \in L^p(0,T; W_0^{1,p}(\Omega))$. Setting

$$v = u - \lambda w, \quad \lambda > 0, \quad w \in L^p(0,T; W_0^{1,p}(\Omega))$$

in the last inequality, we obtain

$$\int_0^T \langle \chi - A(u - \lambda w), w \rangle \, dt \geq 0.$$

Passing here to the limit as $\lambda \to +0$ and taking into account the radial continuity of the p-Laplacian, we obtain the inequality

$$\int_0^T \langle \chi - A(u), w \rangle \, dt \geq 0 \quad \text{for all } w \in L^p(0,T; W_0^{1,p}(\Omega)). \quad (22.73)$$

Assume that $\chi \neq A(u)$; then for certain $w \in L^p(0,T; W_0^{1,p}(\Omega))$, we have the strict inequality

$$\int_0^T \langle \chi - A(u), w \rangle \, dt > 0.$$

Replacing here w by $-w$, we obtain the inequality

$$\int_0^T \langle \chi - A(u), w \rangle \, dt < 0,$$

which contradicts the inequality (22.73). Therefore, $\chi = A(u) = -\Delta_p u$ and hence there exists a weak solution of the original Dirichlet problem.

Remark 22.6. Note that if in (22.59) we take the solution u itself as v, then, on the one hand, we obtain

$$\lim_{m \to +\infty} X_m = 0,$$

and on the other hand, for X_m and $v = u$, the inequality

$$X_m \geq 2^{2-p} \int_0^T \int_\Omega |D_x u_m - D_x u|^p \, dx \, dt$$

holds. Therefore,

$$u_m \to u \quad \text{strongly in } L^p(0, T; W_0^{1,p}(\Omega)) \text{ as } m \to +\infty.$$

Thus, we have proved the following theorem.

Theorem 22.1. *For any $u_0(x) \in W_0^{1,p}(\Omega)$, under the condition $f(x, t) \in C([0, T]; L^2(\Omega))$, there exists a weak solution of the Dirichlet problem (22.1)–(22.3) in the sense of Definition 22.1.*

22.2 Bibliographical Notes

The contents of this lecture is taken from [3, 15, 16, 19, 22, 23, 46].

Lecture 23

Galerkin Method
and Method of Compactness.
Parabolic Equations, II

23.1 Parabolic Equation with the p-Laplacian

Consider the following initial-boundary-value problem for the parabolic equation in a bounded domain Ω with sufficiently smooth boundary $\partial\Omega$:

$$\frac{\partial u}{\partial t} - \operatorname{div}(|D_x u|^{p-2} D_x u) = f(x,t), \qquad (x,t) \in D = \Omega \otimes (0,T),$$

$$u(x,t) = 0, \qquad\qquad\qquad\qquad (x,t) \in \partial\Omega \otimes [0,T], \qquad (23.1)$$

$$u(x,0) = u_0(x) \qquad\qquad\qquad\qquad x \in \Omega, \quad p > 2.$$

Definition 23.1. A *weak solution* of the problem (23.1) is a function $u(t)$ of the class $u(t) \in L^p(0,T; W_0^{1,p}(\Omega))$, $u'(t) \in L^{p'}(0,T; W^{-1,p'}(\Omega))$[1] satisfying the condition

$$\int_0^T \langle L(u), w \rangle \, dt = \int_0^T \langle f, w \rangle \, dt \qquad (23.2)$$

for any $w(t) \in L^p(0,T; W_0^{1,p}(\Omega))$, where

$$L(u) \stackrel{\text{def}}{=} u' - \operatorname{div}(|D_x u|^{p-2} D_x u),$$

$\langle \cdot, \cdot \rangle$ is the duality bracket between the Banach spaces $W_0^{1,p}(\Omega)$ and $W^{-1,p'}(\Omega)$.

Remark 23.1. Introduce the vector space[2]

$$W_{pp'}(0,T) \stackrel{\text{def}}{=} \Big\{ u(t) : \ u(t) \in L^p(0,T; W_0^{1,p}(\Omega)),$$

$$u'(t) \in L^{p'}(0,T; W^{-1,p'}(\Omega)) \Big\}, \quad p' = \frac{p}{p-1}, \quad p > 2, \quad (23.3)$$

which is a Banach space with respect to the norm

$$\|u\|_{pp'} \stackrel{\text{def}}{=} \left(\int_0^T \|u\|_{W_0^{1,p}(\Omega)}^p \, dt \right)^{1/p} + \left(\int_0^T \|u'\|_{W^{-1,p'}(\Omega)}^{p'} \, dt \right)^{1/p'}. \qquad (23.4)$$

---

[1] The derivative u' is treated in the weak sense (see Definition 4.6).
[2] We use the notation from Lecture 4.

Theorem 23.1. *Let functions $f(t)$ and u_0 satisfy the conditions*

$$f(t) \in L^{p'}(0,T; W^{-1,p'}(\Omega)), \quad \frac{1}{p} + \frac{1}{p'} = 1, \quad p > 2, \tag{23.5}$$

and $u_0 \in L^2(\Omega)$. Then there exists a unique weak solutions $u(t)$ of the problem (23.2) in the class

$$u(t) \in L^p(0,T; W_0^{1,p}(\Omega)), \quad u'(t) \in L^{p'}(0,T; W^{-1,p'}(\Omega)). \tag{23.6}$$

Proof. 1. Theorem 21.2 implies that the p-Laplacian acts as follows:

$$A(v) \equiv -\Delta_p v : L^p(0,T; W_0^{1,p}(\Omega)) \to L^{p'}(0,T; W^{-1,p'}(\Omega)), \quad p > 2. \tag{23.7}$$

Then from (23.2) we obtain

$$u'(t) = f(t) + \operatorname{div}\left(|D_x u|^{p-2} D_x u\right) \in L^{p'}(0,T; W^{-1,p'}(\Omega)). \tag{23.8}$$

Therefore, if a weak solution of the problem considered exists, then $u(t) \in W_{pp'}(0,T)$ and hence by Theorem 4.4 we have $W_{pp'}(0,T) \subset C([0,T]; L^2(\Omega))$ and consequently $u(t) \in C([0,T]; L^2(\Omega))$; in particular the initial value $u(0) \in L^2(\Omega)$ is defined.

2. Galerkin approximations. Let $\{w_j\}_{j=1}^{+\infty} \subset W_0^{1,p}(\Omega)$ be a countable set such that finite linear combinations of its elements are dense in $W_0^{1,p}(\Omega)$. Without loss of generality, we may assume that the functional family $\{w_j(x)\}$ is orthonormal:

$$\int_\Omega w_{j_1}(x) w_{j_2}(x)\, dx = \delta_{j_1 j_2}, \tag{23.9}$$

where $\delta_{j_1 j_2}$ is the Kronecker delta.

Introduce the Galerkin approximations $u_m(t)$ for the problem (23.2) as follows:

$$\left(u_m'(t), w_j\right)_2 + \langle A(u_m(t)), w_j \rangle = \langle f(t), w_j \rangle, \quad 1 \le j \le m, \tag{23.10}$$

where

$$u_m(t) := \sum_{k=1}^m c_{mk}(t) w_k, \quad c_{mk}(t) \in C^{(1)}[0, T_m],$$

and

$$u_m(0) = u_{0m} := \sum_{k=1}^m c_{mk}(0) w_k, \quad u_{0m} \to u_0 \quad \text{strongly in } L^2(\Omega) \text{ as } m \to +\infty, \tag{23.11}$$

where $(\cdot, \cdot)_2$ is the scalar product in $L^2(\Omega)$.

The system of ordinary differential equations (23.10) can be rewritten as follows:

$$\frac{dc_{mj}(t)}{dt} = f_j(c_{m1}, \dots, c_{mm}, t) := -\langle A(u_m(t)), w_j \rangle + \langle f(t), w_j \rangle, \quad j = \overline{1, m}; \tag{23.12}$$

the initial conditions have the form

$$c_{mk}(0) := \alpha_{mk}, \quad \sum_{k=1}^m \alpha_{mk} w_k \to u_0 \quad \text{strongly in } L^2(\Omega). \tag{23.13}$$

In Lecture 22, the following assertion was actually proved.

Theorem 23.2. *For any $c_m(0) = (c_{m1}(0), \ldots, c_{mm}(0))$, there exists a positive number $T_0 = T_0(c_{m1}(0), \ldots, c_{mm}(0))$ such that for any $T \in (0, T_0)$, there exists a unique classical solution $c_m(t) \in C^{(1)}[0,T] \otimes \cdots \otimes C^{(1)}[0,T]$ of the Cauchy problem for the system (23.12). Moreover, either $T_0 = +\infty$ or $T_0 < +\infty$, and in the latter case we have*

$$\lim_{t \uparrow T_0} \sqrt{\sum_{k=1}^{m} c_{mk}^2(t)} = +\infty. \tag{23.14}$$

Thus, for any $m \in \mathbb{N}$, there exists $T_0 > 0$ such that there exist Galerkin approximations $u_m(t) \in C^{(1)}([0, T_0); W_0^{1,p}(\Omega))$.

3. A priori estimates. First, we note that

$$\langle A(u), u \rangle = \langle -\Delta_p u, u \rangle = |||D_x u|||_p^p, \tag{23.15}$$

where $\|\cdot\|_p$ is the norm in $L^p(\Omega)$:

$$\|v\|_p \overset{\text{def}}{=} \left(\int_\Omega |v|^p \, dx \right)^{1/p}.$$

The norm in $L^2(\Omega)$ is denoted by $|\cdot|$. Multiplying (23.10) by $c_{mj}(t)$ and summing by $j = \overline{1, m}$, we obtain

$$\frac{1}{2} \frac{d}{dt} |u_m(t)|^2 + \langle A(u_m), u_m \rangle = \langle f(t), u_m \rangle.$$

By the definition of the p-Laplacian, the following relations hold:

$$\langle A(u_m), u_m \rangle = |||D_x u_m|||_p^p, \quad \langle f(t), u_m \rangle \le \|f\|_* |||D_x u_m|||_p.$$

Integrating by time, we arrive at the inequality

$$\frac{1}{2} |u_m(t)|^2 + \int_0^t |||D_x u_m(s)|||_p^p \, ds \le \int_0^t \|f(s)\|_* |||D_x u_m(s)|||_p \, ds + \frac{1}{2} |u_{0m}|^2. \tag{23.16}$$

Using the three-parameter Young inequality (11.18), from the inequality (23.16) we obtain the a priori estimate

$$\frac{1}{2} |u_m(t)|^2 + (1 - \varepsilon) \int_0^t |||D_x u_m(s)|||_p^p \, ds \le c(\varepsilon) \int_0^t \|f(s)\|_*^{p'} \, ds + \frac{1}{2} |u_{0m}|^2 \tag{23.17}$$

for $\varepsilon \in (0, 1)$. By the definition of Galerkin approximations $u_m(t)$, the limit relation (23.11) holds; therefore, in particular, the numerical sequence $\{|u_{0m}|\}$ converges and hence is bounded. Since $f(t) \in L^{p'}(0, T; W^{-1, p'}(\Omega))$, the right-hand side of the inequality (23.17) is bounded by a constant $M_1(T) > 0$ for all $t \in [0, T]$. Thus, we obtain the following a priori estimate:

$$|u_m(t)|^2 \le 2M_1(T), \quad t \in [0, T]. \tag{23.18}$$

Note that, due to (23.9), the following chain of equalities holds:

$$|u_m(t)|^2 = \sum_{j_1=1}^{m} \sum_{j_2=1}^{m} c_{mj_1}(t) c_{mj_2}(t) \int_{\Omega} w_{j_1}(x) w_{j_2}(x) \, dx = \sum_{j=1}^{m} c_{mj}^2(t). \qquad (23.19)$$

From (23.18) and (23.19) we obtain

$$\sum_{j=1}^{m} c_{mj}^2(t) \leq 2M_1(T), \quad t \in [0,T]. \qquad (23.20)$$

By Theorem 23.2, we conclude that a classical solution $c_m(t) \in C^{(1)}[0,T] \otimes \cdots \otimes C^{(1)}[0,T]$ of the Cauchy problem (23.11)–(23.12) exists on the whole segment $[0,T]$ where the function $f(t)$ is defined (see (23.5)).

Thus, it follows from (23.17) that the sequence

$$\{u_m\}_{m=1}^{+\infty} \quad \text{is bounded in } L^\infty(0,T;L^2(\Omega)) \cap L^p(0,T;W_0^{1,p}(\Omega)). \qquad (23.21)$$

By Lemma 21.4 (see (21.7)), for any function $v \in L^p(0,T;W_0^{1,p}(\Omega))$ we have

$$\|A(v)\|_* = \||D_x v|\|_p^{p-1} \quad \Rightarrow \quad \|A(v)\|_*^{p'} = b\||D_x v|\|_p^p$$

$$\Rightarrow \quad \int_0^T \|A(v)\|_*^{p'} \, dt = \int_0^T \||D_x v|\|_p^p \, dt;$$

therefore, the sequence

$$\{A(u_m)\}_{m=1}^{+\infty} \quad \text{is bounded in } L^{p'}(0,T;W^{-1,p'}(\Omega)). \qquad (23.22)$$

4. Passing to the limit. Due to (23.18), (23.21) and (23.22), we can extract a subsequence $\{u_\mu\}$ such that

$$u_\mu \overset{*}{\rightharpoonup} u \qquad *\text{-weakly in } L^\infty(0,T;L^2(\Omega)), \qquad (23.23)$$

$$u_\mu \rightharpoonup u \qquad \text{weakly in } L^p(0,T;W_0^{1,p}(\Omega)), \qquad (23.24)$$

$$u_\mu(T) \rightharpoonup \xi \quad \text{weakly in } L^2(\Omega), \qquad (23.25)$$

$$A(u_\mu) \rightharpoonup \chi \quad \text{weakly in } L^{p'}(0,T;W^{-1,p'}(\Omega)) \qquad (23.26)$$

as $\mu \to +\infty$.

First, we prove (23.23). As is well known,

$$L^\infty(0,T;L^2(\Omega)) = \left(L^1(0,T;L^2(\Omega))\right)^*.$$

Due to (23.21), the sequence $\{u_m\}$ is bounded in $L^\infty(0,T;L^2(\Omega))$. Since the space $L^1(0,T;L^2(\Omega))$ is separable, Theorem 2.2 implies the existence of a subsequence $\{u_\mu\} \subset \{u_m\}$ such that the limit relation (23.23) holds.

Further, we prove that (23.23) implies that for $p > 1$,

$$u_\mu \rightharpoonup u \quad \text{weakly in } L^p(0,T;L^2(\Omega)) \text{ as } \mu \to +\infty. \qquad (23.27)$$

Indeed, the limit relation (23.23) means that

$$\int_0^T f(t) \left[u_\mu(t) - u(t)\right] dt \to 0 \quad \text{as } \mu \to +\infty \qquad (23.28)$$

for any $f(t) \in L^1(0, T; L^2(\Omega))$. Recall the following continuous embedding:

$$L^{p'}(0, T; L^2(\Omega)) \subset L^1(0, T; L^2(\Omega)). \tag{23.29}$$

Therefore, (23.28) holds for all $f(t) \in L^{p'}(0, T; L^2(\Omega))$. For $p > 1$, the Banach space $L^p(0, T; L^2(\Omega))$ is reflexive and hence the notion of $*$-weak convergence in it coincides with the notion of weak convergence. Thus, (23.27) holds.

Now we prove that there exists a function $v(t) \in L^p(0, T; W_0^{1,p}(\Omega))$ such that we can extract from the subsequence $\{u_\mu\} \subset \{u_m\}$ another subsequence (we denote it by the same symbol $\{u_\mu\}$) for which the following limit relation holds:

$$u_\mu \rightharpoonup v \quad \text{weakly in } L^p(0, T; W_0^{1,p}(\Omega)) \text{ as } \mu \to +\infty. \tag{23.30}$$

Remark 23.2. In what follows, we will extract subsequences and preserve the same notation for them without additional prompts.

Indeed, by (23.21), the sequence $\{u_m\}$ is bounded in $L^p(0, T; W_0^{1,p}(\Omega))$ and the Banach space $L^p(0, T; W_0^{1,p}(\Omega))$ $(p > 2)$ is separable and reflexive; therefore (see Theorem 2.1), there exists a subsequence $\{u_\mu\} \subset \{u_m\}$ such that the limit relation (23.30) holds.

Now we prove that $v(t) = u(t)$. The duality bracket between $L^p(0, T; W_0^{1,p}(\Omega))$ and $L^{p'}(0, T; W^{-1,p'}(\Omega))$ have the form

$$\int_0^T \langle f^*(t), w(t) \rangle \, dt \tag{23.31}$$

for all $f^*(t) \in L^{p'}(0, T; W^{-1,p'}(\Omega))$ and $w(t) \in L^p(0, T; W_0^{1,p}(\Omega))$, where $\langle \cdot, \cdot \rangle$ is the duality bracket between $W_0^{1,p}(\Omega)$ and $W^{-1,p'}(\Omega)$. Note that the space $W_0^{1,p}(\Omega)$ is reflexive and

$$W_0^{1,p}(\Omega) \overset{ds}{\subset} L^2(\Omega) \quad \Rightarrow \quad L^2(\Omega) \overset{ds}{\subset} W^{-1,p'}(\Omega).$$

Hence we have the following equality for the duality brackets (see Theorem 1.4):

$$\langle f^*, w \rangle = \int_\Omega f^*(x)w(x) \, dx, \quad f^*(x), w(x) \in L^2(\Omega). \tag{23.32}$$

Thus, for any $f^*(t) \in L^{p'}(0, T; L^2(\Omega))$ and $w(t) \in L^p(0, T; L^2(\Omega))$, we have the equality

$$\int_0^T \langle f^*(t), w(t) \rangle \, dt = \int_0^T \int_\Omega f^*(x, t)w(x, t) \, dx \, dt. \tag{23.33}$$

In particular,

$$\int_0^T \langle f^*(t), u_\mu(t) - v(t) \rangle \, dt = \int_0^T \int_\Omega f^*(x, t) \left[u_\mu(x, t) - v(x, t) \right] \, dx \, dt \tag{23.34}$$

and the left-hand side of this equality tends to zero as $\mu \to +\infty$. Thus,

$$u_\mu \rightharpoonup v \quad \text{weakly in } L^p(0, T; L^2(\Omega)) \text{ as } \mu \to +\infty. \tag{23.35}$$

Due to (23.27), we have

$$u_\mu \rightharpoonup u \quad \text{weakly in } L^p(0, T; L^2(\Omega)) \text{ as } \mu \to +\infty. \tag{23.36}$$

Therefore, $v(t) = u(t)$.

Now we prove (23.25). From (23.18) we conclude that the sequence $\{u_m(T)\}$ is bounded with respect to the norm of the Banach space $L^2(\Omega)$. By Theorem 2.1, there exist an element $\xi(x) \in L^2(\Omega)$ and a subsequence $\{u_\mu\} \subset \{u_n\}$ such that the limit relation (23.25) holds.

Finally, prove (23.26). Due to (23.22), the reflexivity of the Banach space $L^{p'}(0, T; W^{-1,p'}(\Omega))$, and Theorem 2.1, we conclude that there exists a subsequence $\{\Delta_p u_\mu\} \subset \{\Delta_p u_m\}$ such that the limit relation (23.26) holds.

Thus, all limit relations (23.23)–(23.26) are proved. In particular, we have

$$u(t) \in L^p(0, T; W_0^{1,p}(\Omega)), \tag{23.37}$$

$$\xi(x) \in L^2(\Omega), \quad \chi(t) \in L^{p'}(0, T; W^{-1,p'}(\Omega)). \tag{23.38}$$

5. Boundary conditions for $u(t)$ at $t = 0$ and $t = T$. We prolong the functions $u_m(t)$, $u(t)$, $u_m'(t)$, $A(u_m(t))$, f, and χ to $\mathbb{R}^1$ by zero outside $[0, T]$ and denote the corresponding extensions by $\overline{u}_m(t)$, $\overline{u}(t)$, $\overline{u_m'}(t)$ $\overline{A(u_m(t))}$, $\overline{f}$, and $\overline{\chi}$, respectively. We prove the equality

$$(\overline{u}_m'(t), w_j)_2 = (\overline{u_m'}(t), w_j)_2 \quad \text{for } t \neq 0, \, t \neq T, \, j = \overline{1, m}. \tag{23.39}$$

Indeed, for $t \neq 0$ and $t \neq T$ we have

$$\overline{u_m'}(t) = \begin{cases} u_m'(t) & \text{for } t \in (0, T), \\ 0 & \text{for } t \in \mathbb{R}^1 \backslash [0, T] \end{cases} = \{\overline{u}_m'(t)\}.$$

Then from (23.10) we deduce that for $t \neq 0$ and $t \neq T$

$$(\{\overline{u}_m'(t)\}, w_j)_2 + \langle \overline{A(u_m(t))}, w_j \rangle = \langle \overline{f}(t), w_j \rangle, \quad 1 \leq j \leq m. \tag{23.40}$$

Using the well-known result on the relation between classical derivatives with derivatives of generalized functions[3] from $\mathscr{D}'(\mathbb{R}^1)$, we obtain the following equality:

$$\frac{d}{dt}(\overline{u}_m(t), w_j)_2 = (\{\overline{u}_m'(t)\}, w_j)_2 + \left([\overline{u}_m(0 + 0) - \overline{u}_m(0 - 0)], \, w_j\right)_2 \delta(t)$$
$$+ \left([\overline{u}_m(T + 0) - \overline{u}_m(T - 0)], \, w_j\right)\delta(t - T)$$
$$= \left(\{\overline{u}_m'(t)\}, \, w_j\right)_2 + (u_m(0), w_j)_2 \delta(t) - (u_m(T), w_j)_2 \delta(t - T) \tag{23.41}$$

for all $j = \overline{1, m}$; here we used the equalities

$$\overline{u}_m(0 - 0) = \overline{u}_m(T + 0) = 0, \quad \overline{u}_m(0 + 0) = u_m(0), \quad \overline{u}_m(T - 0) = u_m(T),$$

---

[3] Here the derivative d/dt is treated in the sense of generalized functions from $\mathscr{D}'(\mathbb{R}^1)$.

where $\{\overline{u}'_m(t)\}$ coincides with the classical derivative $u'_m(t)$ for $t \neq 0$ and $t \neq T$ and at the points $t = 0$ and $t = T$ it is defined arbitrarily. Then

$$\left(\{\overline{u}'_m(t)\}, \, w_j\right)_2 = \frac{d}{dt}\left(\overline{u}_m(t), w_j\right)_2$$
$$- \left(u_m(0), w_j\right)_2 \delta(t) + \left(u_m(T), w_j\right)_2 \delta(t - T), \quad j = \overline{1, m}. \tag{23.42}$$

Combining (23.40) and (23.42), we obtain the equality

$$\frac{d}{dt}\left(\overline{u}_m(t), w_j\right)_2 + \left\langle \overline{A(u_m(t))}, w_j \right\rangle \tag{23.43}$$

$$= \left\langle \overline{f}(t), w_j \right\rangle + \left(u_m(0), w_j\right)_2 \delta(t) - \left(u_m(T), w_j\right)_2 \delta(t - T), \tag{23.44}$$

which holds for all $t \in \mathbb{R}^1$ in the sense of generalized functions from $\mathscr{D}'(\mathbb{R}^1)$. For example, the following equality holds:

$$\left\langle\!\!\left\langle \frac{d}{dt}\left(\overline{u}_m(t), w_j\right)_2, \, \varphi(t) \right\rangle\!\!\right\rangle = - \left\langle\!\!\left\langle \left(\overline{u}_m(t), w_j\right)_2, \, \varphi'(t) \right\rangle\!\!\right\rangle$$

for all $\varphi(t) \in \mathscr{D}(\mathbb{R}^1)$, where $\langle\!\langle \cdot, \cdot \rangle\!\rangle$ is the duality bracket between $\mathscr{D}(\mathbb{R}^1)$ and $\mathscr{D}'(\mathbb{R}^1)$.

Now, passing to the limit in (23.44)[4] as $m = \mu \to +\infty$ and fixed j, we arrive at the following equality treated in the sense of the space $\mathscr{D}'(\mathbb{R}^1)$:

$$\frac{d}{dt}\left(\overline{u}, w_j\right)_2 + \left\langle \overline{\chi}, w_j \right\rangle = \left\langle \overline{f}, w_j \right\rangle + \left(u_0, w_j\right)_2 \delta(t) - \left(\xi, w_j\right)_2 \delta(t - T) \tag{23.45}$$

for all $j \in \mathbb{N}$. Indeed, by (23.23) we have

$$\left\langle\!\!\left\langle \frac{d}{dt}(\overline{u}_\mu(t), w_j)_2, \, \varphi(t) \right\rangle\!\!\right\rangle = - \left\langle\!\!\left\langle (\overline{u}_\mu(t), w_j)_2, \, \varphi'(t) \right\rangle\!\!\right\rangle$$

$$= - \int_{\mathbb{R}^1} \left(\overline{u}_\mu(t), w_j\right)_2 \varphi'(t) \, dt = - \int_{\mathbb{R}^1} \left(\overline{u}_\mu(t), \varphi'(t)w_j\right)_2 dt$$

$$\to - \int_{\mathbb{R}^1} \left(\overline{u}(t), \varphi'(t)w_j\right)_2 dt = - \int_{\mathbb{R}^1} \left(\overline{u}(t), w_j\right)_2 \varphi'(t) \, dt$$

$$= \left\langle\!\!\left\langle \frac{d}{dt}\left(\overline{u}(t), w_j\right)_2, \, \varphi(t) \right\rangle\!\!\right\rangle \tag{23.46}$$

as $\mu \to +\infty$ for all $\varphi(t) \in \mathscr{D}(\mathbb{R}^1)$. Similarly, due to (23.26), we obtain the following chain of relations:

$$\left\langle\!\!\left\langle \overline{A(u_\mu)}, w_j \right\rangle, \varphi(t) \right\rangle\!\!\right\rangle = \int_{\mathbb{R}^1} \left\langle \overline{A(u_\mu)}, w_j \right\rangle \varphi(t) \, dt = \left\langle \overline{A(u_\mu)}, \varphi(t)w_j \right\rangle dt$$

$$\to \int_{\mathbb{R}^1} \left\langle \overline{\chi}(t), w_j \right\rangle \varphi(t) \, dt = \int_{\mathbb{R}^1} \left\langle \overline{\chi}(t), \varphi(t)w_j \right\rangle dt$$

$$= \left\langle\!\!\left\langle \left\langle \overline{\chi}(t), w_j \right\rangle, \, \varphi(t) \right\rangle\!\!\right\rangle \tag{23.47}$$

---

[4]Equation (23.44) is considered here in the space $\mathscr{D}'(\mathbb{R}^1)$.

as $\mu \to +\infty$ for all $\varphi(t) \in \mathscr{D}(\mathbb{R}^1)$. Moreover, by (23.11) and (23.25), the following limit relations hold for all $\varphi(t) \in \mathscr{D}(\mathbb{R}^1)$:

$$\langle\!\langle (u_\mu(0), w_j)_2 \, \delta(t), \; \varphi(t) \rangle\!\rangle = \varphi(0) \, (u_\mu(0), w_j)_2$$
$$\to \varphi(0) \, (u_0, w_j)_2 = \langle\!\langle (u_0, w_j)_2 \, \delta(t), \; \varphi(t) \rangle\!\rangle \quad (23.48)$$

$$\langle\!\langle (u_\mu(T), w_j)_2 \, \delta(t - T), \; \varphi(t) \rangle\!\rangle = \varphi(T) \, (u_\mu(T), w_j)_2$$
$$\to \varphi(T) \, (\xi, w_j)_2 = \langle\!\langle (\xi, w_j)_2 \, \delta(t - T), \; \varphi(t) \rangle\!\rangle \quad (23.49)$$

as $\mu \to +\infty$.

Now we note that, since $\overline{u}(t) \in L^\infty(\mathbb{R}^1; L^2(\Omega))$ and $w_j \in L^2(\Omega)$, the following chain of equalities holds:

$$\left\langle\!\!\left\langle \frac{d}{dt}(\overline{u}(t), w_j)_2, \; \varphi(t) \right\rangle\!\!\right\rangle = - \left\langle\!\!\left\langle (\overline{u}(t), w_j)_2, \; \varphi'(t) \right\rangle\!\!\right\rangle$$

$$= - \int_{\mathbb{R}^1} (\overline{u}(t), w_j)_2 \varphi'(t) \, dt = - \left(\int_0^T \overline{u}(t)\varphi'(t) \, dt, w_j \right)_2$$

$$= \left(\left\langle\!\!\left\langle \frac{d}{dt}\overline{u}(t), \varphi(t) \right\rangle\!\!\right\rangle_{\mathscr{D}(\mathbb{R}^1; W^{-1,p'}(\Omega))}, \; w_j \right)_2$$

$$= \left\langle \left\langle\!\!\left\langle \frac{d}{dt}\overline{u}(t), \; \varphi(t) \right\rangle\!\!\right\rangle_{\mathscr{D}(\mathbb{R}^1; W^{-1,p'}(\Omega))}, \; w_j \right\rangle, \quad (23.50)$$

where the duality bracket $\langle\!\langle \cdot, \cdot \rangle\!\rangle_{\mathscr{D}(\mathbb{R}^1; W^{-1,p'}(\Omega))}$ means the action of an element of the space $\mathscr{D}'(\mathbb{R}^1; W^{-1,p'}(\Omega))$ on an element of the space $\mathscr{D}(\mathbb{R}^1)$, and the result of this action is an element of the space $W^{-1,p'}(\Omega)$. Therefore, we can rewrite Eq. (23.45) as follows:

$$\left\langle \left\langle\!\!\left\langle \frac{d}{dt}\overline{u}(t) + \overline{\chi}(t) - \overline{f}(t) - u_0\delta(t) + \xi\delta(t - T), \; \varphi(t) \right\rangle\!\!\right\rangle_{\mathscr{D}(\mathbb{R}^1; W^{-1,p'}(\Omega))}, \; w_j \right\rangle = 0 \quad (23.51)$$

for all $j \in \mathbb{N}$ and all $\varphi(t) \in \mathscr{D}(\mathbb{R}^1)$, and hence we obtain the equality

$$\frac{d\overline{u}}{dt} + \overline{\chi} = \overline{f} + u_0\delta(t) - \xi\delta(t - T) \quad (23.52)$$

treated in the sense of the space $\mathscr{D}'(\mathbb{R}^1; W^{-1,p'}(\Omega))$. Note that $\varphi(t) \in \mathscr{D}(0,T) \subset \mathscr{D}(\mathbb{R}^1)$ if we extend the function $\varphi(t)$ by zero outside $(0,T)$. Restricting (23.52) to test functions from $\mathscr{D}(0,T)$, we obtain the following equality treated in the sense of $\mathscr{D}'((0,T); W^{-1,p'}(\Omega))$:

$$\frac{du}{dt} = f(t) - \chi(t). \quad (23.53)$$

The right-hand side of this equality belongs to the space $L^{p'}(0,T; W^{-1,p'}(\Omega))$. For any $\varphi(t) \in \mathscr{D}(0,T)$ we have the following equalities:

$$\int_0^T [f(t) - \chi(t)]\varphi(t) \, dt = \left\langle\!\!\left\langle \frac{du}{dt}, \varphi(t) \right\rangle\!\!\right\rangle = - \int_0^T u(t)\varphi'(t) \, dt \quad (23.54)$$

since
$$f(t) - \chi(t) \in L^{p'}(0, T; W^{-1,p'}(\Omega)) \subset L^1(0, T; W^{-1,p'}(\Omega)),$$
$$u(t) \in L^p(0, T; W_0^{1,p}(\Omega)) \subset L^1(0, T; W^{-1,p'}(\Omega)).$$

Therefore, the function $u(t)$ possesses a weak derivative $u'(t)$ and the following equality holds:
$$u'(t) = f(t) - \chi(t) \in L^{p'}(0, T; W^{-1,p'}(\Omega)). \tag{23.55}$$

Therefore, due to (23.37), we have $u(t) \in W_{pp'}(0, T) \subset C([0, T]; L^2(\Omega))$. Thus, the inclusions $u(0) \in L^2(\Omega)$ and $u(T) \in L^2(\Omega)$ make sense.

Note that, on the one hand, by the definition of the derivative of a $W^{-1,p'}(\Omega)$-valued generalized function and Corollary 4.1 of Theorem 4.6, we have the equality
$$\frac{d\overline{u}(t)}{dt} = \{\overline{u}'(t)\} + u(0)\delta(t) - u(T)\delta(t - T). \tag{23.56}$$

On the other hand, from (23.55) we obtain the equality
$$\{\overline{u}'(t)\} = \begin{cases} f - \chi & \text{for } t \in (0, T), \\ 0 & \text{for } t \in \mathbb{R}^1 \backslash (0, T) \end{cases} = \overline{f} - \overline{\chi}, \tag{23.57}$$

which together with (23.52) implies
$$\frac{d\overline{u}(t)}{dt} = \{\overline{u}'(t)\} + u_0\delta(t) - \xi\delta(t - T). \tag{23.58}$$

Comparing (23.56) with (23.58), we conclude that
$$u(0) = u_0, \quad u(T) = \xi. \tag{23.59}$$

It remains to note that $u_0(x), \xi(x) \in L^2(\Omega) \subset W^{-1,p'}(\Omega)$.

6. Method of monotonicity. To prove the existence of a weak solution of the problem considered, we must show that $\chi = A(u)$. The monotonicity property of the operator A implies that
$$X_\mu := \int_0^T \langle A(u_\mu) - A(v(t)), u_\mu(t) - v(t) \rangle \, dt \geq 0 \tag{23.60}$$

for all $v(t) \in L^p(0, T; W_0^{1,p}(\Omega))$. Multiplying both sides of (23.10) by c_{mj}, summing by $j = \overline{1, m}$, and integrating by $t \in (0, T)$, we obtain the equality
$$\int_0^T \langle A(u_\mu), u_\mu \rangle \, dt = \int_0^T \langle f, u_\mu \rangle \, dt + \frac{1}{2}|u_{0\mu}|^2 - \frac{1}{2}|u_\mu(T)|^2.$$

Therefore,
$$X_\mu = \int_0^T \langle A(u_\mu), u_\mu \rangle \, dt - \int_0^T \langle A(u_\mu), v \rangle \, dt - \int_0^T \langle A(v), u_\mu - v \rangle \, dt$$
$$= \int_0^T \langle f, u_\mu \rangle \, dt + \frac{1}{2}|u_{0\mu}|^2 - \frac{1}{2}|u_\mu(T)|$$
$$- \int_0^T \langle A(u_\mu), v \rangle \, dt - \int_0^T \langle A(v), u_\mu - v \rangle \, dt. \tag{23.61}$$

Since the norm of a reflexive Banach space is weakly lower semicontinuous, by (23.25) and (23.59) we have[5]

$$\liminf_{\mu \to +\infty} |u_\mu(T)| \geq |\xi| = |u(T)| \quad \text{as } \mu \to +\infty.$$

Moreover,

$$\limsup_{\mu \to +\infty} \left(- |u_\mu(T)| \right) = - \liminf_{\mu \to +\infty} |u_\mu(T)| \leq -|u(T)|. \tag{23.62}$$

From (23.11), (23.24), (23.26), (23.60), (23.61), and (23.62) we obtain

$$0 \leq \limsup_{\mu \to +\infty} X_\mu \leq \int_0^T \langle f, u \rangle \, dt + \frac{1}{2}|u_0|^2 - \frac{1}{2}|u(T)|^2$$

$$- \int_0^T \langle \chi, v \rangle \, dt - \int_0^T \langle A(v), u - v \rangle \, dt. \tag{23.63}$$

From (23.55) and Theorem 4.4 we conclude that

$$\langle u' + \chi - f, u \rangle = 0 \quad \Rightarrow \quad \frac{1}{2}\frac{d}{dt}|u|^2 + \langle \chi, u \rangle = \langle f, u \rangle,$$

and

$$\int_0^T \langle f, u \rangle \, dt + \frac{1}{2}|u_0|^2 - \frac{1}{2}|u(T)|^2 = \int_0^T \langle \chi, u \rangle \, dt. \tag{23.64}$$

7. Radial continuity of Δ_p. Comparing Eq. (23.64) with (23.63), we obtain

$$0 \leq \limsup_{\mu \to +\infty} X_\mu \leq \int_0^T \langle \chi, u \rangle \, dt - \int_0^T \langle \chi, v \rangle \, dt - \int_0^T \langle A(v), u - v \rangle \, dt$$

$$= \int_0^T \langle \chi - A(v), u - v \rangle \, dt. \tag{23.65}$$

We set $v = u - \lambda w$, where $\lambda > 0$ and a function $w \in L^p(0, T; W_0^{1,p}(\Omega))$ is arbitrary. Then from (23.65) we have

$$\lambda \int_0^T \langle \chi - A(u - \lambda w), w \rangle \, dt \geq 0, \tag{23.66}$$

which implies

$$\int_0^T \langle \chi - A(u - \lambda w), w \rangle \, dt \geq 0. \tag{23.67}$$

---

[5] Recall that $|\cdot|$ is the norm of the Banach space $L^2(\Omega)$.

Letting λ tend to zero in (23.67), due to the radial continuity of the operator $A(u)$ we obtain

$$\int_0^T \langle \chi - A(u), w \rangle \, dt \geq 0 \tag{23.68}$$

for all $w \in L^p(0, T; W_0^{1,p}(\Omega))$. In particular, the expression on the left-hand side of the formula (23.68) does not change its sign after multiplication of w by (-1). This implies that $\chi = A(u)$.

8. Uniqueness. Now we prove the uniqueness of the weak solution of the problem considered. Let u_1 and u_2 be two solutions of the class $L^p(0, T; W_0^{1,p}(\Omega))$. Then their difference $w = u_1 - u_2$ satisfies the problem

$$w' + A(u_1) - A(u_2) = 0, \quad w(0) = 0,$$

so that

$$\langle w', w \rangle + \langle A(u_1) - A(u_2), u_1 - u_2 \rangle = 0.$$

Due to the monotonicity of the operator A we have

$$\langle w', w \rangle \leq 0.$$

Thus, by Theorem 4.4,

$$\langle w', w \rangle = \frac{1}{2} \frac{d}{dt} |w(t)|^2 \leq 0 \quad \Rightarrow \quad |w|^2(t) \leq |w(0)|^2 = 0 \quad \Rightarrow \quad w(t) = 0.$$

The theorem is proved. $\qquad\square$

23.2 Bibliographical Notes

The contents of this lecture is taken from [15, 16, 46].

Lecture 24

Galerkin Method
and Method of Compactness.
Hyperbolic Equations

24.1 Nonlinear Hyperbolic Equations

Let Ω be a bounded domain in $\mathbb{R}^3$ with smooth boundary $\partial\Omega \in C^{2,\delta}$, $\delta \in (0,1]$.
Consider the following problem in the classical statement:

$$\frac{\partial^2 u}{\partial t^2} - \Delta u + |u|^q u = 0, \qquad\qquad (x,t) \in D = \Omega \otimes (0,T], \qquad (24.1)$$

$$u(x,t) = 0, \qquad\qquad (x,t) \in S = \partial\Omega \otimes [0,T], \qquad (24.2)$$

$$u(x,0) = u_0(x), \quad u'(x,0) = u_1(x), \quad x \in \Omega, \qquad\qquad (24.3)$$

where

$$\Delta u \overset{\text{def}}{=} \frac{\partial^2 u}{\partial x_1^2} + \frac{\partial^2 u}{\partial x_2^2} + \frac{\partial^2 u}{\partial x_3^2}, \quad x = (x_1, x_2, x_3).$$

Consider the generalized statement of the problem (24.1)–(24.3).

Definition 24.1. A *weak solution* of the problem (24.1)–(24.3) is a function
$u(t)$ of the class $u(t) \in L^\infty(0,T; H_0^1(\Omega))$, $u'(t) \in L^\infty(0,T; L^2(\Omega))$, $u''(t) \in L^\infty(0,T; H^{-1}(\Omega))$, satisfying the equality

$$\int_0^T dt \, \langle u'' - \Delta u + |u|^q u, v \rangle = 0 \qquad (24.4)$$

for all $v(t) \in L^1(0,T; H_0^1(\Omega))$ for $q \in (0,4]$, and the initial conditions

$$u(0) = u_0 \in H_0^1(\Omega), \quad u'(0) = u_1 \in L^2(\Omega), \qquad (24.5)$$

where $\langle \cdot, \cdot \rangle$ is the duality bracket between the Hilbert spaces $H_0^1(\Omega)$ and $H^{-1}(\Omega)$.

The problem (24.4)–(24.5) is equivalent to the following problem:

$$\int_0^T \varphi(t) \, \langle u'' - \Delta u + |u|^q u, w \rangle \, dt = 0, \quad \forall w \in H_0^1(\Omega), \quad \forall \varphi(t) \in L^1(0,T), \qquad (24.6)$$

$$u(0) = u_0 \in H_0^1(\Omega), \quad u'(0) = u_1 \in L^2(\Omega). \qquad (24.7)$$

Remark 24.1. Note that if for some $T > 0$ there exists a weak solution in the sense of Definition 24.1, then, in particular,

$$u(t) \in L^p(0,T; H_0^1(\Omega)), \quad u'(t) \in L^p(0,T; L^2(\Omega)), \quad u''(t) \in L^p(0,T; H^{-1}(\Omega))$$

for any $p \geq 1$. Therefore,

$$u(t) \in W_{pp}(0,T) := \left\{ v(t) \in L^p(0,T; H_0^1(\Omega)), \ u'(t) \in L^p(0,T; L^2(\Omega)) \right\},$$

$$u'(t) \in W_{pp}'(0,T) := \left\{ v(t) \in L^p(0,T; L^2(\Omega)), \ v'(t) \in L^p(0,T; H^{-1}(\Omega)) \right\}.$$

Note that the following continuous and dense embeddings holds:

$$H_0^1(\Omega) \overset{ds}{\subset} L^2(\Omega) \overset{ds}{\subset} H^{-1}(\Omega).$$

Applying Lemma 4.6 with

$$X = H_0^1(\Omega), \quad Y = Z = L^2(\Omega),$$

we obtain

$$W_{pp}(0,T) \subset C([0,T]; L^2(\Omega)). \tag{24.8}$$

Similarly, taking

$$X = L^2(\Omega), \quad Y = Z = H^{-1}(\Omega),$$

we obtain

$$W_{pp}'(0,T) \subset C([0,T]; H^{-1}(\Omega)). \tag{24.9}$$

Therefore, the initial conditions (24.5) make sense.

Theorem 24.1. *Let $u_0 \in H_0^1(\Omega)$, $u_1 \in L^2(\Omega)$, and $q \in (0,2]$. Then there exists a unique weak solution of the problem (24.1), (24.2) in the sense of Definition 24.1.*

Proof. 1. Approximate solutions. We consider the following problem on Galerkin approximations for the problem (24.6)–(24.7):

$$\int_0^{T_m} \varphi(t) \langle u_m'' - \Delta u_m + |u_m|^q u_m, w_j \rangle \, dt = 0, \quad j = \overline{1,m} \tag{24.10}$$

for all $\varphi(t) \in L^1(0, T_m)$, where

$$u_m(t) := \sum_{k=1}^m c_{mk}(t) w_k$$

are Galerkin approximations and $\{w_j\} \subset H_0^1(\Omega)$ is a basis consisting of eigenfunctions of the Laplace operator

$$\Delta w_j + \lambda_j w_j = 0, \quad w_j \in H_0^1(\Omega).$$

Without loss of generality, we may assume that this system of eigenfunctions is orthonormal:

$$\int_\Omega w_{j_1}(x)w_{j_2}(x)\,dx = \delta_{j_1 j_2}, \tag{24.11}$$

where $\delta_{j_1 j_2}$ is the Kronecker delta. Since $\{w_j\} \subset H_0^1(\Omega)$, we obtain from (24.11) the following chain of equalities:

$$\int_\Omega (D_x w_{j_1}(x), D_x w_{j_2}(x))\,dx = \langle -\Delta w_{j_1}, w_{j_2}\rangle$$

$$= \lambda_{j_1} \int_\Omega w_{j_1}(x)w_{j_2}(x)\,dx = \lambda_{j_1}\delta_{j_1 j_2}. \tag{24.12}$$

We supplement the system (24.10) with the following initial conditions:

$$c_{mk}(0) = \alpha_{mk}, \quad c'_{mk}(0) = \beta_{mk}, \quad k = \overline{1, m}, \tag{24.13}$$

where $\{\alpha_{mk}\}_{k=1}^m$ and $\{\beta_{mk}\}_{k=1}^m$ are such that

$$u_{m0} := \sum_{k=1}^m \alpha_{mk}w_k \to u_0 \quad \text{strongly in } H_0^1(\Omega) \text{ as } m \to +\infty, \tag{24.14}$$

$$u_{m1} := \sum_{k=1}^m \beta_{mk}w_k \to u_1 \quad \text{strongly in } L^2(\Omega) \text{ as } m \to +\infty. \tag{24.15}$$

We search for a solution of the system (24.10) in the class $c_{mk}(t) \in C^{(2)}[0, T_m]$.

2. Local solvability. Since $C_0^\infty(0, T_m) \subset L^1(0, T_m)$, we set $\varphi(t) \in C_0^\infty(0, T_m)$ in (24.6). In the class $c_{mk}(t) \in C^{(2)}[0, T_m]$ we have

$$\langle u_m'' - \Delta u_m + |u_m|^q u_m, w_j\rangle \in C[0, T_m].$$

Using the fundamental lemma of calculus of variations, we obtain the system of m ordinary second-order differential equations:

$$\langle u_m'' - \Delta u_m + |u_m|^q u_m, w_j\rangle = 0, \quad t \in [0, T_m], \quad j = \overline{1, m}. \tag{24.16}$$

Since $w_j \in H_0^1(\Omega) \overset{ds}{\subset} L^2(\Omega)$, by Theorem 1.4 we have

$$\langle w_k, w_j\rangle = (w_k, w_j)_2, \quad \langle -\Delta w_k, w_j\rangle = (D_x w_k, D_x w_j)_2,$$

where $(\cdot, \cdot)_2$ is the scalar product in $L^2(\Omega)$. Moreover, since $H_0^1(\Omega) \overset{ds}{\subset} L^{q+2}(\Omega)$ for $q \in [0, 4]$, we have by construction

$$u_m \in L^\infty(0, T_m; H_0^1(\Omega)) \subset L^\infty(0, T_m; L^{q+2}(\Omega)), \quad q \in [0, 4].$$

Therefore, we have

$$|u_m|^q u_m \in L^\infty(0, T_m; L^{(q+2)/(q+1)}(\Omega)).$$

On the other hand, $w_j \in H_0^1(\Omega) \subset L^{q+2}(\Omega)$. Therefore, by Theorem 1.4 we have

$$\langle |u_m|^q u_m, w_j\rangle = (|u_m|^q u_m, w_j)_2.$$

Hence, taking into account (24.11) and (24.12), we can rewrite the system (24.16) in the following equivalent form:

$$c''_{mj}(t) + \lambda_j c_{mj}(t) = F_j(c_m), \quad j = \overline{1, m}, \tag{24.17}$$

where

$$F_j(c_m) = F_j(c_{m1}, \ldots, c_{mm}) := -\int_\Omega (|u_m|^q u_m)(x, t) w_j(x)\, dx. \tag{24.18}$$

For arbitrary $c_m^1 = (c_{m1}^1, \ldots, c_{mm}^1)$ and $c_m^2 = (c_{m1}^2, \ldots, c_{mm}^2)$, the following estimate holds:

$$\left| F_j(c_m^1) - F_j(c_m^2) \right| \le \mu_j(R) |c_m^1 - c_m^2|; \tag{24.19}$$

here

$$\mu_j(R) = (q+1) \left(\int_\Omega \left(\sum_{k=1}^m |w_k(x)|^2 \right)^{(q+2)/2} dx \right)^{1/(q+2)} \|w_j\|_{q+2} R^q,$$

$$R = \max\left\{ \|u_m^1\|_{q+2}, \|u_m^2\|_{q+2} \right\},$$

$$u_m^1 := \sum_{k=1}^m c_{mk}^1 w_k, \quad u_m^2 := \sum_{k=1}^m c_{mk}^2 w_k.$$

Indeed, consider the inequality

$$\left| |u_m^1|^q u_m^1 - |u_m^2|^q u_m^2 \right| \le (q+1) \max\left\{ |u_m^1|^q, |u_m^2|^q \right\} |u_m^1 - u_m^2|. \tag{24.20}$$

We have

$$\left| F_j(c_m^1) - F_j(c_m^2) \right| \le (q+1) \int_\Omega \max\left\{ |u_m^1|^q, |u_m^2|^q \right\} |u_m^1 - u_m^2| |w_j|\, dx$$

$$\le (q+1) \max\left\{ \|u_m^1\|_{q+2}^q, \|u_m^2\|_{q+2}^q \right\} \|u_m^1 - u_m^2\|_{q+2} \|w_j\|_{q+2}, \tag{24.21}$$

where we have used the generalized Hölder inequality with

$$q_1 = \frac{q+2}{q}, \quad q_2 = q+2, \quad q_3 = q+2.$$

The following chain of inequalities holds:

$$|u_m^1 - u_m^2| = \left| \sum_{k=1}^m (c_{mk}^1 - c_{mk}^2) w_k \right|$$

$$\le \left(\sum_{k=1}^m |c_{mk}^1 - c_{mk}^2|^2 \right)^{1/2} \left(\sum_{k=1}^m |w_k(x)|^2 \right)^{1/2}$$

$$= |c_m^1 - c_m^2| \left(\sum_{k=1}^m |w_k(x)|^2 \right)^{1/2}, \tag{24.22}$$

where

$$|c_m^1 - c_m^2| := \left(\sum_{k=1}^{m} |c_{mk}^1 - c_{mk}^2|^2 \right)^{1/2}.$$

This implies the estimate

$$\|u_m^1 - u_m^2\|_{q+2} \le \left(\int_{\Omega} \left(\sum_{k=1}^{m} |w_k(x)|^2 \right)^{(q+2)/2} dx \right)^{1/(q+2)} |c_m^1 - c_m^2|. \tag{24.23}$$

The estimates (24.21)–(24.23) imply the required estimate (24.19).

It remains to note that the following inequality holds:

$$\|u_m^l\|_{q+2}^q \le |c_m^l|^q \left(\int_{\Omega} \left(\sum_{k=1}^{m} |w_k(x)|^2 \right)^{(q+2)/2} dx \right)^{q/(q+2)}, \quad l = 1, 2, \tag{24.24}$$

which together with the estimate (24.19) implies the estimate

$$\left| F_j(c_m^1) - F_j(c_m^2) \right| \le \nu_j(R)|c_m^1 - c_m^2|, \tag{24.25}$$

where

$$\nu_j(R) = A_{mj} R^q, \quad R = \max \left\{ |c_m^1|, |c_m^2| \right\}$$

with constants $A_{mj} > 0$, $j = \overline{1, m}$. Therefore, the mapping $F_j(c_m)$ defined by Eq. (24.18) is boundedly Lipschitz continuous and hence for the Cauchy problem for the system (24.17), the following assertion is valid.

Theorem 24.2. *For any initial functions*

$$c_m(0) = \big(c_{m1}(0), \ldots, c_{mm}(0) \big), \quad c_m'(0) = \big(c_{m1}'(0), \ldots, c_{mm}'(0) \big)$$

there exists $T_0 = T_0(c_m(0), c_m'(0)) > 0$ such that for any $T \in (0, T_0)$, there exists a unique classical solution $c_m(t) \in C^{(2)}[0, T] \otimes \cdots \otimes C^{(2)}[0, T]$ of the Cauchy problem for the system (24.17). Moreover, either $T_0 = +\infty$ or $T_0 < +\infty$, and in the latter case, we have

$$\sup_{t \uparrow T_0} \left(\sqrt{\sum_{j=1}^{m} c_{mj}^2(t)} + \sqrt{\sum_{j=1}^{m} (c_{mj}')^2(t)} \right) = +\infty. \tag{24.26}$$

3. A priori estimates. Multiplying Eq. (24.16) corresponding to the subscript j by c_{mj}' and summing by j, we obtain the equality

$$\big(u_m'', u_m' \big)_2 + \big(D_x u_m, D_x u_m' \big)_2 + \big(|u_m|^q u_m, u_m' \big)_2 = 0. \tag{24.27}$$

Therefore,

$$\frac{1}{2} \frac{d}{dt} \left[\|u_m'\|_2^2 + \| |D_x u_m| \|_2^2 \right] + \frac{1}{q+2} \frac{d}{dt} \|u_m\|_{q+2}^{q+2} = 0. \tag{24.28}$$

Integrating (24.28) by time, we obtain

$$\frac{1}{2}\Big[\|u_m'\|_2^2 + \big\||D_x u_m|\big\|_2^2\Big] + \frac{1}{q+2}\|u_m\|_{q+2}^{q+2}$$

$$= \frac{1}{2}\Big[\|u_{m1}\|_2^2 + \big\||D_x u_{m0}|\big\|_2^2\Big] + \frac{1}{q+2}\|u_{m0}\|_{q+2}^{q+2}. \qquad (24.29)$$

Due to (24.14) and (24.15) we have

$$u_{m0} \to u_0 \quad \text{strongly in } H_0^1(\Omega) \hookrightarrow L^{q+2}(\Omega),$$

$$u_{m1} \to u_1 \quad \text{strongly in } L^2(\Omega).$$

This means that the right-hand side of Eq. (24.29) is bounded by a constant $M_1 > 0$ independent of $m \in \mathbb{N}$. Thus, we have

$$\frac{1}{2}\Big[\|u_m'\|_2^2 + \big\||D_x u_m|\big\|_2^2\Big] + \frac{1}{q+2}\|u_m\|_{q+2}^{q+2} \le M_1. \qquad (24.30)$$

It follows from (24.30) that

$$\{u_m\} \text{ is bounded in } L^\infty(0, T; H_0^1(\Omega)), \qquad (24.31)$$

$$\{u_m'\} \text{ is bounded in } L^\infty(0, T; L^2(\Omega)). \qquad (24.32)$$

Taking into account Eqs. (24.11) and (24.12), we obtain the following relations:

$$\|u_m'\|_2^2 = \sum_{j_1=1}^m \sum_{j_2=1}^m c_{mj_1}' c_{mj_2}' \int_\Omega w_{j_1}(x) w_{j_2}(x)\, dx = \sum_{j=1}^m (c_{mj}')^2, \qquad (24.33)$$

$$\big\||D_x u_m|\big\|_2^2 = \sum_{j_1=1}^m \sum_{j_2=1}^m c_{mj_1} c_{mj_2} \int_\Omega (D_x w_{j_1}(x), D_x w_{j_2}(x))\, dx$$

$$= \sum_{j=1}^m \lambda_j c_{mj}^2 \ge \lambda_1 \sum_{j=1}^m c_{mj}^2, \qquad (24.34)$$

where $\lambda_1 > 0$ is the first eigenvalue of the Laplace operator. Thus, taking into account (24.33) and (24.34), from (24.30) we obtain the following *a priori* estimates for any $T > 0$:

$$\sum_{k=1}^m (c_{mk}'(t))^2 \le M_1, \quad \sum_{k=1}^m c_{mk}^2(t) \le \frac{M_1}{\lambda_1}, \quad t \in [0, T]. \qquad (24.35)$$

This and Theorem 24.2 imply that $T_0 = +\infty$. Thus, we have proved that $c_{mk}(t) \in C^{(2)}[0, +\infty)$.

4.   Passing to the limit.   The Banach space $L^\infty(0, T; H_0^1(\Omega))$ is dual to $L^1(0, T; H^{-1}(\Omega))$ and the Banach space $L^\infty(0, T; L^2(\Omega))$ is dual to $L^1(0, T; L^2(\Omega))$. Therefore, due to the *a priori* estimates (24.31) and (24.32) and Theorem 2.2, we can extract from the sequence $\{u_m\}$ a subsequence $\{u_\mu\}$ such that

$$u_\mu \overset{*}{\rightharpoonup} u \quad *\text{-weakly in } L^\infty(0, T; H_0^1(\Omega)), \qquad (24.36)$$

$$u_\mu' \overset{*}{\rightharpoonup} w \quad *\text{-weakly in } L^\infty(0, T; L^2(\Omega)) \qquad (24.37)$$

as $\mu \to +\infty$.

Lemma 24.1. *The $*$-weak convergence of the sequence $\{u_\mu\}$ in $L^\infty(0, T; H_0^1(\Omega))$ implies the $*$-weak convergence in $L^\infty(0, T; L^2(\Omega))$.*

Proof. The duality bracket between $L^1(0, T; H^{-1}(\Omega))$ and $L^\infty(0, T; H_0^1(\Omega))$ has the following explicit form:

$$\int\limits_0^T \langle \varphi(t), f^*(t) \rangle \, dt, \quad f^*(t) \in L^\infty(0, T; H_0^1(\Omega)), \quad \varphi(t) \in L^1(0, T; H^{-1}(\Omega)), \quad (24.38)$$

where $\langle \cdot, \cdot \rangle$ is the duality bracket between $H_0^1(\Omega)$ and $H^{-1}(\Omega)$. Moreover, the space $H_0^1(\Omega)$ is reflexive (since it is a Hilbert space), and the following dense embedding holds:

$$H_0^1(\Omega) \overset{ds}{\subset} L^2(\Omega) \quad \Rightarrow \quad L^2(\Omega) \overset{ds}{\subset} H^{-1}(\Omega),$$

where we used Theorem 1.3. Due to Theorem 1.4, the following equality of the duality brackets is fulfilled:

$$\langle \varphi, f^* \rangle = \int\limits_\Omega \varphi(x) f^*(x) \, dx, \quad f^*(x), \ \varphi(x) \in L^2(\Omega).$$

Therefore, we have

$$\int\limits_0^T \langle \varphi(t), f^*(t) \rangle \, dt = \int\limits_0^T \int\limits_\Omega \varphi(x, t) f^*(x, t) \, dx \, dt \qquad (24.39)$$

for any $f^*(x, t) \in L^\infty(0, T; L^2(\Omega))$ and $\varphi(x, t) \in L^1(0, T; L^2(\Omega))$.

Assume that the limit relation (24.36) holds. Then due to (24.39) we have

$$\int\limits_0^T \int\limits_\Omega \varphi(x, t) \left[u_\mu(x, t) - u(x, t) \right] dx \, dt = \int\limits_0^T \langle \varphi(t), u_\mu(t) - u(t) \rangle \, dt \to +0 \qquad (24.40)$$

as $\mu \to +\infty$ for any $\varphi(x, t) \in L^1(0, T; L^2(\Omega))$. Thus, we obtain the limit relation

$$u_\mu \overset{*}{\rightharpoonup} u \quad \text{*-weakly in } L^\infty(0, T; L^2(\Omega)) \text{ as } \mu \to +\infty.$$

The lemma is proved. $\qquad\qquad\qquad\qquad\qquad\qquad\qquad\qquad\qquad\qquad\qquad\qquad\qquad\quad$ $\square$

Lemma 24.2. *The limit functions in* (24.36) *and* (24.37) *are related by the formula* $w(t) = u'(t)$.

Proof. Since $\{u_\mu(t)\} \subset L^\infty(0, T; H_0^1(\Omega)) \subset L^\infty(0, T; L^2(\Omega))$, we have the following equality treated in the sense of $L^2(\Omega)$-valued distributions:

$$\langle\!\langle u_\mu'(t), \varphi(t) \rangle\!\rangle = -\langle\!\langle u_\mu(t), \varphi'(t) \rangle\!\rangle, \quad \varphi(t) \in \mathscr{D}(0, T). \qquad (24.41)$$

By the limit relations (24.36) and (24.37) and Lemma 24.1, we conclude that

$$\langle\!\langle u_\mu'(t), \varphi(t) \rangle\!\rangle = \int\limits_0^T u_\mu'(t) \varphi(t) \, dt \rightharpoonup \int\limits_0^T w(t) \varphi(t) \, dt, \qquad (24.42)$$

$$\langle\!\langle u_\mu(t), \varphi'(t)\rangle\!\rangle = \int\limits_0^T u_\mu(t)\varphi'(t)\,dt \;\rightharpoonup\; \int\limits_0^T u(t)\varphi'(t)\,dt \qquad (24.43)$$

weakly in $L^2(\Omega)$ as $\mu \to +\infty$ for all $\varphi(t) \in \mathscr{D}(0,T)$. Thus, from (24.41)–(24.43) we obtain the following equality:

$$\int\limits_0^T u(t)\varphi'(t)\,dt = -\int\limits_0^T w(t)\varphi(t)\,dt, \quad \varphi(t) \in \mathscr{D}(0,T). \qquad (24.44)$$

Therefore, $w(t) = u'(t)$ in the weak sense. The lemma is proved. $\qquad\square$

Note that it follows from (24.30) that the sequences $\{|D_x u_m|\}$ and $\{u_m\}$ are bounded in $L^\infty(0,T;L^2(\Omega)) \subset L^2(0,T;L^2(\Omega)) = L^2(D)$, $D := \Omega \otimes (0,T)$ and the sequence $\{u'_m\}$ is bounded in $L^\infty(0,T;L^2(\Omega)) \subset L^2(0,T;L^2(\Omega)) = L^2(D)$. Therefore, the sequence $\{u_m\}$ lies in a bounded set in $H^1(D)$. However, it is well known that the embedding of $H^1(D)$ into $L^{q+2}(D)$ is completely continuous for $q \in [0,4)$; therefore, it is totally continuous, since $N = 3$. *Here we used the method of compactness.*

Our further reasoning is as follows. Since the sequence $\{u_m\}$ is bounded in $H^1(D)$, we can extract a subsequence $\{u_\mu\} \subset \{u_m\}$ such that

$$u_\mu \rightharpoonup v \quad \text{weakly in } H^1(D) \text{ as } \mu \to +\infty. \qquad (24.45)$$

Next, we apply the following assertion.

Lemma 24.3. *The limit functions in* (24.36) *and* (24.45) *coincide.*

Proof. Similarly to the proof of Lemma 24.1, we can show that the limit relation (24.36) implies[1] that

$$u_\mu \rightharpoonup u \quad \text{weakly in } L^2(D) \text{ as } \mu \to +\infty; \qquad (24.46)$$

obviously, (24.45) implies that

$$u_\mu \to v \quad \text{strongly in } L^2(D) \text{ as } \mu \to +\infty \qquad (24.47)$$

and hence this convergence is also valid in the weak sense. The assertion of the lemma follows from (24.46) and (24.47). $\qquad\square$

Due to the totally continuous embedding $H^1(D) \hookrightarrow L^{q+2}(D)$, $q \in [0,4)$, we obtain from (24.45) that

$$u_\mu \to u \quad \text{strongly in } L^{q+2}(D) \text{ as } \mu \to +\infty,$$

and hence a certain subsequence converges to the same function u almost everywhere.

---

[1] See the proof of Lemma 22.1.

Thus, we can assume that the subsequence $\{u_\mu\} \subset \{u_m\}$ satisfies the condition

$$u_\mu \to u \quad \text{strongly in } L^{q+2}(D) \text{ and almost everywhere for } (x,t) \in D \text{ as } \mu \to +\infty.$$
(24.48)

Note that the operator

$$N_f(u) := |u|^q u : L^{q+2}(\Omega) \to L^{(q+2)/(q+1)}(\Omega)$$
(24.49)

is a Nemytsky operator satisfying the growth condition

$$\||u|^q u| \le |u|^{q+1}.$$

By the Krasnosel'sky theorem on the continuity of Nemytsky operators (see Theorem 6.4), we obtain from (24.48) the following limit relation:

$$|u_\mu|^q u_\mu \to |u|^q u \quad \text{strongly in } L^{(q+2)/(q+1)}(D) \text{ as } \mu \to +\infty.$$
(24.50)

Note that the *a priori* estimate (24.30) implies the inequality

$$\|u_m\|_{q+2}^{q+2} \le (q+2)M_1 < +\infty,$$
(24.51)

and hence the estimate

$$\left\| |u_m|^q u_m \right\|_{(q+2)/(q+1)} \le \left[(q+2)M_1 \right]^{(q+1)/(q+2)} < +\infty.$$
(24.52)

Therefore, the sequence $\{|u_m|^q u_m\}$ is bounded in $L^\infty(0,T; L^{(q+2)/(q+1)}(\Omega))$. Choosing an appropriate subsequence, we can also assume that

$$|u_\mu|^q u_\mu \overset{*}{\rightharpoonup} g \quad *\text{-weakly in } L^\infty(0,T; L^{(q+2)/(q+1)}(\Omega)) \text{ as } \mu \to +\infty.$$
(24.53)

Lemma 24.4. *The equality $g = |u|^q u$ is fulfilled.*

Proof. Indeed, in particular, from (24.53) we conclude[2] that

$$|u_\mu|^q u_\mu \rightharpoonup g \quad \text{weakly in } L^{(q+2)/(q+1)}(0,T; L^{(q+2)/(q+1)}(\Omega)).$$
(24.54)

From (24.50) we see that

$$|u_\mu|^q u_\mu \rightharpoonup |u|^q u \quad \text{weakly in } L^{(q+2)/(q+1)}(D) \text{ as } \mu \to +\infty.$$
(24.55)

From (24.54) and (24.55) the assertion of the lemma follows. □

On the one hand, due to the limit relation (24.37) and Lemma 24.2, we arrive at the following chain of relations:

$$\langle u_\mu'', w_j \rangle = \frac{d^2}{dt^2} \langle u_\mu, w_j \rangle \overset{*}{\to} \frac{d^2}{dt^2} \langle u, w_j \rangle \quad \text{as } \mu \to +\infty$$
(24.56)

$*$-weakly in the sense of $\mathscr{D}'(0,T)$. On the other hand, we have

$$\Delta u_\mu - |u_\mu|^q u_\mu \overset{*}{\rightharpoonup} \Delta u - |u|^q u$$
(24.57)

---

[2]See the proof of Lemma 22.1.

∗-weakly in $L^\infty(0, T; H^{-1}(\Omega))$ as $\mu \to +\infty$. Indeed, due to (24.36),

$$\int_0^T \langle f(t), u_\mu(t) - u(t) \rangle \, dt \to 0 \quad \text{as } \mu \to +\infty \qquad (24.58)$$

for any $f(t) \in L^1(0, T; H^{-1}(\Omega))$. Let $\varphi(t) \in L^1(0, T; H_0^1(\Omega))$; then[3] $\Delta\varphi(t) \in L^1(0, T; H^{-1}(\Omega))$, and the following chain of equalities holds:

$$\int_0^T \langle \Delta\varphi(t), u_\mu(t) - u(t) \rangle \, dt = - \int_0^T \langle \Delta u_\mu(t) - \Delta u(t), \varphi(t) \rangle \, dt. \qquad (24.59)$$

By (24.58), for any function $\varphi(t) \in L^1(0, T; H_0^1(\Omega))$, the left-hand side of Eq. (24.59) tends to zero. Therefore,

$$\Delta u_\mu \overset{*}{\rightharpoonup} \Delta u \quad \text{∗-weakly in } L^\infty(0, T; H^{-1}(\Omega)) \text{ as } \mu \to +\infty. \qquad (24.60)$$

From (24.53) and Lemma 24.4 we conclude that

$$|u_\mu|^q u_\mu \overset{*}{\rightharpoonup} |u|^q u \quad \text{∗-weakly in } L^\infty(0, T; L^{(q+2)/(q+1)}(\Omega)). \qquad (24.61)$$

Using the continuous embedding

$$L^{(q+2)/(q+1)}(\Omega) \subset H^{-1}(\Omega),$$

we can prove that[4]

$$|u_\mu|^q u_\mu \overset{*}{\rightharpoonup} |u|^q u \quad \text{∗-weakly in } L^\infty(0, T; H^{-1}(\Omega)). \qquad (24.62)$$

Therefore, (24.57) follows now from (24.60) and (24.62).

In particular, using (24.60) and (24.61), we obtain the following limit relations:

$$\langle -\Delta u_\mu, w_j \rangle \overset{*}{\rightharpoonup} \langle -\Delta u, w_j \rangle \quad \text{∗-weakly in } L^\infty(0, T), \qquad (24.63)$$

$$\langle |u_\mu|^q u_\mu, w_j \rangle \overset{*}{\rightharpoonup} \langle |u|^q u, w_j \rangle \quad \text{∗-weakly in } L^\infty(0, T) \qquad (24.64)$$

as $\mu \to +\infty$.

Thus, from (24.16), (24.56), (24.63), and (24.64), passing to the limit as $\mu \to +\infty$, we arrive at the following equality:

$$\frac{d^2}{dt^2} \langle u, w_j \rangle = \langle \Delta u - |u|^q u, w_j \rangle \in L^\infty(0, T), \quad j \in \mathbb{N}. \qquad (24.65)$$

Note that for any function $\varphi(t) \in \mathscr{D}(0, T)$, the following equality holds:

$$\left\langle\!\!\left\langle \frac{d^2}{dt^2} \langle u, w_j \rangle, \varphi(t) \right\rangle\!\!\right\rangle = \left\langle\!\!\left\langle \langle u, w_j \rangle, \varphi''(t) \right\rangle\!\!\right\rangle$$

$$= \left\langle \left\langle\!\!\left\langle u, \varphi''(t) \right\rangle\!\!\right\rangle, w_j \right\rangle = \left\langle \left\langle\!\!\left\langle \frac{d^2 u}{dt^2}, \varphi(t) \right\rangle\!\!\right\rangle, w_j \right\rangle. \qquad (24.66)$$

---

[3]The fact that the operator Δ transforms strongly measurable functions from $L^1(0, T; H_0^1(\Omega))$ into strongly measurable functions from $L^1(0, T; H^{-1}(\Omega))$ can be proved as a similar fact for the operator Δ_p (see the proof of Theorem 21.1).

[4]The proof is similar to the proof of Lemma 22.1.

Then from (24.65) and (24.66) we obtain the following equality:

$$\left\langle \left\langle\!\!\left\langle \frac{d^2 u}{dt^2} - \Delta u + |u|^q u, \; \varphi(t) \right\rangle\!\!\right\rangle, \; w_j \right\rangle = 0, \quad j \in \mathbb{N}. \tag{24.67}$$

Since $\{w_j\}_{j=1}^{+\infty}$ is a in $H_0^1(\Omega)$, we conclude that the following equality holds:

$$\left\langle \left\langle\!\!\left\langle \frac{d^2 u}{dt^2} - \Delta u + |u|^q u, \; \varphi(t) \right\rangle\!\!\right\rangle, \; w \right\rangle = 0, \quad w \in H_0^1(\Omega). \tag{24.68}$$

Therefore, due to the arbitrariness of $\varphi(t) \in \mathscr{D}(0,T)$, we obtain the following equality in the sense of the space $\mathscr{D}'(0,T; H^{-1}(\Omega))$:

$$\frac{d^2 u}{dt^2} = \Delta u - |u|^q u \in L^\infty(0,T; H^{-1}(\Omega)). \tag{24.69}$$

Here $\langle\!\langle \cdot, \cdot \rangle\!\rangle$ is the duality bracket between $\mathscr{D}(0,T)$ and $\mathscr{D}'(0,T; \mathbb{B})$ for any Banach space $\mathbb{B}$ (in particular, for $\mathbb{B} = \mathbb{R}^1$ and $\mathbb{B} = H^{-1}(\Omega)$). Note that

$$\frac{du}{dt} = u'(t) \in L^\infty(0,T; L^2(\Omega)) \subset L^\infty(0,T; H^{-1}(\Omega)). \tag{24.70}$$

Thus, for any $\varphi(t) \in \mathscr{D}(0,T)$, we have the following equalities in the sense of $\mathscr{D}'(0,T; H^{-1}(\Omega))$:

$$\int_0^T (\Delta u - |u|^q u)\varphi(t)\,dt = \langle\!\langle \Delta u - |u|^q u, \varphi(t) \rangle\!\rangle = \left\langle\!\!\left\langle \frac{d^2 u(t)}{dt^2}, \varphi(t) \right\rangle\!\!\right\rangle$$

$$= -\left\langle\!\!\left\langle \frac{du(t)}{dt}, \varphi'(t) \right\rangle\!\!\right\rangle = -\langle\!\langle u'(t), \varphi'(t) \rangle\!\rangle$$

$$= -\int_0^T u'(t)\varphi'(t)\,dt. \tag{24.71}$$

Thus, from (24.69)–(24.71) we obtain

$$\frac{d^2 u}{dt^2} = u''(t) = \Delta u - |u|^q u \in L^\infty(0,T; H^{-1}(\Omega)). \tag{24.72}$$

Moreover, due to (24.57), (24.16), and (24.72), we arrive at the following limit relations:

$$\langle u_\mu''(t), w_j \rangle = \langle\!\langle \Delta u_\mu - |u_\mu|^q u_\mu, w_j \rangle\!\rangle \overset{*}{\rightharpoonup} \langle\!\langle \Delta u - |u|^q u, w_j \rangle\!\rangle = \langle u''(t), w_j \rangle$$

$$*\text{-weakly in } L^\infty(0,T) \text{ as } \mu \to +\infty \text{ for all } j \in \mathbb{N}. \tag{24.73}$$

5. Initial conditions. It remains to prove that the function $u(t)$ constructed satisfies the initial conditions (24.7).

By construction, we have

$$u_\mu(0) = u_{\mu 0} \to u_0 \quad \text{strongly in } H_0^1(\Omega) \text{ as } \mu \to +\infty. \tag{24.74}$$

On the other hand, due to (24.36), (24.37), and Lemma 24.2 we obtain that

$$u_\mu(0) \rightharpoonup u(0) \quad \text{weakly in } L^2(\Omega) \text{ as } \mu \to +\infty. \tag{24.75}$$

Indeed, consider the following function:

$$\varphi_\mu(t) := \int_\Omega (u_\mu(x)(t) - u(x)(t))\, w(x)\, dx$$

for arbitrary fixed $w(x) \in L^2(\Omega)$. By (24.48) we have

$$\varphi_\mu(t) \to 0 \quad \text{for almost all } t \in [0, T]. \tag{24.76}$$

Indeed, the following estimate holds:

$$|\varphi_\mu(t)| \le \|u_\mu(t) - u(t)\|_{L^{q+2}(\Omega)} \|w\|_{L^{(q+2)/(q+1)}(\Omega)}. \tag{24.77}$$

This implies the following estimate:

$$\int_0^T |\varphi_\mu(t)|^{(q+2)/(q+1)}\, dt \le \left(\int_0^T \|u_\mu(t) - u(t)\|_{L^{q+2}(\Omega)}^{q+2}\, dt \right)^{1/(q+1)}$$
$$\times T^{q/(q+1)} \|w\|_{L^{(q+2)/(q+1)}(\Omega)}^{(q+2)/(q+1)}. \tag{24.78}$$

The right-hand side of the last expression tends to zero as $\mu \to +\infty$ by (24.48). Therefore, choosing an appropriate subsequence $\{u_\mu\}$, we arrive at the limit relation

$$\varphi_\mu(t) \to 0 \quad \text{as } \mu \to +\infty \text{ for almost all } t \in [0, T].$$

On the other hand, due to (24.36), (24.37), and Lemma 24.2 we have

$$u(t) \in L^\infty(0, T; H_0^1(\Omega)), \quad u'(t) \in L^\infty(0, T; L^2(\Omega)).$$

The embedding (24.8) implies that

$$u_\mu(t),\ u(t) \in C([0, T]; L^2(\Omega)) \quad \Rightarrow \quad \varphi_\mu(t) \in C([0, T]).$$

Then for any $\varepsilon > 0$, there exist numbers $\delta = \delta(\varepsilon) > 0$ and $n_0 = n_0(\varepsilon) \in \mathbb{N}$ and a set $E \subset [0, T]$ of zero Lebesgue measure, $\mu(E) = 0$, such that for all $\mu \ge n_0$, all $0 < t < \delta(\varepsilon)$, and $t \in [0, T] \backslash E$, the following chain of inequalities holds:

$$|\varphi_\mu(0)| \le |\varphi_\mu(0) - \varphi_\mu(t)| + |\varphi_\mu(t)| < \frac{\varepsilon}{2} + \frac{\varepsilon}{2} = \varepsilon.$$

Therefore,

$$u_\mu(0) \rightharpoonup u(0) \quad \text{weakly in } L^2(\Omega) \text{ as } \mu \to +\infty.$$

Hence from (24.74) and (24.75) we conclude that the initial condition $u(0) = u_0$ holds (note that $u_0(x) \in H_0^1(\Omega)$ by the assumption).

Now we prove that $u'(0) = u_1$. Indeed, on the one hand, due to (24.73) we have

$$\langle u_\mu'', w_j \rangle \overset{*}{\rightharpoonup} \langle u'', w_j \rangle \quad \text{*-weakly in } L^\infty(0, T) \text{ as } \mu \to +\infty. \tag{24.79}$$

Consider the function

$$\psi_{j,\mu}(t) := \langle u_\mu', w_j \rangle, \quad w_j = w_j(x) \in H_0^1(\Omega). \tag{24.80}$$

The following equality holds:

$$\psi_{j,\mu}(t) = \psi_{j,\mu}(0) + \int_0^t \langle u''_\mu(s), w_j \rangle \, ds = \langle u_{\mu 1}, w_j \rangle + \int_0^T \langle u''_\mu(s), w_j \rangle \chi(s) \, ds, \quad (24.81)$$

where

$$\chi(s) = \begin{cases} 1 & \text{for } s \in [0, t], \\ 0 & \text{for } s \in (t, T]; \end{cases}$$

clearly, $\chi(s) \in L^1(0, T)$. From (24.15), (24.79), and (24.81), we obtain the following limit relation as $\mu \to +\infty$:

$$\psi_{j,\mu}(t) \to \psi_j(t) := \langle u_1, w_j \rangle + \int_0^T \langle u''(s), w_j \rangle \chi(s) \, ds$$

$$= \langle u_1, w_j \rangle + \int_0^t \langle u''(s), w_j \rangle \, ds \quad \text{for all } t \in [0, T]. \quad (24.82)$$

We prove that $\psi_j(t) = \langle u'(t), w_j \rangle$. Indeed, on the one hand, by (24.37) and Lemma 24.2, we have the chain of relations

$$\int_0^t \psi_{j,\mu}(s) \, ds = \int_0^T \psi_{j,\mu}(s) \chi(s) \, ds \to \int_0^T \langle u'(t), w_j \rangle \chi(s) \, ds = \int_0^t \langle u'(t), w_j \rangle \, ds$$

$$(24.83)$$

as $\mu \to +\infty$. On the other hand, we have

$$\int_0^t \psi_{j,\mu}(s) \, ds \to \int_0^t \psi_j(s) \, ds \quad \text{as } \mu \to +\infty \text{ for all } t \in [0, T]. \quad (24.84)$$

Indeed, note that

$$\langle u''_\mu(s), w_j \rangle = \left(D_x u_\mu, D_x w_j \right)_2 - \left(|u_\mu|^q u_\mu, w_j \right)_2. \quad (24.85)$$

Using the *a priori* estimates (24.30) and (24.52), we can prove the inequalities

$$|\langle u''_\mu(s), w_j \rangle| \leq |||D_x u_\mu|||_2 |||D_x w_j|||_2$$
$$+ |||u_\mu|^q u_\mu||_{(q+2)/(q+1)} ||w_j||_{q+2} \leq M_3(T) < +\infty, \quad (24.86)$$

where the constant $M_3(T) > 0$ is independent of $\mu \in \mathbb{N}$. Therefore, we have the estimate

$$\left| \int_0^t \langle u''_\mu(s), w_j \rangle \, ds \right| \leq T M_3(T). \quad (24.87)$$

Moreover, it follows from (24.73) that

$$\int_0^T \chi(s) \langle u''_\mu(s), w_j \rangle \, ds \to \int_0^T \chi(s) \langle u''(s), w_j \rangle \, ds \quad \text{as } \mu \to +\infty. \quad (24.88)$$

Therefore, due to (24.82), (24.87), and (24.88), we can apply the Lebesgue theorem on the passing to the limit in the Lebesgue integral and prove the validity of (24.84).

Comparing (24.83) with (24.84), we obtain the equality

$$\int_0^t \langle u'(t), w_j \rangle \, ds = \langle u_1, w_j \rangle t + \int_0^t \int_0^s \langle u''(\sigma), w_j \rangle \, d\sigma \, ds.$$

By the Lebesgue theorem on the derivative of an absolutely continuous function, from the last equality we obtain

$$\langle u'(t), w_j \rangle = \langle u_1, w_j \rangle + \int_0^t \langle u''(s), w_j \rangle \, ds = \psi_j(t) \qquad (24.89)$$

for almost all $t \in [0, T]$.

Now, due to (24.82), we have

$$\langle u'_\mu, w_j \rangle \to \langle u', w_j \rangle \quad \text{as } \mu \to +\infty \qquad (24.90)$$

for almost all $t \in [0, T]$. Moreover, from (24.36), Lemma 24.2, and (24.72) we have

$$u'(t) \in L^\infty(0, T; L^2(\Omega)), \quad u''(t) \in L^\infty(0, T; H^{-1}(\Omega)).$$

Due to the embedding (24.9), we obtain $\langle u', w_j \rangle \in C[0, T]$. Consider the function:

$$h_{j,\mu}(t) := \langle u'_\mu(t) - u'(t), w_j \rangle \in C[0, T].$$

By (24.90),

$$h_{j,\mu}(t) \to 0 \quad \text{for almost all } t \in [0, T].$$

Therefore, for any $\varepsilon > 0$, there exist numbers $\delta = \delta(\varepsilon) > 0$ and $n_0 = n_0(\varepsilon) \in \mathbb{N}$ and a set $E \subset [0, T]$ of zero Lebesgue measure $\mu(E) = 0$ such that for all $\mu \geq n_0$ and all $0 < t < \delta(\varepsilon)$ and $t \in [0, T] \backslash E$, the following chain of inequalities holds:

$$|h_{j,\mu}(0)| \leq |h_{j,\mu}(0) - h_{j,\mu}(t)| + |h_{j,\mu}(t)| < \frac{\varepsilon}{2} + \frac{\varepsilon}{2} = \varepsilon.$$

Thus, we conclude that

$$\langle u'_\mu(0), w_j \rangle \to \langle u'(0), w_j \rangle \quad \text{as } \mu \to +\infty. \qquad (24.91)$$

Due to (24.15) we have

$$\langle u'_\mu(0), w_j \rangle = \langle u_{1\mu}, w_j \rangle \to \langle u_1, w_j \rangle \quad \text{as } \mu \to +\infty. \qquad (24.92)$$

Therefore,

$$\langle u'(0), w_j \rangle = \langle u_1, w_j \rangle \quad \text{for all } j \in \mathbb{N} \quad \Rightarrow \quad u'(0) = u_1.$$

6. Uniqueness. Finally, we prove the uniqueness of the solution obtained.

Lemma 24.5. *Let the conditions of Theorem 24.1 hold and, moreover, let $q \in (0, 2]$. Then the solution u obtained in Theorem 24.1 is unique.*

Proof. 1. Let u_1 and u_2 be two weak solutions of the problem in the sense of Definition 24.1 and let $w = u_1 - u_2$. For $s \in (0, T)$, we set

$$\psi(t) := \begin{cases} -\displaystyle\int_t^s w(\sigma)\, d\sigma & \text{for } t \leq s, \\[3mm] 0 & \text{for } t > s. \end{cases}$$

We have $\psi(t) = w_1(t) - w_1(s)$ for $t \leq s$, where

$$w_1(t) := \int_0^t w(\sigma)\, d\sigma.$$

2. Setting $v = \psi(t)$, from (24.4) we obtain the equality

$$\int_0^T dt\, \langle w'' - \Delta w + |u_1|^q u_1 - |u_2|^q u_2, \psi(t) \rangle = 0, \tag{24.93}$$

where $w(0) = 0$ and $w'(0) = 0$. Integrating by parts, we obtain

$$\int_0^s (w''(t), \psi(t))_2\, dt = (w'(t), \psi(t))_2 \Big|_{t=0}^{t=s} - \int_0^s (w', \psi')_2\, dt = -\int_0^s (w', \psi')_2\, dt,$$

since $\psi(s) = 0$ and $w'(0) = 0$. Therefore, from (24.93) we obtain the equality

$$-\int_0^s (w', \psi')_2\, dt + \int_0^s (D_x w, D_x \psi)_2\, dt = -\int_0^s (|u_1|^q u_1 - |u_2|^q u_2, \psi)_2\, dt; \tag{24.94}$$

since $\psi'(t) = w(t)$, by Lemma 4.4 we have

$$-\int_0^s (w', \psi')_2\, dt = -\int_0^s (w', w)_2\, dt = -\frac{1}{2}\|w\|_2^2(s),$$

since $w(0) = 0$.

3. Since $w_1(t) \in C^{(1)}([0, T]; H_0^1(\Omega))$, $w(t) = \psi'(t)$, and $\psi(0) = -w_1(s)$, the following chain of equalities holds:

$$\int_0^s \left(D_x w(t), D_x \psi(t) \right)_2 dt$$

$$= \int_0^s \left(D_x \psi'(t), D_x \psi(t) \right)_2 dt = \frac{1}{2} \int_0^s \frac{d}{dt} \left(D_x \psi(t), D_x \psi(t) \right)_2 dt$$

$$= \frac{1}{2} \left(D_x \psi(s), D_x \psi(s) \right)_2 - \frac{1}{2} \left(D_x \psi(0), D_x \psi(0) \right)_2 = -\frac{1}{2} \||D_x w_1(s)|\|_2^2.$$

Then from (24.94) we obtain the equality

$$\frac{1}{2}\|w\|_2^2(s) + \frac{1}{2}\|D_x w_1\|_2^2(s) = \int_0^s \left(|u_1|^q u_1 - |u_2|^q u_2, \psi \right)_2 dt. \tag{24.95}$$

4. Consider the expression in the right-hand side separately:

$$I = \int_0^s \left(|u_1|^q u_1 - |u_2|^q u_2, \, \psi\right)_2 dt = \int_0^s \left(|u_1|^q u_1 - |u_2|^q u_2, \, w_1(t) - w_1(s)\right)_2 dt$$

$$\leq (q+1) \int_0^s \int_\Omega dx \, |w(t)| \left[|w_1(t)| + |w_1(s)|\right] \max\left\{|u_1|^q, |u_2|^q\right\} dt. \tag{24.96}$$

Now we apply the generalized Hölder inequality with the following parameters:

$$p_1 = 2, \quad p_2 = r, \quad p_3 = N, \quad \frac{1}{p_1} + \frac{1}{p_2} + \frac{1}{p_3} = 1 \quad \Rightarrow \quad r = \frac{2N}{N-2}, \quad N = 3.$$

We have the continuous embedding $H_0^1(\Omega) \subset L^r(\Omega)$. Moreover, we have

$$qN \leq \frac{2N}{N-2} \quad \Rightarrow \quad q \leq \frac{2}{N-2} \quad \text{for } N = 3;$$

therefore, we have

$$\left\||u_k|^q\right\|_N = \|u_k\|_{qN}^q \leq c_1 \left\||D_x u_k|\right\|_2^q \leq c_2, \quad k = 1, 2.$$

Thus, from (24.96) we obtain the inequality

$$I \leq c_3 \int_0^s dt \, \|w\|_2(t) \left[\|w_1\|_r(t) + \|w_1\|_r(s)\right] \times \max\left\{\left\||u_1|^q\right\|_N(t), \, \left\||u_2|^q\right\|_N(t)\right\}$$

$$\leq c_4 \int_0^s \|w\|_2(t) \left[\||D_x w_1|\|_2(t) + \||D_x w_1|\|_2(s)\right] dt. \tag{24.97}$$

For any $\varepsilon > 0$, the following inequalities hold:

$$\|w\|_2(t) \||D_x w_1|\|_2(t) \leq \frac{1}{2} \|w\|_2^2(t) + \frac{1}{2} \||D_x w_1|\|_2^2(t),$$

$$\||D_x w_1|\|_2(s) \|w\|_2(t) \leq \frac{\varepsilon}{2T} \||D_x w_1|\|_2^2(s) + \frac{T}{2\varepsilon} \|w\|_2^2(t).$$

Taking these inequalities into account, we obtain from (24.95) the inequality

$$\frac{1}{2} \|w\|_2^2(s) + \frac{1}{2} \||D_x w_1|\|_2^2(s) \leq \frac{c_4}{2} \int_0^s \left[\|w\|_2^2(t) + \||D_x w_1|\|_2^2(t)\right] dt$$

$$+ \frac{c_4}{2} \varepsilon \||D_x w_1|\|_2^2(s) + \frac{T}{2\varepsilon} c_4 \int_0^s \|w\|_2^2(t) \, dt.$$

Setting here $\varepsilon = 1/(2c_4)$, we arrive at the inequality

$$\|w\|_2^2(s) + \||D_x w_1|\|_2^2(s) \leq c_5(T) \int_0^s dt \left[\|w\|_2^2(t) + \||D_x w_1|\|_2^2(t)\right], \quad s \in [0, T].$$

Applying the Grönwall–Bellman lemma (see [11]), we conclude that $u_1 = u_2$ almost everywhere. The lemma is proved. □

Theorem 24.2 is completely proved. □

24.2 Bibliographical Notes

The contents of this lecture is taken from [13, 15, 16, 19, 20, 44, 46].

PART 5

Methods Based on the Maximum Principle

Lecture 25

Method of Upper and Lower Solutions for Elliptic Equations

25.1 Upper and Lower Solutions: Definitions

Let $\Omega \subset \mathbb{R}^N$ be a bounded domain with smooth boundary $\partial\Omega$. Moreover, let a function $f(x, u) \in C(\overline{\Omega}, \mathbb{R}^1; \mathbb{R}^1)$ be differentiable with respect to $u \in \mathbb{R}^1$ for all $x \in \Omega$ and the following inequalities be fulfilled:

$$|f(x_1, u) - f(x_2, u)| \le K_1(M)\, |x_1 - x_2|^\alpha, \quad x_1, x_2 \in \Omega, \qquad (25.1)$$

$$\max\left\{|f(x, u)|,\ |f'_u(x, u)|\right\} \le K_2(M), \quad x \in \Omega, \qquad (25.2)$$

for all $|u| \le M < +\infty$, where $\alpha \in (0, 1]$, $0 < K_1, K_2 < +\infty$, and the constants K_1 and K_2 are independent of u.

Consider the following Dirichlet problem:

$$-\Delta u = f(x, u) \quad \text{for } x \in \Omega, \qquad (25.3)$$

$$u(x) = 0 \qquad \text{for } x \in \partial\Omega. \qquad (25.4)$$

We consider classical solutions of this problem: $u(x) \in C^{(2)}(\Omega) \cap C(\overline{\Omega})$.

Definition 25.1. A function $\underline{U}(x) \in C^{(2)}(\Omega) \cap C(\overline{\Omega})$ satisfying the problem

$$-\Delta \underline{U}(x) \le f(x, \underline{U}(x)) \quad \text{for } x \in \Omega, \qquad (25.5)$$

$$\underline{U}(x) \le 0 \qquad \text{for } x \in \partial\Omega \qquad (25.6)$$

is called a *lower solution* of the problem (25.3).

Definition 25.2. A function $\overline{U}(x) \in C^{(2)}(\Omega) \cap C(\overline{\Omega})$ satisfying the problem

$$-\Delta \overline{U}(x) \ge f(x, \overline{U}(x)), \quad \text{for } x \in \Omega, \qquad (25.7)$$

$$\overline{U}(x) \ge 0 \qquad \text{for } x \in \partial\Omega \qquad (25.8)$$

is called an *upper solution* of the problem (25.3).

25.2 Main Theorem

Theorem 25.1. *Let $\underline{U}(x)$ and $\overline{U}(x)$ be lower and upper solutions of the problem (25.3) such that*

$$\underline{U}(x) \le \overline{U}(x) \quad \forall x \in \Omega$$

(Note that there exist examples in which this inequality is violated at all points of the domain.) *Then the following assertions hold:*

(i) *there exists a solution $u(x) \in C^{2+\alpha}(\overline{\Omega})$ of the problem (25.3) satisfying the inequalities*

$$\underline{U}(x) \leq u(x) \leq \overline{U}(x), \quad x \in \Omega; \tag{25.9}$$

(ii) *there exist minimal and maximal solutions $\overline{u}(x) \in C^{2+\alpha}(\overline{\Omega})$ and $\underline{u}(x) \in C^{2+\alpha}(\overline{\Omega})$ of the problem (25.3) such that*

$$\underline{U}(x) \leq \underline{u}(x) \leq \overline{u}(x) \leq \overline{U}(x), \quad x \in \Omega. \tag{25.10}$$

Proof. 1. Consider the function

$$g(x, u) \overset{\text{def}}{=\!=} f(x, u) + \lambda u, \quad \lambda \geq 0. \tag{25.11}$$

We choose $\lambda \geq 0$ so large that the mapping $u \mapsto g(x, u)$ be increasing on the segment $u \in [\underline{U}(x); \overline{U}(x)]$ for all $x \in \Omega$. For this end, it suffices to take

$$\lambda \geq \max\left\{ - f_u(x, u); \ x \in \overline{\Omega}, \ u \in [\underline{U}(x); \overline{U}(x)] \right\}. \tag{25.12}$$

In particular, this condition is fulfilled for the function $f(x, u) = |u|^q u$, $q > 0$, and an arbitrary function $f(u)$ such that $|f'(t)| \leq K_2$.

2. Instead of the original problem (25.3), we consider the following equivalent problem:

$$-\Delta u + \lambda u = g(x, u) \quad \text{for } x \in \Omega, \tag{25.13}$$
$$u(x) = 0 \qquad \text{for } \partial\Omega. \tag{25.14}$$

Finally, consider the following sequence of recurrent linear problems $(n \geq 1)$:

$$-\Delta u_n + \lambda u_n = g(x, u_{n-1}) \quad \text{for } x \in \Omega, \tag{25.15}$$
$$u_n(x) = 0 \qquad \text{for } \partial\Omega; \tag{25.16}$$

moreover, $u_0(x) = \overline{U}(x) \in C^{(2)}(\Omega) \cap C(\overline{\Omega})$. In this case, the solution $u_1(x)$ of the linear problem (25.15) for $n = 1$ belongs to the class $C^{(2)}(\Omega) \cap C^{\alpha}(\overline{\Omega})$ and the subsequent iterations $u_n(x)$, $n \geq 2$, belong to the class $C^{2+\alpha}(\overline{\Omega})$. Indeed, due to (25.1) and (25.2), we have the following chain of inequalities:

$$\left| g(x_1, u(x_1)) - g(x_2, u(x_2)) \right|$$

$$\leq \lambda|u(x_1) - u(x_2)| + \left| f(x_1, u(x_1)) - f(x_2, u(x_2)) \right|$$

$$\leq \lambda|u(x_1) - u(x_2)| + \left| f(x_1, u(x_1)) - f(x_2, u(x_1)) \right|$$

$$+ \left| f(x_2, u(x_1)) - f(x_2, u(x_2)) \right|$$

$$\leq \lambda|u(x_1) - u(x_2)| + K_1|x_1 - x_2|^{\alpha} + K_2|x_1 - x_2|$$

for all $|u(x)| \leq M < +\infty$. Below we prove that this condition for the iteration sequence is fulfilled. Therefore, on one side, if $u(x) \in C^{\alpha}(\overline{\Omega})$, then the right-hand side $g(x, u(x))$ lies in the class $C^{\alpha}(\overline{\Omega})$. On the other hand, the first approximation $u_1(x)$ lies in the class $C^{(1)}(\overline{\Omega}) \subset C^{\alpha}(\overline{\Omega})$. Therefore, using the Schauder theory (see, e.g. [37]), we conclude that $u_n(x) \in C^{2+\alpha}(\overline{\Omega})$, $n \geq 2$.

Lemma 25.1. *The following chain of inequalities holds:*

$$\underline{U}(x) \leq \cdots \leq u_{n+1}(x) \leq u_n(x) \leq \cdots \leq u_0(x) = \overline{U}(x), \quad x \in \Omega. \tag{25.17}$$

Proof. The proof is based on the weak maximum principle. First, we prove that

$$u_1(x) \leq u_0(x) = \overline{U}(x).$$

By the definition of $u_1(x)$ and $\overline{U}(x)$, we obtain the inequalities

$$-\Delta\left(\overline{U}(x) - u_1(x)\right) + \lambda\left(\overline{U}(x) - u_1(x)\right) \geq g(x, \overline{U}(x)) - g(x, u_0(x))$$
$$= g(x, \overline{U}(x)) - g(x, \overline{U}(x)) = 0 \quad (25.18)$$

for $x \in \Omega$ and

$$\overline{U}(x) - u_1(x) \geq 0, \quad x \in \partial\Omega. \quad (25.19)$$

By the weak maximum principle, we have

$$\overline{U}(x) \geq u_1(x), \quad x \in \Omega. \quad (25.20)$$

Similarly, we have

$$-\Delta\left(\underline{U}(x) - u_1(x)\right) + \lambda\left(\underline{U}(x) - u_1(x)\right) \leq g(x, \underline{U}(x)) - g(x, \overline{U}) \leq 0, \quad (25.21)$$
$$\underline{U}(x) - u_1(x) \leq 0, \quad x \in \partial\Omega. \quad (25.22)$$

By the weak maximum principle, we have

$$u_1(x) \geq \underline{U}(x), \quad x \in \Omega. \quad (25.23)$$

Now we use induction. Assume that

$$\underline{U}(x) \leq \cdots \leq u_n(x) \leq u_{n-1}(x) \leq \cdots \leq u_1(x) \leq u_0(x) = \overline{U}(x), \quad x \in \Omega, \quad (25.24)$$

and prove that

$$\underline{U}(x) \leq u_{n+1}(x) \leq u_n(x). \quad (25.25)$$

Indeed, using the recurrent formulas $u_{n+1}(x)$ and $u_n(x)$ for $n \geq 1$, we obtain

$$-\Delta(u_n(x) - u_{n+1}(x)) + \lambda\left(u_n(x) - u_{n+1}(x)\right)$$
$$= g(x, u_{n-1}) - g(x, u_n) \geq 0, \quad x \in \Omega, \quad (25.26)$$

and

$$u_n(x) - u_{n+1}(x) = 0 \geq 0, \quad x \in \partial\Omega. \quad (25.27)$$

From the weak maximum principle we obtain

$$u_n(x) \geq u_{n+1}(x), \quad x \in \Omega. \quad (25.28)$$

Now we note that by the definition of a lower solution we have

$$-\Delta\underline{U}(x) + \lambda\underline{U}(x) \leq g(x, \underline{U}), \quad x \in \Omega, \quad (25.29)$$
$$\underline{U}(x) \leq 0 \quad\quad\quad\quad\quad x \in \partial\Omega. \quad (25.30)$$

The recurrent definition of $u_{n+1}(x)$ implies

$$-\Delta\left(u_{n+1}(x) - \underline{U}(x)\right) + \lambda\left(u_{n+1}(x) - \underline{U}(x)\right)$$
$$\geq g(x, u_n) - g(x, \underline{U}) \geq 0, \quad x \in \Omega, \quad (25.31)$$

and

$$u_{n+1}(x) - \underline{U}(x) \geq 0, \quad x \in \partial\Omega. \quad (25.32)$$

Again, by the weak maximum principle we have

$$\underline{U}(x) \leq u_{n+1}(x), \quad x \in \Omega. \qquad\qquad \square$$

3. By Lemma 25.1, there exists a function $u(x)$ such that for all $x \in \Omega$, the following limit relation holds:

$$u_n(x) \searrow u(x) \quad \text{as } n \to +\infty. \tag{25.33}$$

Moreover, we have

$$\underline{U}(x) \le u_n(x) \le \overline{U}(x), \quad x \in \Omega. \tag{25.34}$$

Since $\underline{U}(x), \overline{U}(x) \in C(\overline{\Omega})$, we have[1]

$$|u_n(x)|_{0;D} \le M < +\infty. \tag{25.35}$$

We must pass to the limit as $n \to +\infty$ in the recurrent equality (25.15) and prove that $u(x)$ is a solution of the limit equality. Consider the functional sequence

$$g_n(x) \overset{\text{def}}{=\!=} g(x, u_n(x));$$

it satisfies the following inequalities:

$$g(x, \underline{U}(x)) \le g_{n+1}(x) \le g_n(x) \le g(x, \overline{U}(x)) \implies |g_n(x)|_{0;D} \le M_1 < +\infty. \tag{25.36}$$

Indeed, due to (25.2),

$$|g(x, u)| \le |\lambda| M + K_2(M) < +\infty \quad \text{for } |u| \le M.$$

Since the function $g(x, z)$ is differentiable with respect to $z \in \mathbb{R}^1$, by Lagrange's formula for $n \ge 1$ we have[2]

$$\left| f(x_1, u_n(x_1)) - f(x_2, u_n(x_2)) \right|$$
$$\le \left| f(x_1, u_n(x_1)) - f(x_2, u_n(x_1)) \right| + \left| f(x_2, u_n(x_1)) - f(x_2, u_n(x_2)) \right|$$
$$\le K_1(M)|x_1 - x_2|^\alpha$$
$$\quad + \max \left\{ |f_z(x_2, u_n(x_1))|, \; |f_z(x_2, u_n(x_2))| \right\} |u_n(x_1) - u_n(x_2)|$$
$$\le \left(K_1(M) + K_2(M)[u_n]_{\alpha;D} \right) |x_1 - x_2|^\alpha. \tag{25.37}$$

Therefore, from (25.37) we obtain the inequality

$$[f(x, u_n)]_{\alpha;D} \le K_1(M) + K_2(M)[u_n]_{\alpha;D}. \tag{25.38}$$

From the Schauder *a priori* estimate applied to Eq. (25.15) for $n \ge 2$, we obtain the inequality

$$|u_n|_{2+\alpha;D} \le K \big| g(x, u_{n-1}(x)) \big|_{\alpha;D}. \tag{25.39}$$

---

[1] Here and below, we use the following notation:

$$|u|_{0;D} := \sup_{x \in D} |u(x)|, \quad [u]_{\alpha;D} := \sup_{x_1, x_2 \in D} \frac{|u(x_1) - u(x_2)|}{|x_1 - x_2|^\alpha}, \quad |u|_{\alpha;D} := |u|_{0;D} + [u]_{\alpha;D}.$$

[2] If $f(x, u) = f(u)$, then under the conditions specified, we immediately obtain $f(u(x)) \in C^\alpha(\overline{\Omega})$ and hence we may apply the classical (i.e., $C^{2+\alpha}$) Schauder *a priori* estimate.

From (25.36)–(25.39) we get the following chain of inequalities:

$$\left|g(x, u_{n-1}(x))\right|_{\alpha;D} = \left|g(x, u_{n-1}(x))\right|_{0;D} + \left[g(x, u_{n-1}(x))\right]_{\alpha;D}$$
$$\leq M_1 + \lambda\left[u_{n-1}\right]_{\alpha;D} + \left[f(x, u_{n-1})\right]_{\alpha;D}$$
$$\leq M_1 + K_1 + \left(\lambda + K_2\right)\left[u_{n-1}\right]_{\alpha;D}. \tag{25.40}$$

From the chain of inequalities (25.40) and the Schauder *a priori* estimate (25.39) we obtain the inequality

$$\left|u_n\right|_{2+\alpha;D} \leq M_2 + M_3\left[u_{n-1}\right]_{\alpha;D}. \tag{25.41}$$

It remains to apply the interpolation inequality[3]

$$\left[u_{n-1}\right]_{\alpha;D} \leq \varepsilon\left|u_{n-1}\right|_{2+\alpha;D} + c_1(\varepsilon)\left|u_{n-1}\right|_{0;D}, \quad c_1 > 0, \tag{25.42}$$

which is valid for arbitrarily small $\varepsilon > 0$, and we arrive at the following iteration inequality:

$$\left|u_n\right|_{2+\alpha;D} \leq M_2 + M_3\varepsilon\left|u_{n-1}\right|_{2+\alpha;D} + M_3 c_1(\varepsilon)\left|u_{n-1}\right|_{0;D}, \quad n \geq 2. \tag{25.43}$$

Due to the estimate (25.35), we can rewrite the inequality (25.43) in the following form:

$$\left|u_n\right|_{2+\alpha;D} \leq M_4(\varepsilon) + M_3\varepsilon\left|u_{n-1}\right|_{2+\alpha;D}, \quad M_4 := M_2 + M_3 c_1 M, \tag{25.44}$$

for $n \geq 2$. Under the condition $0 < M_3\varepsilon \leq 1/2$, from the recurrent inequality (25.44) we obtain the *a priori* estimate

$$\left|u_{n+2}\right|_{2+\alpha;D} \leq M_4(\varepsilon) \sum_{m=1}^{+\infty} \left(M_3\varepsilon\right)^m + \left|u_2\right|_{2+\alpha;D}, \quad n \in \mathbb{N} \cup \{0\}. \tag{25.45}$$

Thus, we arrive at the *a priori* estimate

$$\left|u_n\right|_{2+\alpha;D} \leq M_5 < +\infty, \quad n \geq 2, \tag{25.46}$$

where a positive constant M_5 is independent of n.

4. Due to a well-known result[4] the estimate (25.46) implies the existence of a subsequence $\{u_{n_k}\} \subset \{u_n\}$ such that

$$u_{n_k}(x) \to u(x) \in C^{2+\alpha}(\overline{\Omega}) \quad \text{strongly in } C^{(2)}(\overline{\Omega}) \text{ as } n_k \to +\infty. \tag{25.47}$$

Consider the corresponding problem

$$-\Delta u_{n_k} + \lambda u_{n_k} = g(x, u_{n_k-1}) \quad \text{for } x \in \Omega, \tag{25.48}$$
$$u_{n_k}(x) = 0 \quad \text{for } x \in \partial\Omega. \tag{25.49}$$

Due to (25.47), we can pass to the limit as $n_k \to +\infty$ taking into account the fact that the subsequence $\{u_{n_k-1}\} \subset \{u_n\}$ satisfies the limit relation

$$g(x, u_{n_k-1}) \searrow g(x, u) \quad \text{as } n_k \to +\infty \text{ for all } x \in \Omega.$$

---

[3] For details, see [37].
[4] See [43].

Then we obtain the following equalities for $u(x) \in C^{2+\alpha}(\overline{\Omega})$:

$$-\Delta u(x) + \lambda u(x) = g(x, u) = \lambda u(x) + f(x, u(x)) \qquad \text{for } x \in \Omega,$$
$$u(x) = 0 \qquad \text{for } x \in \partial\Omega.$$

5. Thus, we have proved the existence of a solution $\overline{u}(x) \in C^{2+\alpha}(\Omega)$ satisfying the inequality

$$\underline{U}(x) \leq \overline{u}(x) \leq \overline{U}(x), \qquad x \in \Omega.$$

Applying the scheme used for the choice of $u_0(x) = \underline{U}(x)$ and realizing the iterative process $\{u^m(x)\}$, we similarly prove the existence of a solution $\underline{u}(x) \in C^{2+\alpha}(\Omega)$ satisfying the inequality

$$\underline{U}(x) \leq \underline{u}(x) \leq \overline{U}(x), \qquad x \in \Omega,$$

and the limit relation

$$u^m(x) \nearrow \underline{u}(x) \quad \text{as } m \to +\infty.$$

Based on the iterative process, we can prove that for all $m, n \in \mathbb{N}$, the corresponding iterative sequences $\{u_n(x)\}$ and $\{u^m(x)\}$ are related by the inequality

$$\underline{U}(x) \leq \cdots \leq u^m(x) \leq u^{m+1}(x) \leq \cdots \leq \underline{u}(x)$$
$$\leq \overline{u}(x) \leq \cdots \leq u_{n+1}(x) \leq u_n(x) \leq \cdots \leq \overline{U}(x), \qquad x \in \Omega. \quad (25.50)$$

6. Let $u(x) \in C^{(2)}(\Omega) \cap C(\overline{\Omega})$ be an arbitrary solution of the Dirichlet problem (25.3). Then, obviously, $u(x)$ is a *lower solution* of the Dirichlet problem. Therefore, if we set $\underline{U}(x) = u(x)$ and realize the iterative process $\{u_n(x)\}$, we obtain

$$u(x) = \underline{U}(x) \leq \overline{u}(x), \qquad x \in \Omega. \tag{25.51}$$

Now we note that, on the other hand, $u(x)$ is an *upper solution* $\overline{U}(x)$. Therefore, if we set $\overline{U}(x) = u(x)$ for the iterative process $\{u^m(x)\}$, then we arrive at the inequality

$$\underline{u}(x) \leq \overline{U}(x) = u(x), \qquad x \in \Omega. \tag{25.52}$$

Thus, from the inequalities (25.50), (25.51), and (25.52), we obtain the inequalities[5]

$$\underline{u}(x) \leq u(x) \leq \overline{u}(x) \quad \text{for all } x \in \Omega.$$

Therefore, $\underline{u}(x)$ is a minimal solution and $\overline{u}(x)$ is a maximal solution of the Dirichlet problem (25.3). The theorem is proved. □

25.3　Bibliographical Notes

The contents of this lecture is taken from [69].

---

[5]Note that this result makes sense since the original Dirichlet problem has a countable set of linearly independent solutions in $H_0^1(\Omega)$ (see Lecture 16).

Method of Upper and Lower Solutions for Parabolic Equations

26.1 Interpolation Inequality

Let $D = \Omega \otimes (0, T)$, where $\Omega \subset \mathbb{R}^N$ is a bounded domain with smooth boundary $\partial \Omega$ and $T > 0$ is finite. For simplicity, we consider the case of a convex domain $\Omega \subset \mathbb{R}^N$. Below, we denote by $\partial' D$ the parabolic boundary of the domain D[1]

$$\partial' D \overset{\text{def}}{=\!=} B \cup S, \quad B \overset{\text{def}}{=\!=} \Omega \otimes \{t = 0\}, \quad S \overset{\text{def}}{=\!=} \partial \Omega \otimes [0, T],$$

$$B_T \overset{\text{def}}{=\!=} \Omega \otimes \{t = T\}, \quad \partial B_T \overset{\text{def}}{=\!=} \partial \Omega \otimes \{t = T\}.$$

We recall the following notation:

$$|u|_{2+\alpha, 1+\alpha/2; D} \overset{\text{def}}{=\!=} |u|_{0;D} + \sum_{i=1}^{N} |u_{x_i}|_{0;D} + |u_t|_{0;D}$$

$$+ \sum_{i,j=1,1}^{N,N} |u_{x_i x_j}|_{0;D} + [u]_{2+\alpha; 1+\alpha/2; D} < +\infty, \tag{26.1}$$

where

$$[u]_{2+\alpha, 1+\alpha/2; D} \overset{\text{def}}{=\!=} [u_t]_{\alpha, \alpha/2; D} + \sum_{i,j=1,1}^{N,N} [u_{x_i x_j}]_{\alpha, \alpha/2; D}, \tag{26.2}$$

$$[u]_{\alpha, \alpha/2; D} \overset{\text{def}}{=\!=} \sup_{\substack{z_1 \neq z_2, \\ z_1, z_2 \in D}} \frac{|u(z_1) - u(z_2)|}{\rho^\alpha(z_1, z_2)}, \tag{26.3}$$

$$|u|_{\alpha, \alpha/2; D} \overset{\text{def}}{=\!=} |u|_{0;D} + [u]_{\alpha, \alpha/2; D}, \quad |u|_{0;D} = \sup_{(x,t) \in D} |u(x,t)| \tag{26.4}$$

for $\alpha \in (0, 1]$; $\rho(z_1, z_2) = |x_1 - x_2| + |t_1 - t_2|^{1/2}$ is the so-called parabolic distance.

We denote by $C^{\alpha, \alpha/2}(\overline{D})$ the Banach space of all functions $u(x, t) \in C(\overline{D})$ with finite norm $|u|_{\alpha, \alpha/2; D} < +\infty$ and by $C^{2+\alpha, 1+\alpha/2}(\overline{D})$ the Banach space of real-valued functions $u(x, t) \in C_{x,t}^{2,1}(\overline{D})$ with finite norm $|u|_{2+\alpha, 1+\alpha/2; D} < +\infty$.

In what follows, we need several estimates of $|h|_{\alpha, \alpha/2; D}$ for functions $h(x, t, u) \in C^{\alpha, \alpha/2, \alpha}(\overline{D} \otimes \mathbb{R}^1)$ under the condition $u(x, t) \in C^{2+\alpha, 1+\alpha/2}(\overline{D})$.

---

[1]Recall that the complete boundary is $\partial D = B_T \cup B \cup S$.

First, we note that following chain of inequalities holds:

$$|h|_{\alpha,\alpha/2;D} = |h|_{0;D}$$

$$+ \sup_{z_1 \neq z_2,\ z_1,z_2 \in D} \frac{|h(x_1,t_1,u_1) - h(x_2,t_2,u_2)|}{\rho^\alpha\big((z_1,u_1),(z_2,u_2)\big)} \frac{\rho^\alpha\big((z_1,u_1),(z_2,u_2)\big)}{\rho^\alpha(z_1,z_2)}$$

$$\leq |h|_{0;D} + K_1\left(1 + |u_t|_{0;D} + \sum_{i=1}^{N} |u_{x_i}|_{0;D}\right)$$

$$\leq |h|_{0;D} + K_1 + \varepsilon|u|_{2+\alpha,1+\alpha/2;D} + c_1(\varepsilon)|u|_{0;D}, \qquad (26.5)$$

where

$$\rho(z_1,z_2) \overset{\text{def}}{=\!=} |x_1 - x_2| + |t_1 - t_2|^{1/2},$$

$$\rho\big((z_1,u_1),(z_2,u_2)\big) \overset{\text{def}}{=\!=} \rho(z_1,z_2) + |u_1 - u_2|,$$

$u_1 := u(x_1,t_1)$, $u_2 := u(x_2,t_2)$, $\varepsilon > 0$. Indeed,

$$\frac{|x_1 - x_2| + |t_1 - t_2|^{1/2} + |u_1(x_1,t_1) - u_2(x_2,t_2)|}{|x_1 - x_2| + |t_1 - t_2|^{1/2}}$$

$$\leq 1 + \frac{|u(x_1,t_1) - u(x_2,t_2)|}{|x_1 - x_2| + |t_1 - t_2|^{1/2}},$$

$$|u(x_1,t_1) - u(x_2,t_2)| \leq |u(x_1,t_1) - u(x_2,t_1)| + |u(x_2,t_1) - u(x_2,t_2)|$$

$$\leq \left|\int_0^1 \frac{\partial u\big(sx_1 + (1-s)x_2, t_1\big)}{\partial s}\, ds\right| + \left|\int_0^1 \frac{\partial u\big(x_2, st_1 + (1-s)t_2\big)}{\partial s}\, ds\right|$$

$$\leq \sup_{(x,t)\in D} |\nabla_x u(x,t)| \cdot |x_1 - x_2| + \sup_{(x,t)\in D} |u_t(x,t)| \cdot |t_1 - t_2|$$

$$\leq K\left[\sum_{i=1}^{N} |u_{x_i}|_{0;D} + |u_t|_{0;D}\right]\big[|x_1 - x_2| + |t_1 - t_2|\big],$$

$$\frac{|x_1 - x_2| + |t_1 - t_2|^{1/2} + |u_1(x_1,t_1) - u_2(x_2,t_2)|}{|x_1 - x_2| + |t_1 - t_2|^{1/2}}$$

$$\leq 1 + K\left[\sum_{i=1}^{N} |u_{x_i}|_{0;D} + |u_t|_{0;D}\right]\big[1 + |t_1 - t_2|^{1/2}\big]$$

$$\leq K_1\left(1 + \sum_{i=1}^{N} |u_{x_i}|_{0;D} + |u_t|_{0;D}\right), \qquad K_1 = K_1(T).$$

Next, we use the well-known interpolation inequality

$$\sum_{j=1}^{N} |u_{x_j x_j}|_{0;D} + |u_t|_{0;D} + \sum_{j=1}^{N} |u_{x_j}|_{\alpha,\alpha/2;D} + |u|_{\alpha,\alpha/2;D} \qquad (26.6)$$

$$\leq \varepsilon|u|_{2+\alpha,1+\alpha/2;D} + c(\varepsilon)|u|_{0;D}, \qquad (26.7)$$

which is valid for all $u(x,t) \in C^{2+\alpha,1+\alpha/2}(\overline{D})$ and all $\varepsilon > 0$ (see [37]).

Remark 26.1. Instead of the condition $h(x, t, u) \in C^{\alpha, \alpha/2, \alpha}(\overline{D} \otimes \mathbb{R}^1)$, we can impose the following conditions:

$$h(x, t, u) \in C_{x,t,u}^{0,0,1}(\overline{D} \otimes \mathbb{R}^1), \tag{26.8}$$

$$\left| h(x_1, t_1, u) - h(x_2, t_2, u) \right| \leq K_1 \left[|x_1 - x_2| + |t_1 - t_2|^{1/2} \right]^\alpha, \quad |u| \leq M_1, \tag{26.9}$$

$$\max \left\{ \left| h(x, t, u) \right|, \left| h_u'(x, t, u) \right| \right\} \leq K_2, \quad |u| \leq M_1, \tag{26.10}$$

where K_1 and K_2 are positive constants independent of (x_1, t_1, u) and (x_2, t_2, u) and of (x, t, u), respectively.

26.2 Definitions of Upper and Lower Solutions

In this section, we apply the method of upper and lower solutions to the proof of the existence of a classical solution $u(x, t) \in C_{x,t}^{2,1}(\overline{D})$ to the first boundary-value problem for a semilinear parabolic equation in $D \in \Omega \otimes (0, T)$, where $\Omega \subset \mathbb{R}^N$ is a bounded domain with smooth boundary $\partial \Omega$.

Consider the following first boundary-value problem:

$$\frac{\partial u}{\partial t} - Lu(x, t) = f(x, t, u) \quad \text{in } (x, t) \in D \cup B_T, \tag{26.11}$$

$$u(x, t) = g(x, t) \qquad \text{in } (x, t) \in \partial' D, \tag{26.12}$$

where

$$Lu(x, t) \overset{\text{def}}{=\!=} \sum_{i,j=1,1}^{N,N} a_{ij}(x, t) \frac{\partial^2 u(x, t)}{\partial x_i \partial x_j} + \sum_{i=1}^N b_i(x, t) \frac{\partial u(x, t)}{\partial x_i} + c(x, t) u(x, t), \tag{26.13}$$

$$a_{ij}(x, t), \; b_i(x, t), \; c(x, t) \in C^{\alpha, \alpha/2}(\overline{D}), \quad a_{ij} = a_{ji}, \quad c(x, t) \leq 0. \tag{26.14}$$

Assume that there exist constants $\lambda > 0$ and $\Lambda > 0$ such that

$$\lambda |\xi|^2 \leq \sum_{i,j=1,1}^{N,N} a_{ij}(x, t) \xi_i \xi_j \leq \Lambda |\xi|^2 \tag{26.15}$$

for all $\xi \in \mathbb{R}^N$ and $(x, t) \in D$. Moreover, we assume that the function $f(x, t, u)$ belongs to the class $C^{\alpha, \alpha/2, \alpha}(\overline{D} \otimes \mathbb{R}^1)$ and the function $g(x, t)$ can be extended to a function of the class $C^{2+\alpha, 1+\alpha/2}(\overline{D})$ for some $\alpha \in (0, 1]$. Obviously, function $g(x, t)$ must satisfy the equation $g_t - Lg = f(x, t, g)$ on the parabolic boundary $\partial' D$.

Note that due to the classical Schauder *a priori* estimate (see, e.g., [37]), solutions of the problem

$$2\frac{\partial u}{\partial t} - Lu(x, t) = \hat{f}(x, t) \in C^{\alpha, \alpha/2}(\overline{D}), \tag{26.16}$$

$$u(x, t) = g(x, t) \in C^{2+\alpha, 1+\alpha/2}(\overline{D}) \tag{26.17}$$

satisfy the following *a priori* estimate :

$$|u|_{2+\alpha, 1+\alpha/2; D} \leq K(N, D, \alpha) \left(|\hat{f}|_{\alpha, \alpha/2} + |g|_{2+\alpha, 1+\alpha/2} \right) \tag{26.18}$$

(here we assume that there exists a extension of the function $g(x,t)$ of the required class from the parabolic boundary $\partial' D$ to the closure $\overline{D}$.)

Definition 26.1. A function $\overline{U}(x,t) \in C^{2,1}_{x,t}(D \cup B_T) \cap C(\overline{D})$ is called an *upper solution* of the first boundary-value problem (26.11), (26.12) if

$$\frac{\partial \overline{U}(x,t)}{\partial t} - L\overline{U}(x,t) \geq f(x,t,\overline{U}) \quad \text{for } (x,t) \in D, \tag{26.19}$$

$$\overline{U}(x,t) \geq g(x,t) \qquad\qquad\qquad \text{for } (x,t) \in \partial' D. \tag{26.20}$$

Definition 26.2. A function $\underline{U}(x,t) \in C^{2,1}_{x,t}(D \cup B_T) \cap C(\overline{D})$ is called a *lower solution* of the first boundary-value problem (26.11), (26.12) if

$$\frac{\partial \underline{U}(x,t)}{\partial t} - L\underline{U}(x,t) \leq f(x,t,\underline{U}) \quad \text{for } (x,t) \in D, \tag{26.21}$$

$$\underline{U}(x,t) \leq g(x,t) \qquad\qquad\qquad \text{for } (x,t) \in \partial' D. \tag{26.22}$$

In the sequel, we assume that the upper and lower solutions satisfy the inequality

$$\overline{U}(x,t) \geq \underline{U}(x,t) \quad \text{for } (x,t) \in \overline{D}; \tag{26.23}$$

note that there exist examples in which this inequality is violated. In the case where the inequality (26.23) holds, we say that the functions $\overline{U}(x,t)$ and $\underline{U}(x,t)$ are ordered.

Definition 26.3. For any ordered pair $\overline{U}(x,t)$ and $\underline{U}(x,t)$, we introduce the set

$$\langle \underline{U}, \overline{U} \rangle \overset{\text{def}}{=} \left\{ u(x,t) \in C(\overline{D}) : \underline{U}(x,t) \leq u(x,t) \leq \overline{U}(x,t), \ (x,t) \in \overline{D} \right\}. \tag{26.24}$$

Assume that a nonlinear function $f(x,t,u)$ satisfy the one-sided Lipschitz condition:

$$f(x,t,u_1) - f(x,t,u_2) \geq -\lambda(u_1 - u_2), \quad \underline{U} \leq u_2 \leq u_1 \leq \overline{U}, \tag{26.25}$$

where $\lambda \geq 0$ is a constant.

Example 26.1. For example, the function

$$f(x,t,u) = f_0(x,t)|u|^{p-2}u, \quad f_0(x,t) \geq 0, \quad p > 1,$$

satisfies the condition (26.25) with $\lambda = 0$.

Note that due to the condition (26.25),

$$F(x,t,u) \overset{\text{def}}{=} \lambda u + f(x,t,u) \tag{26.26}$$

is a monotonically nondecreasing function of u for all $(x,t) \in D$ and all $u \in \langle \underline{U}, \overline{U} \rangle$. We rewrite Eq. (26.11) in the following equivalent form:

$$\frac{\partial u}{\partial t} - Lu(x,t) + \lambda u = F(x,t,u), \quad (x,t) \in D \cup B_T. \tag{26.27}$$

26.3 Iterative Scheme

Consider the following iterative scheme:

$$\frac{\partial u_k}{\partial t} - L u_k(x,t) + \lambda u_k = F(x,t,u_{k-1}) \quad \text{in } (x,t) \in D \cup B_T, \qquad (26.28)$$

$$u_k(x,t) = g(x,t) \qquad\qquad\qquad\quad \text{on } (x,t) \in \partial' D, \qquad (26.29)$$

where $u_0(x,t)$ is an arbitrary function of the class $C^{2,1}_{x,t}(D \cup B_T) \cap C(\overline{D})$.

In the linear $C^{2+\alpha,1+\alpha/2}$-theory of parabolic equations, the following facts were proved (see, e.g., [37]): the first iteration $u_1(x,t)$ belongs to the class $C^{\alpha,\alpha/2}(\overline{D})$, whereas for the subsequent iterations we have

$$u_k(x,t) \in C^{2+\alpha,1+\alpha/2}(\overline{D}), \quad k = 2,3,\dots.$$

Now, as the initial approximation $u_0(x,t) \in C^{2,1}_{x,t}(D) \cap C(\overline{D})$ we take

$$\text{either } u_0(x,t) = \overline{U}(x,t) \text{ or } u_0(x,t) = \underline{U}(x,t), \qquad (26.30)$$

which satisfy the condition (26.23). We denote by $\overline{u}_k(x,t)$ and $\underline{u}_k(x,t)$, $k \in \mathbb{N}$, the iterative sequences of solutions of the problem (26.28), (26.29) with the initial conditions $u_0(x,t) = \overline{U}(x,t)$ and $u_0(x,t) = \underline{U}(x,t)$, respectively.

Lemma 26.1. *Let $\overline{U}(x,t)$ and $\underline{U}(x,t)$ be ordered upper and lower solutions of the first boundary-value problem (26.11), (26.12) and let a function $f(x,t,u)$ satisfy the condition (26.25). Then the sequences $\{\overline{u}_k(x,t)\}$ and $\{\underline{u}_k(x,t)\}$ possess the following monotonicity property:*

$$\underline{U}(x,t) = \underline{u}_0(x,t) \le \underline{u}_k(x,t) \le \underline{u}_{k+1}(x,t)$$
$$\le \overline{u}_{k+1}(x,t) \le \overline{u}_k(x,t) \le \overline{u}_0(x,t) = \overline{U}(x,t) \qquad (26.31)$$

for all $k \in \mathbb{N}$.

Proof. 1. Let $w(x,t) = \overline{u}_0(x,t) - \overline{u}_1(x,t) = \overline{U}(x,t) - \overline{u}_1(x,t)$, $(x,t) \in \overline{D}$. Then $w(x,t) \in C^{2,1}_{x,t}(D \cup B_T) \cap C(\overline{D})$ is a solution of the problem

$$\frac{\partial w}{\partial t} - L w + \lambda w \ge F(x,t,\overline{U}) - F(x,t,\overline{U}) = 0 \quad \text{for } (x,t) \in D \cup B_T,$$

$$w(x,t) \ge g(x,t) - g(x,t) = 0 \qquad\qquad \text{for } (x,t) \in \partial' D.$$

The weak maximum principle yields the inequality $w(x,t) \ge 0$ on $\overline{D}$, i.e.,

$$\overline{u}_1(x,t) \le \overline{u}_0(x,t) = \overline{U}(x,t), \quad (x,t) \in \overline{D}.$$

Similarly, one can obtain the inequality

$$\underline{u}_1(x,t) \ge \underline{u}_0(x,t) = \underline{U}(x,t), \quad (x,t) \in \overline{D}.$$

2. Now let $w_1(x,t) = \overline{u}_1(x,t) - \underline{u}_1(x,t)$, $(x,t) \in \overline{D}$. Then the function $w_1(x,t) \in C^{2,1}_{x,t}(D \cup B_T) \cap C(\overline{D})$ satisfies the relations

$$\frac{\partial w_1}{\partial t} - L w_1 + \lambda w_1 = F(x,t,\overline{U}) - F(x,t,\underline{U}) \ge 0 \quad \text{for } (x,t) \in D \cup B_T,$$

$$w_1(x,t) \ge g(x,t) - g(x,t) = 0 \qquad\qquad \text{for } (x,t) \in \partial' D.$$

Due to the weak maximum principle we have $w_1(x,t) \geq 0$, $(x,t) \in \overline{D}$. Therefore,

$$\underline{u}_0(x,t) = \underline{U}(x,t) \leq \underline{u}_1(x,t) \leq \overline{u}_1(x,t) \leq \overline{U}(x,t) = \overline{u}_0(x,t), \quad (x,t) \in \overline{D}.$$

3. Assume that $\underline{u}_{k-1}(x,t) \leq \underline{u}_k(x,t) \leq \overline{u}_k(x,t) \leq \overline{u}_{k-1}(x,t)$, $(x,t) \in \overline{D}$. Then the function

$$w_k(x,t) = \overline{u}_k(x,t) - \overline{u}_{k+1}(x,t), \quad (x,t) \in \overline{D},$$

satisfies the relations

$$\frac{\partial w_k}{\partial t} - Lw_k + \lambda w_k \geq F(x,t,\overline{u}_{k-1}) - F(x,t,\overline{u}_k) \geq 0 \quad \text{for } (x,t) \in D,$$

$$w_k(x,t) = g(x,t) - g(x,t) = 0 \qquad\qquad\qquad \text{for } (x,t) \in \partial'D.$$

Due to the weak maximum principle we see $w_k(x,t) \geq 0$ in $\overline{D}$, i.e.,

$$\overline{u}_{k+1}(x,t) \leq \overline{u}_k(x,t), \quad (x,t) \in \overline{D}.$$

Similarly, one can prove that

$$\underline{u}_{k+1}(x,t) \leq \underline{u}_k(x,t), \quad \underline{u}_{k+1}(x,t) \leq \overline{u}_{k+1}(x,t), \quad (x,t) \in \overline{D}. \qquad \square$$

Now we prove the main existence theorem. First, we note that the iterative sequence $\{\overline{u}_k(x,t)\}$ is a monotonically nonincreasing and lower bounded lower solution $\underline{U}(x,t)$ whereas the iterative sequence $\{\underline{u}_k(x,t)\}$ is a monotonically nondecreasing and upper bounded upper solution $\overline{U}(x,t)$. Therefore, there exist limits

$$\lim_{k\to+\infty} \overline{u}_k(x,t) = \overline{u}(x,t), \quad \lim_{k\to+\infty} \underline{u}_k(x,t) = \underline{u}(x,t), \quad (x,t) \in \overline{D}, \qquad (26.32)$$

and, moreover,

$$\underline{U}(x,t) \leq \underline{u}(x,t) \leq \overline{u}(x,t) \leq \overline{U}(x,t), \quad (x,t) \in \overline{D}.$$

We prove below that both functions $\overline{u}(x,t)$ and $\underline{u}(x,t)$ are solutions of the first boundary-value problem (26.11), (26.12). Moreover, if there exists a constant $\mu \geq 0$ such that

$$f(x,t,u_1) - f(x,t,u_2) \geq -\mu(u_1 - u_2), \quad \underline{U} \leq u_1 \leq u_2 \leq \overline{U}, \qquad (26.33)$$

then the solution of the first boundary-value problem is unique.

26.4   Main Theorem

Theorem 26.1. *Let $\overline{U}(x,t)$ and $\underline{U}(x,t)$ be ordered upper and lower solutions of the first boundary-value problem (26.11), (26.12) and let a function $f(x,t,u)$ satisfy the conditions (26.8)–(26.10) and (26.25). Then the following assertions hold:*

(i) *the sequence $\{\overline{u}_k(x,t)\}$ converges monotonically from above to a solution $\overline{u}(x,t) \in C^{2+\alpha,1+\alpha/2}(\overline{D})$ of the first boundary-value problem (26.11), (26.12), whereas the sequence $\{\underline{u}_k(x,t)\}$ converges monotonically from below to a solution $\underline{u}(x,t) \in C^{2+\alpha,1+\alpha/2}(\overline{D})$ of the same first boundary-value problem; moreover,*

$$\underline{u}(x,t) \leq \overline{u}(x,t), \quad (x,t) \in \overline{D}; \qquad (26.34)$$

(ii) *any solution* $u^*(x,t) \in \langle \underline{U}, \overline{U} \rangle$ *of the first boundary-value problem* (26.11), (26.12) *satisfies the inequality*

$$\underline{u}(x,t) \le u^*(x,t) \le \overline{u}(x,t), \quad (x,t) \in \overline{D}; \tag{26.35}$$

(iii) *if, in addition, the condition* (26.33) *is fulfilled, then* $\overline{u}(x,t) = \underline{u}(x,t)$ *is a unique solution in* $\langle \underline{U}, \overline{U} \rangle$.

Proof. 1. Since the sequence $\{\overline{u}_k(x,t)\}$ monotonically converges to $\overline{u}(x,t)$ from above, due to the monotonicity of the function $F(x,t,u)$, the sequence $\{F(x,t,\overline{u}_k)\}$ monotonically converges to $F(\overline{u},x,t)$ from above for all $(x,t) \in \overline{D}$. Similarly, the sequence $\{F(\underline{u}_k,x,t)\}$ monotonically converges to $F(\underline{u},x,t)$ from below. We see earlier that

$$\underline{u}_1(x,t), \ \overline{u}_1(x,t) \in C^{\alpha,\alpha/2}(\overline{D}), \quad \underline{u}_k(x,t), \ \overline{u}_k(x,t) \in C^{2+\alpha,1+\alpha/2}(\overline{D}), \quad k = 2,3,\dots.$$

Owing to classical Schauder *a priori* estimates for both iterative schemes (26.28)–(26.30), we have the following estimates:

$$\big|\overline{u}_k\big|_{2+\alpha,1+\alpha/2;D} \le K\Big(\big|g\big|_{2+\alpha,1+\alpha/2;D} + \big|F(x,t,\overline{u}_{k-1})\big|_{\alpha,\alpha/2;D}\Big), \tag{26.36}$$

$$\big|\underline{u}_k\big|_{2+\alpha,1+\alpha/2;D} \le K\Big(\big|g\big|_{2+\alpha,1+\alpha/2;D} + \big|F(x,t,\underline{u}_{k-1})\big|_{\alpha,\alpha/2;D}\Big), \tag{26.37}$$

where $k \ge 2$ and the constant $K = K(N,D,\alpha) > 0$ is independent of $k \in \mathbb{N}$. Due to the inequality (26.5), we obtain the following chain of inequalities (note that here we denote both iterative sequences by $\{u_k(x,t)\}$):

$$\begin{aligned}
\big|F(x,t,u_{k-1})\big|_{\alpha,\alpha/2;D} &\le \lambda\big|u_{k-1}\big|_{\alpha,\alpha/2;D} + \big|f(x,t,u_{k-1})\big|_{\alpha,\alpha/2;D} \\
&\le \varepsilon\big|u_{k-1}\big|_{2+\alpha,1+\alpha/2;D} + c_2(\varepsilon)\big|u_{k-1}\big|_{0;D} + \big|f(x,t,u_{k-1})\big|_{0;D} \\
&\quad + K_1 + \varepsilon\big|u_{k-1}\big|_{2+\alpha,1+\alpha/2;D} + c_1(\varepsilon)\big|u_{k-1}\big|_{0;D} \\
&= 2\varepsilon\big|u_{k-1}\big|_{2+\alpha,1+\alpha/2;D} \\
&\quad + K_1 + c_3(\varepsilon)\big|u_{k-1}\big|_{0;D} + \big|f(x,t,u_{k-1})\big|_{0;D}. \tag{26.38}
\end{aligned}$$

Note that

$$\big|u_{k-1}\big|_{0;D} \le \sup_{x \in D}\Big\{\big|\underline{U}(x)\big|, \ \big|\overline{U}(x)\big|\Big\} = c_4$$

and due to (26.10), we arrive at the inequality

$$\big|f(x,t,u_{k-1})\big|_{0;D} \le c_5.$$

Using the inequalities (26.36), (26.37), and (26.38), we obtain the iterative inequality

$$\big|u_k\big|_{2+\alpha,1+\alpha/2;D} \le 2\varepsilon K\big|u_{k-1}\big|_{2+\alpha,1+\alpha/2;D} + K_2(\varepsilon). \tag{26.39}$$

For brevity, we will write ε instead of $2\varepsilon K$; then from the inequality (26.39) we obtain the inequality

$$\big|u_k\big|_{2+\alpha,1+\alpha/2;D} \le \varepsilon\big|u_{k-1}\big|_{2+\alpha,1+\alpha/2;D} + K_3(\varepsilon), \quad k \ge 3. \tag{26.40}$$

Separately, we consider the iterative inequality

$$z_{k+1} \leq \varepsilon z_k + d, \quad \varepsilon \in (0,1), \quad k \geq 2.$$

It implies the estimate

$$z_k \leq z_2 + d \sum_{n=1}^{+\infty} \varepsilon^n = K_4(\varepsilon) < +\infty, \quad \text{for } k \geq 2.$$

Therefore, both sequences $\{\underline{u}_k(x,t)\}$ and $\{\overline{u}_k(x,t)\}$ are uniformly bounded in the space $C^{2+\alpha,1+\alpha/2}(\overline{D})$ with respect to $k \geq 2$. Therefore, there exist subsequences (see [43]) $\{\overline{u}_\mu\}$ and $\{\underline{u}_\mu\}$ such that

$$\overline{u}_\mu \to \overline{u} \quad \text{strongly in } C_{x,t}^{2,1}(\overline{D}) \text{ as } \mu \to +\infty,$$
$$\underline{u}_\mu \to \underline{u} \quad \text{strongly in } C_{x,t}^{2,1}(\overline{D}) \text{ as } \mu \to +\infty;$$

moreover, the limit functions are such that

$$\underline{u}(x,t), \ \overline{u}(x,t) \in C^{2+\alpha,1+\alpha/2}(\overline{D}).$$

It remains to pass to the limit as $\mu \to +\infty$ in Eq. (26.28) and (26.29) for the corresponding subsequences. The property (i) is proved.

2. Note that if $u^*(x,t) \in \langle \underline{U}, \overline{U} \rangle$ is a solution of the class $C_{x,t}^{2,1}(D \cup B_T) \cap C(\overline{D})$, then, obviously, $u^*(x,t)$ is an upper and lower solution simultaneously. Therefore, for the ordered pairs $\langle u^*, \overline{U} \rangle$ and $\langle \underline{U}, u^* \rangle$, the iterative scheme leads us to the conclusion that

$$u^*(x,t) \leq \overline{u}(x,t) \quad \text{for all } (x,t) \in \overline{D},$$
$$u^*(x,t) \geq \underline{u}(x,t) \quad \text{for all } (x,t) \in \overline{D}.$$

Thus, the property (ii) is also proved.

3. To prove the property (iii), it suffices to verify that

$$\overline{u}(x,t) \leq \underline{u}(x,t) \quad \text{for all } (x,t) \in \overline{D}. \tag{26.41}$$

Indeed, the function

$$w(x,t) = \underline{u}(x,t) - \overline{u}(x,t), \quad (x,t) \in \overline{D},$$

satisfies the relations

$$\frac{\partial w}{\partial t} - Lw = f(x,t,\underline{u}) - f(x,t,\overline{u}) \geq -\mu w \quad \text{for } (x,t) \in D\cup,$$
$$w(x,t) = g(x,t) - g(x,t) = 0 \quad\quad\quad\quad \text{for } (x,t) \in \partial'D.$$

Therefore, due to the weak maximum principle, we have $w(x,t) \geq 0$ in $\overline{D}$. Thus, the property (iii) is established. Theorem 26.1 is completely proved. $\qquad\square$

26.5   Bibliographical Notes

The contents of this lecture is taken from [69].

Method of Upper and Lower Weak Solutions

In this lecture, we consider the method of weak upper and lower solutions for the following nonlinear inhomogeneous p-Laplacian equation in a bounded domain $\Omega \subset \mathbb{R}^3$ with smooth boundary $\partial\Omega$:

$$-\Delta_p u = f(x, u), \ x \in \Omega, \quad u = 0, \ x \in \partial\Omega, \quad 2 < p < 3, \tag{27.1}$$

where $f(x, u) : \overline{\Omega} \otimes \mathbb{R}^1 \to \mathbb{R}^1$ is a continuous function and

$$|f(x, u)| \le a_1 + b_1 |u|^{q+1}, \ q > 0, \quad q + 2 \le p^* = \frac{3p}{3 - p}$$

with constants $a_1 \ge 0$ and $b_1 > 0$. Moreover, we assume that

$$\text{either } f_u'(x, u) \ge 0 \text{ or } |f_u'(x, u)| \le a \tag{27.2}$$

for all $x \in \Omega$ and $u \in \mathbb{R}^1$. Recall that

$$\Delta_p u := \operatorname{div}(|D_x u|^{p-2} D_x u).$$

Lemma 27.1 (see [69, Proposition 1.3.10]). If $f(x) \in W_0^{1,p}(\Omega)$, then $(f(x))^+ := \max\{f(x), 0\} \in W_0^{1,p}(\Omega)$ and the following equality holds:

$$D_x(f(x))^+ = \begin{cases} D_x f(x) & \text{for } f(x) \ge 0, \\ 0 & \text{for } f(x) < 0. \end{cases} \tag{27.3}$$

Definition 27.1.

(i) A function $\overline{U}(x) \in W^{1,p}(\Omega) \cap C(\overline{\Omega})$ is called a *weak upper solution* of the problem (27.1) if

$$\int_\Omega \left(|D_x \overline{U}(x)|^{p-2} D_x \overline{U}(x), \ D_x v \right) dx \ge \int_\Omega f(x, \overline{U}(x)) v(x) \, dx \tag{27.4}$$

for any function $v(x) \in W_0^{1,p}(\Omega)$, $v(x) \ge 0$ almost everywhere.

(ii) A function $\underline{U}(x) \in W^{1,p}(\Omega) \cap C(\overline{\Omega})$ is called a *weak lower solution* of the problem (27.1) if

$$\int_\Omega \left(|D_x \underline{U}(x)|^{p-2} D_x \underline{U}(x), \ D_x v(x) \right) dx \le \int_\Omega f(x, \underline{U}(x)) v(x) \, dx \tag{27.5}$$

for any function $v(x) \in W_0^{1,p}(\Omega)$, $v(x) \ge 0$ almost everywhere.

(iii) A function $u(x) \in W_0^{1,p}(\Omega) \cap C(\overline{\Omega})$ is called a *weak solution* of the problem (27.1) if

$$\int_\Omega \left(|D_x u(x)|^{p-2} D_x u(x), \; D_x v(x) \right) dx = \int_\Omega f(x, u(x)) v(x) \, dx \qquad (27.6)$$

for any function $v(x) \in W_0^{1,p}(\Omega)$.

Remark 27.1. If $\overline{U}(x), \underline{U}(x) \in C^2(\Omega) \cap C(\overline{\Omega})$, then from (27.4) and (27.5) we obtain

$$-\Delta_p \overline{U}(x) \geq f(x, \overline{U}(x)), \quad -\Delta_p \underline{U}(x) \leq f(x, \underline{U}(x)) \quad \text{in } \Omega,$$

which corresponds to the classical definitions of upper and lower solution. The idea of the proof of this assertion is as follows. Assume that at a certain point $x_0 \in \Omega$, we have $-\Delta_p \overline{U} < f(x, \overline{U})$; then, due to the continuity of this expression, this inequality is valid in some closed neighborhood of this point. It suffices to take $v(x) \geq 0$ whose support lies in this closed neighborhood and obtain a contradiction with the definition of a weak upper solution $\overline{U}(x)$.

Theorem 27.1. *Assume that there exist upper $\overline{U}(x)$ and lower $\underline{U}(x)$ solutions of the problem* (27.1) *such that*

$$\underline{U}(x) \leq 0, \quad \overline{U}(x) \geq 0 \quad \text{on } \partial\Omega, \qquad \underline{U}(x) \leq \overline{U}(x) \quad \text{a.e. in } \Omega. \qquad (27.7)$$

Then there exists a weak solution $u(x)$ of the problem (27.1) *such that*

$$\underline{U}(x) \leq u(x) \leq \overline{U}(x) \quad \text{a.e. in } \Omega.$$

Proof. 1. We fix a sufficiently large number $\lambda > 0$ such that the mapping

$$z \to f(x, z) + \lambda z \qquad (27.8)$$

does not decrease for all $x \in \Omega$; this is possible due to (27.2).

Now we set $u_0(x) = \underline{U}(x)$ and construct the sequence $u_k(x)$, $k = 0, 1, 2, \ldots$, by the following recurrent rule: $u_{k+1}(x) \in W_0^{1,p}(\Omega)$ is a unique weak solution of the boundary-value problem

$$-\Delta_p u_{k+1} + \lambda u_{k+1} = f(x, u_k) + \lambda u_k \text{ in } \Omega, \quad u_{k+1} = 0 \text{ on } \partial\Omega; \qquad (27.9)$$

such a solution exists due to the Browder–Minty theorem (see Theorem 20.1).

2. We prove that

$$\underline{U} = u_0 \leq u_1 \leq \cdots \leq u_k \leq u_{k+1} \cdots \quad \text{a.e. in } \Omega. \qquad (27.10)$$

First, we note that, due to (27.9),

$$\int_\Omega \left(\left(|D_x u_1|^{p-2} D_x u_1, D_x v \right) + \lambda u_1 v \right) dx = \int_\Omega \left(f(x, u_0) + \lambda u_0 \right) v \, dx \qquad (27.11)$$

for any $v \in W_0^{1,p}(\Omega)$. Subtracting (27.11) from (27.5), we obtain the following inequality:

$$\int_\Omega \left[\left(|D_x u_0|^{p-2} D_x u_0 - |D_x u_1|^{p-2} D_x u_1, D_x v \right) + \lambda(u_0 - u_1, v) \right] dx \le 0, \quad u_0 = \underline{U}.$$

Setting

$$v = (u_0 - u_1)^+ \in W_0^{1,p}(\Omega), \quad v \ge 0 \text{ a.e.}$$

(note that $v(x) \in W_0^{1,p}(\Omega)$ due to Lemma 27.1 and the facts $u_1(x) \in W_0^1(\Omega)$ and $u_0(x) = \underline{U}(x) \le 0$ for almost all $x \in \partial\Omega$), we obtain

$$\int_\Omega \left(|D_x u_0|^{p-2} D_x u_0 - |D_x u_1|^{p-2} D_x u_1, \right.$$

$$\left. D_x(u_0 - u_1)^+ + \lambda(u_0 - u_1)(u_0 - u_1)^+ \right) dx \le 0. \tag{27.12}$$

However,

$$D_x(u_0 - u_1)^+ = \begin{cases} D_x(u_0 - u_1) & \text{a.e. on } \{u_0 \ge u_1\}, \\ 0 & \text{a.e. on } \{u_1 \ge u_0\}. \end{cases}$$

Now, applying the inequality (21.6), we arrive at the inequality

$$\int_{u_0 \ge u_1} \left(|D_x u_0|^{p-2} D_x u_0 - |D_x u_1|^{p-2} D_x u_1, \ D_x u_0 - D_x u_1 \right) dx$$

$$\ge 2^{2-p} \int_{u_0 \ge u_1} |D_x u_0 - D_x u_1|^p dx.$$

Therefore,

$$\int_{u_0 \ge u_1} \left[2^{2-p} |D_x(u_0 - u_1)|^p + \lambda(u_0 - u_1)^2 \right] dx \le 0;$$

hence $u_0(x) \le u_1(x)$ almost everywhere on Ω. Now, by induction, we assume that

$$u_{k-1}(x) \le u_k(x) \quad \text{a.e. in } \Omega. \tag{27.13}$$

From (27.9) we have

$$\int_\Omega \left[\left(|D_x u_{k+1}|^{p-2} D_x u_{k+1}, \ D_x v \right) + \lambda u_{k+1} v \right] dx = \int_\Omega (f(x, u_k) + \lambda u_k) v \, dx,$$

$$\tag{27.14}$$

$$\int_\Omega \left[\left(|D_x u_k|^{p-2} D_x u_k, \ D_x v \right) + \lambda u_k v \right] dx = \int_\Omega (f(x, u_{k-1}) + \lambda u_{k-1}) v \, dx \tag{27.15}$$

for any $v \in W_0^{1,p}(\Omega)$. Subtracting and setting

$$v = (u_k - u_{k+1})^+ \in W_0^{1,p}(\Omega),$$

we have

$$\int\limits_{u_k \geq u_{k+1}} \left[2^{2-p} |D_x(u_k - u_{k+1})|^p + \lambda (u_k - u_{k+1})^2 \right] dx$$

$$\leq \int\limits_{\Omega} \left[\big(f(x, u_{k-1}) + \lambda u_{k-1} \big) - \big(f(x, u_k) + \lambda u_k \big) \right] \big(u_k - u_{k+1} \big)^+ dx \leq 0.$$

The last inequality is valid due to (27.13) and (27.8). Therefore, $u_k \leq u_{k+1}$ almost everywhere in Ω, as was asserted.

3. Now we prove that

$$u_k \leq \overline{U} \quad \text{a.e. in } \Omega, \quad k = 0, 1, 2, \ldots. \tag{27.16}$$

For $k = 0$, (27.16) is valid due to (27.7). Let for $k \in \mathbb{N}$,

$$u_k \leq \overline{U} \quad \text{a.e. in } \Omega. \tag{27.17}$$

We prove that then $u_k \leq \overline{U}$ almost everywhere in Ω. Subtracting (27.4) from (27.14) and setting $v = \big(u_{k+1} - \overline{U} \big)^+$, we obtain

$$\int\limits_{u_{k+1} \geq \overline{U}} \left[2^{2-p} |D_x(u_{k+1} - \overline{U})|^p + \lambda (u_{k+1} - \overline{U})^2 \right] dx$$

$$\leq \int\limits_{\Omega} \left[\big(f(x, u_k) + \lambda u_k \big) - \big(f(x, \overline{U}) + \lambda \overline{U} \big) \right] \big(u_{k+1} - \overline{U} \big)^+ dx \leq 0$$

due to (27.17) and (27.8). Thus, $u_{k+1} \leq \overline{U}$ almost everywhere in Ω.

4. Owing to (27.10) and (27.16), we have

$$\underline{U} \leq \cdots \leq u_k \leq u_{k+1} \leq \cdots \leq \overline{U} \quad \text{a.e. in } \Omega. \tag{27.18}$$

Therefore, the limit

$$u(x) := \lim_{k \to +\infty} u_k(x) \tag{27.19}$$

exists for almost all $x \in \Omega$. Moreover,

$$u_k \to u \quad \text{strongly in } L^{q+2}(\Omega) \subset L^2(\Omega), \quad q \geq 0 \tag{27.20}$$

due to the dominated convergence theorem and (27.18). Indeed, we have

$$|u_k(x) - u(x)| \leq 2V(x) < +\infty,$$

$$V(x) \overset{\text{def}}{=} \max \left\{ |\underline{U}(x)|, \; |\overline{U}(x)| \right\} \in C(\overline{\Omega}).$$

This together with (27.19) implies the required assertion.

Next, due to the fact that $f(x, u)$ is a Carathéodory function with the growth condition (27.2), the corresponding Nemytsky operator (see Lecture 6)

$$N_f(u) : L^{q+2}(\Omega) \to L^{(q+2)/(q+1)}(\Omega)$$

is continuous, i.e., in particular, by (27.20)

$$\|N_f(u_n) - N_f(u)\|_{(q+2)/(q+1)} \to +0 \quad \text{as } n \to +\infty. \tag{27.21}$$

Multiplying (27.9) by $u_{k+1} \in W_0^{1,p}(\Omega)$ using the scalar product generated by the duality bracket

$$\langle \cdot, \cdot \rangle : W^{-1,p'}(\Omega) \otimes W_0^{1,p}(\Omega) \to \mathbb{R}^1,$$

we obtain the equality

$$\langle -\Delta_p u_{k+1} + \lambda u_{k+1}, u_{k+1} \rangle = \langle f(x, u_k) + \lambda u_k, u_{k+1} \rangle.$$

Integrating by parts, we have

$$\||D_x u_{k+1}|\|_p^p + \lambda \|u_{k+1}\|_2^2 = \int_\Omega f(x, u_k) u_{k+1}\, dx + \lambda \int_\Omega u_k u_{k+1}\, dx. \tag{27.22}$$

The following chain of inequalities holds:

$$\left| \int_\Omega f(x, u_k) u_{k+1}\, dx \right| \le a_1 \int_\Omega |u_{k+1}|\, dx + b_1 \int_\Omega |u_k|^{q+1}|u_{k+1}|\, dx$$

$$\le \varepsilon \|u_{k+1}\|_2^2 + c_1(\varepsilon) + b_1 \|u_k\|_{q+2}^{q+1} \|u_{k+1}\|_{q+2}. \tag{27.23}$$

Note that $\underline{U}(x) \le u_k(x) \le \overline{U}(x)$ a.e. in Ω and, moreover,

$$\underline{U}(x), \overline{U}(x) \in W^{1,p}(\Omega) \subset L^{q+2}(\Omega).$$

Therefore,

$$|u_k(x)| \le V(x) \stackrel{\text{def}}{=} \max\left\{ |\underline{U}(x)|, |\overline{U}(x)| \right\} \in L^{q+2}(\Omega) \subset L^2(\Omega)$$

for almost all $x \in \Omega$. Then we have

$$b_1 \|u_k\|_{q+2}^{q+1} \|u_{k+1}\|_{q+2} \le b_1 \|V(x)\|_{q+2}^{q+1} K_{fr} \||D_x u_{k+1}|\|_p, \tag{27.24}$$

where K_{fr} is the Friedrichs constant. Applying the three-parameter Young inequality with small $\varepsilon > 0$ to the right-hand side of (27.24), we arrive at the inequality

$$\left| \int_\Omega f(x, u_k) u_{k+1}\, dx \right| \le c_2(\varepsilon) + \varepsilon \|u_{k+1}\|_2^2 + \varepsilon \||D_x u_{k+1}|\|_p^p. \tag{27.25}$$

Finally, the following obvious inequality holds:

$$\lambda \left| \int_\Omega u_k u_{k+1}\, dx \right| \le \varepsilon \|u_{k+1}\|_2^2 + c(\varepsilon) \|u_k\|_2^2 \le \varepsilon \|u_{k+1}\|_2^2 + c(\varepsilon) \|V(x)\|_2^2. \tag{27.26}$$

Due to (27.23), (27.25), and (27.26), from (27.22) we obtain the inequality

$$(1 - \varepsilon)\||D_x u_{k+1}|\|_p^p + (\lambda - 2\varepsilon)\|u_{k+1}\|_2^2 \le c_3(\varepsilon) \tag{27.27}$$

for $\varepsilon \in (0, \min\{1, \lambda/2\})$. From the estimate (27.27) we conclude that the sequence $\{u_k\}$ is uniformly (with respect to $k \in \mathbb{N}$) bounded in $W_0^{1,p}(\Omega)$. Since the Banach space $W_0^{1,p}(\Omega)$ is reflexive (and hence weakly closed), there exists a subsequence $\{u_{k_j}\} \subset \{u_k\}$ such that

$$u_{k_j}(x) \rightharpoonup u(x) \quad \text{weakly in } W_0^{1,p}(\Omega) \text{ as } k_j \to +\infty. \tag{27.28}$$

Note that, due to (27.9), the following equality holds:

$$\langle -\Delta_p u_{k_j}, u_{k_j} - u \rangle = \langle -\lambda u_{k_j} + \lambda u_{k_j-1} + f(x, u_{k_j-1}), u_{k_j} - u \rangle$$

$$= \int_\Omega \left(-\lambda u_{k_j}(x) + \lambda u_{k_j-1}(x) + f\left(x, u_{k_j-1}(x)\right) \right)$$

$$\times \left(u_{k_j}(x) - u(x) \right) dx := I. \tag{27.29}$$

For the integral I, the following estimate holds:

$$|I| \le \left\| u_{k_j} - u \right\|_{q+2} \left\| -\lambda u_{k_j} + \lambda u_{k_j-1} + f(x, u_{k_j-1}) \right\|_{(q+2)/(q+1)}$$

$$\le B(\lambda) \| u_{k_j} - u \|_{q+2} \to +0 \tag{27.30}$$

as $k_j \to +\infty$. Here we used the continuous embedding

$$L^{q+2}(\Omega) \subset L^{(q+2)/(q+1)}(\Omega)$$

and the limit properties (27.20) and (27.21). Indeed,

$$\| u_{k_j} \|_{(q+2)/(q+1)} \to \| u \|_{(q+2)/(q+1)} \qquad \text{as } k_j \to +\infty,$$

$$\| N_f(u_{k_j-1}) \|_{(q+2)/(q+1)} \to \| N_f(u) \|_{(q+2)/(q+1)} \quad \text{as } k_j \to +\infty$$

and hence the numerical sequence

$$\| u_{k_j} \|_{(q+2)/(q+1)}, \quad \| u_{k_j-1} \|_{(q+2)/(q+1)}, \quad \| f(x, u_{k_j-1}) \|_{(q+2)/(q+1)}$$

are bounded.

Thus, from (27.29) and (27.30) we obtain that

$$\lim_{k_j \to +\infty} \langle -\Delta u_{k_j}, u_{k_j} - u \rangle = 0. \tag{27.31}$$

Due to the S^+-property of the p-Laplacian (see Lecture 19) and the limit properties (27.28) and (27.31), we conclude that

$$u_{k_j}(x) \to u(x) \quad \text{strongly in } W_0^{1,p}(\Omega) \text{ as } k_j \to +\infty. \tag{27.32}$$

Therefore, by Lemma 21.5, we obtain

$$\Delta_p u_{k_j}(x) \to \Delta_p u(x) \quad \text{strongly in } W^{-1,p'}(\Omega) \text{ as } k_j \to +\infty. \tag{27.33}$$

5. Finally, we verify that u is a weak solution of the problem (27.1). We fix $v(x) \in W_0^{1,p}(\Omega) \subset L^{q+2}(\Omega)$. Then from (27.9) we have

$$\int_\Omega \left[\left(|D_x u_{k_j}|^{p-2} D_x u_{k_j}, \, D_x v \right) + \lambda u_{k_j} v \right] dx = \int_\Omega \left(f(x, u_{k_j-1}) + \lambda u_{k_j-1} \right) v \, dx. \tag{27.34}$$

Letting k_j tend to $+\infty$, due to (27.20), (27.21), and (27.33), we obtain the limit equality

$$\int_\Omega \left[\left(|D_x u|^{p-2} D_x u, \, D_x v \right) + \lambda u v \right] dx = \int_\Omega \left(f(x, u) + \lambda u \right) v \, dx.$$

Cancelling the term involving λ, we arrive at the required equality

$$\int_\Omega \left(|D_x u|^{p-2} D_x u, \, D_x v \right) dx = \int_\Omega f(x, u) v \, dx \quad \text{for all } v(x) \in W_0^{1,p}(\Omega).$$

The theorem is proved. $\qquad\qquad\qquad\qquad\qquad\qquad\qquad\qquad\qquad\qquad\qquad$ $\square$

Bibliographical Notes

The contents of this lecture is taken from [19].

PART 6

Schauder Principle
and Contraction Mapping Theorem

Topological Schauder Principle

28.1 Contraction Mapping Theorem

The method based on the contraction mapping theorem is one of the most popular methods of nonlinear analysis. Recall the related notions and facts.

Definition 28.1. A point $f \in \operatorname{dom} A$ is called a *fixed point* of an operator A if $f = Af$.

Recall the following definitions.

Definition 28.2. We say that an operator $A : \mathbb{B} \to \mathbb{B}$ satisfies the *Lipschitz condition* on $D \subset \mathbb{B}$ if there exists $0 < q < +\infty$ such that
$$\|Af - Ag\| \leq q\,\|f - g\| \quad \text{for all } f, g \in D.$$
In this case, the number $q > 0$ is called the *Lipschitz constant* and the operator A is said to be *Lipschitz continuous*.

Definition 28.3. An operator A satisfying the Lipschitz condition with constant $q \in (0, 1)$ is called a *contraction mapping* or a *contractive operator*.

Lemma 28.1. *In the inequality*
$$\|f_{n+1} - f_n\| \leq q\|f_n - f_{n-1}\| \tag{28.1}$$
holds for $n \geq 1$, where $q \in (0, 1)$, then for any $n \geq 1$
$$\|f_{n+k} - f_n\| \leq q^n(1 - q)^{-1}\|f_1 - f_0\| \quad \text{for } k \geq 1. \tag{28.2}$$
Thus, the sequence $\{f_n\}$ is a Cauchy sequence in the Banach space $\mathbb{B}$.

Proof. By induction, from (28.1) we have
$$\|f_{n+1} - f_n\| \leq q^n\|f_1 - f_0\|, \quad n \geq 1.$$
Therefore, for $k \geq 1$
$$\|f_{n+k} - f_n\| = \left\|\sum_{j=1}^{k}(f_{n+j} - f_{n+j-1})\right\| \leq \sum_{j=1}^{k}\|f_{n+j} - f_{n+j-1}\|$$
$$\leq \|f_1 - f_0\|\sum_{j=1}^{k}q^{n+j-1} \leq q^n(1 - q)^{-1}\|f_1 - f_0\|.$$

Since $q < 1$, the right-hand side tends to zero as $n \to +\infty$. Therefore, $\{f_n\}$ is a Cauchy sequence.   □

Theorem 28.1 (contraction mapping theorem). *Assume that an operator A maps a closed subset D of a Banach space $\mathbb{B}$ into D and is a contraction mapping on D. Then A possesses in D a unique fixed point, say f. Further, for any initial value $f_0 \in D$, consecutive approximations $f_{n+1} = Af_n$, $n \geq 0$, converge to f and the following estimate of the convergence rate is valid:*

$$\|f - f_n\| \leq q^n (1 - q)^{-1} \|Af_0 - f_0\|. \tag{28.3}$$

Proof. 1. First, we note that all terms of the iterative sequence $\{f_n\}$ belong to D. Indeed, since $f_0 \in D$, we conclude that $f_1 = Af_0 \in D$. Assume that $f_n \in D$; then, obviously, $f_{n+1} = Af_n \in D$. Therefore, by induction we have $\{f_n\} \subset D$.

Since A is a contractive operator on $D \subset \mathbb{B}$, we have

$$\|f_{n+1} - f_n\| = \|Af_n - Af_{n-1}\| \leq q \|f_n - f_{n-1}\|.$$

This implies the inequality

$$\|f_{n+1} - f_n\| \leq q \|f_n - f_{n-1}\| \quad \text{for } n \geq 1 \text{ and } q \in (0, 1). \tag{28.4}$$

Lemma 28.1 implies that for $n > m$

$$\|f_n - f_m\| \leq q^m (1 - q)^{-1} \|Af_0 - f_0\| \quad \text{for } k \geq 1, \tag{28.5}$$

since $f_1 = Af_0$. This proves that the sequence $\{f_n\} \subset D$ constructed by $f_0 \in D$ is a Cauchy sequence in the Banach space $\mathbb{B}$.

2. By (28.5) the sequence $\{f_n\}$ is a cauchy sequence in $\mathbb{B}$ and hence strongly converges in $\mathbb{B}$ to some $\overline{f} \in \mathbb{B}$. Due to the closedness of $D \subset \mathbb{B}$, we conclude that $\overline{f} \in D$.

Due to the continuity of A on the closed set $D \subset \mathbb{B}$ (which follows from the contractive property of the operator A on D), we obtain the following chain of limit equalities:

$$A\overline{f} = \lim_{n \to +\infty} Af_n = \lim_{n \to +\infty} f_{n+1} = \overline{f}, \tag{28.6}$$

i.e., $\overline{f}$ is a fixed point. Note that the limits in (28.6) are treated in the sense of the strong convergence in $\mathbb{B}$.

3. To prove the uniqueness, we assume that $\overline{g}$ is another fixed point of A. Then

$$\|\overline{f} - \overline{g}\| = \|A\overline{f} - A\overline{g}\| \leq q \|\overline{f} - \overline{g}\|.$$

Since $0 < q < 1$, this means that $\overline{f} = \overline{g}$.

4. The estimate of the convergence rate is obtained from (28.2) by the passing to the limit as $k \to +\infty$. The theorem is proved.   □

28.2 Schauder Fixed-Point Theorem

First, we recall the famous Brouwer fixed-point theorem in a finite-dimensional space.

Theorem 28.2 (Brouwer fixed-point theorem). *Let D be a bounded, closed, convex subset of a finite-dimensional normed vector space. If A is a continuous mapping of D into itself, then A has a fixed point in D.*

Definition 28.4. Let $\mathbb{B}$ be a Banach space and M be a finite set

$$M := \{x_i \in \mathbb{B} : \ i = 1, \ldots, n\}.$$

The set of all linear combinations of the form

$$\left\{ \sum_{i=1}^{n} \lambda_i x_i : \ \sum_{i=1}^{n} \lambda_i = 1, \ \lambda_i \geq 0, \ x_i \in M, \ i = \overline{1, n} \right\}$$

is called a *convex hull* $\mathrm{Co}(M)$ of the set M.

The Brouwer theorem allows one to prove various fixed-point theorems for non-linear operators in infinite-dimensional Banach space.

Theorem 28.3 (Schauder fixed-point theorem or Schauder principle). *Let an operator A maps a closed, bounded, convex set D of a Banach space $\mathbb{B}$ into itself. If A is completely continuous[1] on D, then it has a fixed point in D.*

Proof. 1. Assume that the operator A has no fixed points on D. Then there exists $\varepsilon_0 > 0$ such that

$$\|A(x) - x\| \geq \varepsilon_0 \quad \forall x \in D. \tag{28.7}$$

Indeed, in the opposite case, there exists a sequence $\{x_n\} \subset D$ such that

$$\|A(x_n) - x_n\| \to 0, \quad n \to +\infty. \tag{28.8}$$

Then, due to the compactness of $\overline{A(D)}$ in $\mathbb{B}$, from the sequence $\{A(x_n)\} \subset \overline{A(D)}$ we can extract a subsequence $\{A(x_{n'})\}$ such that

$$A(x_{n'}) \to x_0 \quad \text{strongly in } \mathbb{B} \text{ as } n' \to +\infty.$$

Moreover, $x_0 \in \overline{A(D)}$ due to the closedness of $\overline{A(D)}$. Note that the following inequality holds:

$$\|x_{n'} - x_0\| \leq \|x_{n'} - A(x_{n'})\| + \|A(x_{n'}) - x_0\|.$$

This inequality and (28.8) imply that

$$x_{n'} \to x_0 \quad \text{strongly in } \mathbb{B} \text{ as } n' \to +\infty.$$

Moreover, $x_0 \in D$ since D is closed. Setting $n = n'$ in (28.8) and passing to the limit as $n' \to +\infty$, due to the continuity of the operator A we obtain $A(x_0) = x_0$,

---

[1] I.e., compact and continuous.

which contradicts the assumption on the absence of fixed points of the operator A on D. Thus, the inequality (28.7) holds.

2. Further, we assume that $\theta \in D$ (recall that $\theta \in \mathbb{B}$ is the zero element in $\mathbb{B}$). This condition is not restrictive. Indeed, let $y_0 \in D$; we consider the set $D_0 := D - y_0$ and the operator

$$A_0 x := A(x + y_0) - y_0.$$

We prove that D_0 is closed. Let $\{x_n\} \subset D_0$ and

$$x_n \to x_0 \quad \text{strongly in } \mathbb{B} \text{ as } n \to +\infty.$$

Then $\{x_n + y_0\} \subset D$ and

$$x_n + y_0 \to x_0 + y_0 \quad \text{as } n \to +\infty.$$

The set D is closed in $\mathbb{B}$ and hence $x_0 + y_0 \in D$. Therefore, $x_0 \in D_0 = D - y_0$. Thus, the set D_0 is closed.

Now we prove that D_0 is convex. Let $x_1, x_2 \in D_0$. Then $z_1 = x_1 + y_0 \in D$ and $z_2 = x_2 + y_0 \in D$. The set D is convex; therefore,

$$tz_1 + (1-t)z_2 \in D \quad \text{for all } t \in [0,1]$$
$$\Rightarrow t(x_1 + y_0) + (1-t)(x_2 + y_0) = tx_1 + (1-t)x_2 + y_0 \in D$$
$$\Rightarrow tx_1 + (1-t)x_2 \in D - y_0 = D_0 \quad \text{for all } t \in [0,1].$$

Therefore, the set D_0 is convex.

Prove that A_0 is a completely continuous operator. Let $\{x_n\} \subset D_0$ and

$$x_n \to x_0 \quad \text{strongly in } \mathbb{B} \text{ as } n \to +\infty;$$

then

$$x_n + y_0 \to x_0 + y_0 \quad \text{strongly in } \mathbb{B} \text{ as } n \to +\infty.$$

Due to the continuity of the operator A, we have

$$A(x_n + y_0) \to A(x_0 + y_0) \quad \text{strongly in } \mathbb{B} \text{ as } n \to +\infty.$$

Therefore,

$$A_0(x_n) = A(x_n + y_0) - y_0 \to A(x_0 + y_0) - y_0 = A_0(x_0) \qquad (28.9)$$

strongly in $\mathbb{B}$ as $n \to +\infty$. Prove the complete continuity of the operator A_0 on D_0. Let $\{x_n\} \subset D_0$ be a bounded sequence. Then $\{x_n + y_0\} \subset D$ is bounded and, due to the complete continuity of the operator A, there exists a subsequence $\{x_{n_k} + y_0\} \subset D$ such that

$$A(x_{n_k} + y_0) \to v \quad \text{strongly in } \mathbb{B} \text{ as } n_k \to +\infty;$$

then we have

$$A_0(x_{n_k}) = A(x_{n_k} + y_0) - y_0 \to v - y_0 \qquad (28.10)$$

strongly in $\mathbb{B}$ as $n \to +\infty$. From the limit relations (28.9) and (28.10) we conclude that the operator A_0 is completely continuous on D_0.

Now we note that if $x_0 \in D_0$ is a fixed point of the operator A_0 on a set D_0, then the point $x_0 + y_0 \in D$ is a fixed point of the operator A. Indeed, the following chain of equalities holds:

$$x_0 = A_0(x_0) = A(x_0 + y_0) - y_0 \quad \Leftrightarrow \quad A(x_0 + y_0) = x_0 + y_0. \tag{28.11}$$

Therefore, without loss of generality, we may assume that $\theta \in D$.

3. Fix an arbitrary number $\varepsilon \in (0, \varepsilon_0)$. Let

$$M_\varepsilon := \{y_i \in \overline{A(D)}, \ i = 1, \ldots, n\}$$

be a finite ε-net of the compact set $\overline{A(D)}$. In the set M_ε, consider a maximal linearly independent system of elements; for example, we can assume that it consists of elements of the set

$$N_\varepsilon := \{y_i, \ i = 1, \ldots, m\}, \quad m \leq n.$$

Consider the m-dimensional Banach space $\mathbb{B}_m$ spanned by elements of the set N_ε; obviously, it is a subspace of the Banach space $\mathbb{B}$. Let

$$K_\varepsilon := \overline{\mathrm{Co}(\theta \cup M_\varepsilon)}$$

be the closure of the convex hull of the set consisting of the zero element $\theta \in \mathbb{B}$ and the points of the finite ε-net M_ε. Obviously, $K_\varepsilon \subset \mathbb{B}_m$. Since by the condition of the theorem $\overline{A(D)} \subset D$ and D is convex, closed, and contains θ, we see that $K_\varepsilon \subset D$.

4. Consider the operator $A_\varepsilon : D \to D$ defined as follows:

$$A_\varepsilon(x) := \frac{\sum\limits_{i=1}^{n} \mu_i(x) y_i}{\sum\limits_{i=1}^{n} \mu_i(x)} \quad \text{for all } x \in D, \tag{28.12}$$

where

$$\mu_i(x) := \begin{cases} 0 & \text{if } \|A(x) - y_i\| \geq \varepsilon, \\ \varepsilon - \|A(x) - y_i\| & \text{if } \|A(x) - y_i\| < \varepsilon. \end{cases} \tag{28.13}$$

The operator A_ε is often called the Schauder ε-projector.

5. Now we consider the restriction of the operator A_ε to the set K_ε. Prove that A_ε maps K_ε into itself and A_ε is continuous on K_ε. The first assertion follows from the facts that the operator is defined on $D \supset K_\varepsilon$ and

$$A_\varepsilon(x) = \sum\limits_{i=1}^{n} \lambda_i(x) y_i \subset \overline{\mathrm{Co}(\theta \cup M_\varepsilon)} = K_\varepsilon \quad \text{for all } x \in K_\varepsilon$$

(we take into account the convexity of K_ε), where

$$\lambda_i(x) := \frac{\mu_i(x)}{\sum\limits_{j=1}^{n} \mu_j(x)} \geq 0, \quad \sum\limits_{i=1}^{n} \lambda_i(x) = 1 \quad \text{for } x \in K_\varepsilon.$$

Prove the second assertion. Let $\{x_k\} \subset K_\varepsilon$ be an arbitrary sequence such that

$$x_k \to x \quad \text{strongly in } \mathbb{B} \text{ as } k \to +\infty.$$

Due to the closedness of K_ε, we have $x \in K_\varepsilon$. By the definition of A_ε, we have the equality

$$A_\varepsilon(x_k) - A_\varepsilon(x) = \sum_{i=1}^{n} \Big[\lambda_i(x_k) - \lambda_i(x)\Big] y_i, \quad \lambda_i(x) := \frac{\mu_i(x)}{\sum\limits_{j=1}^{n} \mu_j(x)}. \tag{28.14}$$

Since $K_\varepsilon \subset \overline{A(D)}$, for any $x \in K_\varepsilon$ there exists a number $j_0 \in \overline{1,n}$ such that $\mu_{j_0}(x) > 0$. Note that the norm possesses the following property:

$$\big|\|A(x) - y_i\| - \|A(x_k) - y_i\|\big| \le \|A(x) - A(x_k)\| \to +0 \quad \text{as } n \to +\infty,$$

since the operator A is continuous on D. This implies that[2]

$$\mu_i(x_k) \to \mu_i(x) \quad \text{as } k \to +\infty, \ i = \overline{1,n}. \tag{28.15}$$

The following chain of inequalities holds:

$$\sum_{i=1}^{n} |\lambda_i(x) - \lambda_i(x_k)|$$

$$= \frac{1}{\sum\limits_{j=1}^{n} \mu_j(x) \sum\limits_{j=1}^{n} \mu_j(x_k)} \sum_{i=1}^{n} \left| \mu_i(x) \sum_{j=1}^{n} \mu_j(x_k) - \mu_i(x_k) \sum_{j=1}^{n} \mu_j(x) \right|$$

$$\le \frac{\sum\limits_{i=1}^{n} |\mu_i(x) - \mu_i(x_k)|}{\sum\limits_{j=1}^{n} \mu_j(x)} + \frac{\sum\limits_{j=1}^{n} |\mu_j(x_k) - \mu_j(x)|}{\sum\limits_{j=1}^{n} \mu_j(x)}$$

$$= 2\frac{\sum\limits_{i=1}^{n} |\mu_i(x) - \mu_i(x_k)|}{\sum\limits_{j=1}^{n} \mu_j(x)} \to +0 \quad \text{as } k \to +\infty. \tag{28.16}$$

Therefore, from (28.14) and (28.16) we obtain the limit relation

$$\|A_\varepsilon(x_k) - A_\varepsilon(x)\| \le \sum_{i=1}^{n} |\lambda_i(x_k) - \lambda_i(x)| \, \|y_i\|$$

$$\le \max_{i=\overline{1,n}} \|y_i\| \sum_{i=1}^{n} |\lambda_i(x_k) - \lambda_i(x)| \to +0 \quad \text{as } k \to +\infty.$$

The continuity of A_ε on K_ε is proved.

Thus, we may apply the Brouwer fixed-point theorem (Theorem 28.2) to the restriction of the operator A_ε to the closed, convex, bounded set K_ε; we conclude that there exists a fixed point $x_\varepsilon \in K_\varepsilon$ of the operator A_ε, i.e., $A_\varepsilon(x_\varepsilon) = x_\varepsilon$.

---

[2]Recall that the real-valued function $x^+ := \max\{x, 0\}$ is continuous for $x \in \mathbb{R}^1$.

6. Note that the operator A_ε possesses the following property:

$$\|A(x) - A_\varepsilon(x)\| \leq \varepsilon \tag{28.17}$$

for all $x \in D$, i.e., the operator A_ε approximates the operator A on D with accuracy ε. Indeed,

$$A(x) - A_\varepsilon(x) = \frac{\sum\limits_{i=1}^{n} \mu_i(x) A(x)}{\sum\limits_{i=1}^{n} \mu_i(x)} - \frac{\sum\limits_{i=1}^{n} \mu_i(x) y_i}{\sum\limits_{i=1}^{n} \mu_i(x)} = \frac{\sum\limits_{i=1}^{n} \mu_i(x)(A(x) - y_i)}{\sum\limits_{i=1}^{n} \mu_i(x)}.$$

This implies that

$$\|A(x) - A_\varepsilon(x)\| \leq \frac{\sum\limits_{i=1}^{n} \mu_i(x)\|A(x) - y_i\|}{\sum\limits_{i=1}^{n} \mu_i(x)},$$

where summing in the numerator and the denominator is performed only by subscripts i for which $\|A(x) - y_i\| < \varepsilon$, since

$$\mu_i(x) = 0 \quad \text{for} \quad \|A(x) - y_i\| \geq \varepsilon.$$

Therefore,

$$\|A(x) - A_\varepsilon(x)\| < \frac{\sum\limits_{i=1}^{n} \mu_i(x)\varepsilon}{\sum\limits_{i=1}^{n} \mu_i(x)} = \varepsilon.$$

7. Due to (28.17) we have

$$\|A(x_\varepsilon) - x_\varepsilon\| = \|A(x_\varepsilon) - A_\varepsilon(x_\varepsilon)\| < \varepsilon.$$

This contradicts the inequality (28.7) since we have chosen $\varepsilon \in (0, \varepsilon_0)$. Therefore, the assumption on the absence of fixed points of the operator A on D is invalid. The Schauder theorem is proved. □

The following consequence of the Schauder principle is very important in the theory of nonlinear boundary-value problems.

Corollary 28.1. *Let A be a completely continuous mapping of a Banach space $\mathbb{B}$ into itself. Assume that there exists a constant $M > 0$ such that for all pairs $(x, \alpha) \in \mathbb{B} \times [0, 1]$ satisfying the equation*

$$x = \alpha A(x), \tag{28.18}$$

the inequality

$$\|x\| < M \tag{28.19}$$

holds, where the constant M is independent of the choice of x and α. Then the operator A has a fixed point.

Proof. 1. Without loss of generality, we may assume that $M = 1$. Indeed, performing the substitution $x = My$, we transform Eq. (28.18) to the form $y = \alpha \hat{A}(y)$, where $\hat{A}(y) := A(My)/M$, and the inequality (28.19) to the form $\|y\| < 1$. If the operator A is completely continuous, then the operator $\hat{A}$ is also completely continuous, and the converse assertion is also valid.

Introduce the mapping

$$A^*(x) := \begin{cases} A(x) & \text{if } \|A(x)\| \le 1, \\ A(x)/\|A(x)\| & \text{if } \|A(x)\| \ge 1. \end{cases}$$

We prove that this mapping transforms the unit ball $D_1 = \{x \in \mathbb{B} : \|x\| \le 1\}$ into itself. Indeed, let $x \in D_1$; then the following two cases are possible:

 (i) $\|A(x)\| \le 1$,
 (ii) $\|A(x)\| > 1$.

In both cases, we have $\|A^*x\| \le 1$.

2. Now we obtain an estimate of the norm of the difference $A^*(x_1) - A^*(x_2)$ (for the norm $\|\cdot\|$ of the Banach space $\mathbb{B}$) in the case where $x_1, x_2 \in D_1$. The following two cases are possible:

 (a) $\|A(x_1)\| \le 1$ and $\|A(x_2)\| \le 1$;
 (b) $\|A(x_1)\| \ge 1$ and $\|A(x_2)\| \ge 1$.

In the case (a), we obtain the equality

$$\|A^*(x_1) - A^*(x_2)\| = \|A(x_1) - A(x_2)\|. \tag{28.20}$$

In the case (b), we have

$$\begin{aligned}
\|A^*&(x_1) - A^*(x_2)\| \\
&\le \left\| \frac{A(x_1)}{\|A(x_1)\|} - \frac{A(x_2)}{\|A(x_2)\|} \right\| \\
&\le \frac{1}{\|A(x_1)\|\,\|A(x_2)\|} \Big\| \|A(x_2)\|A(x_1) - \|A(x_1)\|A(x_2) \Big\| \\
&= \frac{1}{\|A(x_1)\|\,\|A(x_2)\|} \Big\| \|A(x_2)\| \big[A(x_1) - A(x_2) \big] + \big[\|A(x_2)\| - \|A(x_1)\| \big] A(x_2) \Big\| \\
&\le \frac{1}{\|A(x_1)\|} \|A(x_1) - A(x_2)\| + \frac{1}{\|A(x_1)\|} \big| \|A(x_2)\| - \|A(x_1)\| \big| \\
&\le 2\|A(x_1) - A(x_2)\|, \tag{28.21}
\end{aligned}$$

where we used the inequality

$$\big| \|A(x_2)\| - \|A(x_1)\| \big| \le \|A(x_1) - A(x_2)\|.$$

The inequalities (28.20) and (28.21) imply that if the operator A is continuous and completely continuous, then the operator A^* also possesses these properties.

First, we prove the continuity. Let $\{x_n\} \subset D_1$ and

$$x_n \to x \quad \text{strongly in } \mathbb{B} \text{ as } n \to +\infty.$$

Due to the closedness of D_1, we have $x \in D_1$. Due to the continuity of A, we have the limit property

$$A(x_n) \to A(x) \quad \text{strongly in } \mathbb{B} \text{ as } n \to +\infty. \tag{28.22}$$

The following three cases are possible:

$$\|A(x)\| < 1, \quad \|A(x)\| > 1, \quad \text{or} \quad \|A(x)\| = 1. \tag{28.23}$$

In the first case, we have

$$\|A(x_n)\| < 1 \quad \text{for } n \geq n_0$$
$$\Rightarrow \quad \|A^*(x)_n - A^*(x)\| = \|A(x_n) - A(x)\| \to +0 \quad \text{as } n \to +\infty.$$

In the second case, the proof is similar, but one must use the estimate (28.21) and conclude that

$$\|A^*(x_n) - A^*(x)\| \leq 2\|A(x_n) - A(x)\| \to +0 \quad \text{as } n \to +\infty. \tag{28.24}$$

In the third case, either the estimate (28.20) or the estimate (28.21) holds, depending on the location of the point $A(x_n)$: inside or outside of the ball. In any case, the rougher estimate (28.21) holds, and we again arrive at the limit property (28.24). The continuity is proved.

Similarly, we can prove the complete continuity of the operator A^* on D_1. Indeed, let $\{x_n\} \subset D_1$ be an arbitrary sequence. Due to the compactness of A on D_1, there exists a subsequence $\{x_{n_m}\} \subset \{x_n\}$ such that

$$A(x_{n_m}) \to v \quad \text{strongly in } \mathbb{B} \text{ as } m \to +\infty. \tag{28.25}$$

We consider the following three cases:

$$\|v\| < 1, \quad \|v\| > 1, \quad \text{and} \quad \|v\| = 1.$$

In the first case, we use the estimate (28.20) and obtain, as above, the limit property

$$\|A^*(x_{n_m}) - v\| = \|A(x_{n_m}) - v\| \to +0 \quad \text{as } m \to +\infty.$$

In the second and third cases, we use the estimate (28.21) and obtain the limit property

$$\left\| A^*(x_{n_m}) - \frac{v}{\|v\|} \right\| \leq 2\|A(x_{n_m}) - v\| \to +0 \quad \text{as } m \to +\infty.$$

3. Therefore, due to the Schauder principle (see Theorem 28.3), we conclude that the operator A^* has a fixed point x_0. We prove that the point x_0 is a fixed point of the mapping A.

Indeed, assume that $\|Ax_0\| \geq 1$. Then $x_0 = A^*(x_0) = \alpha A(x_0)$, where $\alpha = 1/\|A(x_0)\|$, and hence $\|x_0\| = \|A^*(x_0)\| = 1$: this contradicts the inequality (28.19) with the constant $M = 1$.

Therefore, the assumption $\|A(x_0)\| \geq 1$ is invalid, i.e., $\|A(x_0)\| < 1$. Then $x_0 = A^*(x_0) = A(x_0)$. Corollary 28.1 is proved. $\qquad \square$

28.3   Quasilinear Equation with p-Laplacian

Consider the following classical problem for $p \geq 2$:

$$\Delta_p u \stackrel{\text{def}}{=} \text{div}(|D_x u|^{p-2} D_x u) = -f(x, u), \quad x \in \Omega, \qquad (28.26)$$

$$u(x) = 0, \qquad\qquad\qquad\qquad\qquad x \in \partial\Omega, \qquad (28.27)$$

where $\Omega \subset \mathbb{R}^N$ is a bounded domain with smooth boundary $\partial\Omega \in C^{2,\delta}$, $\delta \in (0, 1]$. Introduce the notation

$$p^* := \begin{cases} Np/(N-p) & \text{for } p < N, \\ \infty & \text{for } p \geq N. \end{cases}$$

Assume that $f : \Omega \times \mathbb{R}^1 \to \mathbb{R}^1$ is a Carathéodory function (see Lecture 6) satisfying the growth condition

$$|f(x, s)| \leq c|s|^{q-1} + b(x), \quad x \in \Omega, \quad s \in \mathbb{R}^1, \quad q \in (1, p^*), \qquad (28.28)$$

where $c > 0$ is a constant, $b(x) \in L^{q'}(\Omega)$, $b(x) \geq 0$ almost everywhere in Ω, and

$$\frac{1}{q} + \frac{1}{q'} = 1.$$

The condition $q \in (1, p^*)$ guarantees the complete continuity of the embedding $W_0^{1,p}(\Omega) \hookrightarrow\hookrightarrow L^q(\Omega)$.

To the Carathéodory function $f(x, u)$, we assign the Nemytsky operator

$$N_f(u)(x) \stackrel{\text{def}}{=} f(x, u(x)), \quad N_f : L^q(\Omega) \to L^{q'}(\Omega);$$

by the Krasnosel'sky theorem (see Theorem 6.2), it is a continuous and bounded operator. Introduce the embedding operator

$$J : W_0^{1,p}(\Omega) \to L^q(\Omega); \qquad (28.29)$$

its transposed operator (see Theorem 1.3)

$$J^t : L^{q'}(\Omega) \to W^{-1,p'}(\Omega) \qquad (28.30)$$

is continuous. Consider the operator

$$\hat{N}_f(u) := J^t N_f(Ju) : W_0^{1,p}(\Omega) \to W^{-1,p'}(\Omega). \qquad (28.31)$$

Lemma 28.2. *The operator $\hat{N}_f$ defined by Eq. (28.31) is completely continuous and bounded.*

Proof. The continuity and boundedness follow from the continuity and boundedness of all three operators generating the operator $\hat{N}_f(u)$.

Prove the complete continuity. Let $\{u_n\} \subset W_0^{1,p}(\Omega)$ be a bounded sequence. Due to the complete continuity of the operator J, there exists a subsequence $\{u_{n_k}\} \subset \{u_n\}$ such that

$$J u_{n_k} \to v \quad \text{strongly in } L^q(\Omega) \text{ as } n_k \to +\infty. \qquad (28.32)$$

Then, due to the continuity of the operators $N_f(u)$ and J^t, we conclude that

$$\hat{N}_f(u_{n_k}) \to J^t N_f(v) \quad \text{strongly in } W^{-1,p'}(\Omega) \text{ as } n_k \to +\infty. \qquad (28.33)$$

The lemma is proved. $\square$

Definition 28.5. A function $u \in W_0^{1,p}(\Omega)$ satisfying the equation

$$\langle -\Delta_p u, v \rangle = \left\langle \hat{N}_f(u), v \right\rangle \quad \text{for all } v \in W_0^{1,p}(\Omega), \tag{28.34}$$

where $\langle \cdot, \cdot \rangle$ is the duality bracket between the Banach spaces $W_0^{1,p}(\Omega)$ and $W^{-1,p'}(\Omega)$, is called a *weak solution* of the problem (28.26), (28.27).

By Lemma 21.6, the operator

$$(-\Delta_p)^{-1} : W^{-1,p'}(\Omega) \to W_0^{1,p}(\Omega)$$

is bounded and continuous. Therefore, (28.34) can be equivalently rewritten in the form

$$u = (-\Delta_p)^{-1} \hat{N}_f u \tag{28.35}$$

with the operator

$$A \overset{\text{def}}{=} (-\Delta_p)^{-1} \hat{N}_f : W_0^{1,p}(\Omega) \to W_0^{1,p}(\Omega). \tag{28.36}$$

Lemma 28.3. *The operator* $A : W_0^{1,p}(\Omega) \to W_0^{1,p}(\Omega)$ *is completely continuous and bounded.*

Proof. By Lemma 28.2, the operator $\hat{N}_f(u) : W_0^{1,p}(\Omega) \to W^{-1,p'}(\Omega)$ is completely continuous and bounded. The operator

$$(-\Delta_p)^{-1} : W^{-1,p'}(\Omega) \to W_0^{1,p}(\Omega)$$

is continuous and bounded due to Lemma 21.6. As in the proof of Lemma 28.2, we can verify that the composition

$$(-\Delta_p)^{-1} \hat{N}_f(u) : W_0^{1,p}(\Omega) \to W_0^{1,p}(\Omega)$$

is a completely continuous and bounded operator. The lemma is proved. $\qquad\square$

Now we prove that the set

$$S := \left\{ u \in W_0^{1,p}(\Omega) \ \middle| \ u = \alpha A(u) \text{ for the pair } (u, \alpha) \in W_0^{1,p}(\Omega) \otimes [0,1] \right\}$$

is bounded in $W_0^{1,p}(\Omega)$. Indeed, the following chain of equalities is valid for arbitrary $u \in W_0^{1,p}(\Omega)$:

$$\|A(u)\|_{W_0^{1,p}(\Omega)}^p = \|D_x A(u)\|_p^p = \langle (-\Delta_p) A(u), A(u) \rangle = \left\langle \hat{N}_f(u), A(u) \right\rangle$$

$$= \int_\Omega f(x, u(x)) A(u) \, dx \leq \int_\Omega \left(c|u|^{q-1} + b(x) \right) |A(u)| \, dx.$$

Moreover, for $u \in S$, i.e., $u = \alpha A(u)$ with certain $\alpha \in [0,1]$ and $u(x) \in W_0^{1,p}(\Omega)$, we have the chain of inequalities

$$\|A(u)\|_{W_0^{1,p}(\Omega)}^p \leq c\alpha^{q-1} \|A(u)\|_q^q + \|b\|_{q'} \|A(u)\|_q$$

$$\leq cc_1^q \alpha^{q-1} \|A(u)\|_{W_0^{1,p}(\Omega)}^q + c_1 \|b\|_{q'} \|A(u)\|_{W_0^{1,p}(\Omega)}$$

$$\leq cc_1^q \|A(u)\|_{W_0^{1,p}(\Omega)}^q + c_1 \|b\|_{q'} \|A(u)\|_{W_0^{1,p}(\Omega)},$$

where c_1 is the best constant of the embedding $W_0^{1,p}(\Omega) \to L^q(\Omega)$. Therefore, for each $u \in S$, the inequality

$$\|A(u)\|_{W_0^{1,p}(\Omega)}^p \leq K_1 \|A(u)\|_{W_0^{1,p}(\Omega)}^q + K_2 \|A(u)\|_{W_0^{1,p}(\Omega)} \tag{28.37}$$

holds, where $K_1, K_2 \geq 0$ are some constants. Note that the inequality (28.37) for $q \in (1, p)$ implies the existence of a constant $M > 0$ such that

$$\|A(u)\|_{W_0^{1,p}(\Omega)} < M.$$

Indeed, due to the three-parameter Young inequality (see (11.18)), we have

$$K_1 \cdot a^q \leq \varepsilon a^p + c_2(\varepsilon, K_1), \tag{28.38}$$

where

$$p_1 := \frac{p}{q} > 1, \quad p_2 := \frac{p}{p-q}, \quad \frac{1}{p_1} + \frac{1}{p_2} = 1.$$

Moreover, again by the three-parameter Young inequality, we have

$$K_2 \cdot a \leq \varepsilon a^p + c_3(\varepsilon, K_2). \tag{28.39}$$

From (28.37), (28.38), and (28.39), we obtain the inequality

$$\|A(u)\|_{W_0^{1,p}(\Omega)}^p \leq 2\varepsilon \|A(u)\|_{W_0^{1,p}(\Omega)}^p + c_4(\varepsilon, K_1, K_2),$$

where $a := \|A(u)\|_{W_0^{1,p}(\Omega)}$, and set $\varepsilon = 1/4$.

This implies the boundedness of S since

$$\|u\|_{W_0^{1,p}(\Omega)} = \alpha \|A(u)\|_{W_0^{1,p}(\Omega)} < M.$$

Note that always $p < p^*$. Therefore, imposing the condition $q < p$ (this is necessary in the proof of the estimate (28.38)), we automatically obtain that $q < p^*$ (this provides the compactness of the embedding $W_0^{1,p}(\Omega) \subset L^q(\Omega)$).

Thus, due to Corollary 28.1 of the Schauder theorem, we arrive at the following theorem.

Theorem 28.4. *If a Carathéodory function $f : \Omega \times \mathbb{R}^1 \to \mathbb{R}^1$ satisfies (28.28) with $q \in (1, p)$, $p \geq 2$, then the operator $(-\Delta_p)^{-1} N_f$ has a fixed point in $W_0^{1,p}(\Omega)$ or, equivalently, the problem (28.34) has a solution. Moreover, all solutions of this problem form a bounded set in $W_0^{1,p}(\Omega)$.*

Remark 28.1. The last assertion of this theorem follows from the boundedness of the set S.

28.4   Bibliographical Notes

The contents of this lecture is taken from [13, 20, 21, 29, 30, 36, 46, 66].

Picard Theorem: Simplest Case

29.1 Autonomous Equation with Globally Lipschitzian Right-Hand Side

Theorem 29.1 (Picard theorem). *Let $\mathbb{B}$ be a Banach space with norm $\|\cdot\|$ and a function $\Phi : \mathbb{B} \to \mathbb{B}$ be defined on the whole space $\mathbb{B}$ and by Lipschitz continuous, i.e., there exists a number $L > 0$ such that for all $x_1, x_2 \in \mathbb{B}$, the following inequality holds*[1]:

$$\|\Phi(x_1) - \Phi(x_2)\| \leq L \|x_1 - x_2\|.$$

Then for any $t_0 \in \mathbb{R}$ and $x_0 \in \mathbb{B}$, the Cauchy problem

$$\frac{d}{dt}x = \Phi(x), \ t \geq t_0, \quad x(t_0) = x_0 \tag{29.1}$$

is globally uniquely solvable, i.e.,

(i) *it has a solution $x(t) \in C^1([t_0, +\infty); \mathbb{B})$;*

(ii) *if $\tilde{x}(t)$ is another solution of the Cauchy problem (29.1) on the segment $\mathcal{T} = [t_0, T]$ $(t_0 < T < +\infty)$ or on the interval $\mathcal{T} = [t_0, T)$ $(t_0 < T \leq +\infty)$, then it coincides with $x(t)$ on $\mathcal{T} \cap [t_0, +\infty)$.*

First, we prove the following two lemmas.

Lemma 29.1. *Fix a number $h \leq 1/(2L)$. For any $t_1 \geq t_0$ and $x_1 \in \mathbb{B}$, there exists a unique solution of the Cauchy problem*

$$\frac{d}{dt}x = \Phi(x), \ t \in [t_1, t_1 + h], \quad x(t_1) = x_1. \tag{29.2}$$

Proof. Consider the following abstract Volterra integral equation:

$$x(t) = x_1 + \int_{t_1}^{t} \Phi(x(\tau))\, d\tau. \tag{29.3}$$

Prove that the following two assertions (A) and (B) are equivalent:

---

[1]Obviously, a Lipschitz continuous function is continuous; we will use this fact below.

(A) $x(t) \in C^1([t_1, t_1 + h]; \mathbb{B})$ and $x(t)$ is a solution of the Cauchy problem (29.2);
(B) $x(t) \in C([t_1, t_1 + h]; \mathbb{B})$ and $x(t)$ is a solution of the integral equation (29.3).

(A)$\Longrightarrow$(B). Note that if the derivative of the function $x(t)$ is continuous on the segment $[t_1, t_1 + h]$, then (due to Eq. (29.2)) the right-hand side of the equation (the composite function $x \mapsto \Phi(x(t))$) is also continuous. Therefore, both sides of Eq. (29.2) can be integrated by t from t_1 to an arbitrary point on the segment $[t_1, t_1 + h]$:

$$\int_{t_1}^{t} x'(\tau)\, d\tau = \int_{t_1}^{t} \Phi(x(\tau))\, d\tau, \quad t \in [t_1, t_1 + h].$$

Applying the fundamental theorem of calculus (the Newton–Leibniz formula) to the left-hand side, we obtain from (29.2) the relation

$$x(t) - x(t_1) = \int_{t_1}^{t} \Phi(x(\tau))\, d\tau, \quad t \in [t_1, t_1 + h],$$

which coincides with (29.3) if we take into account the equality $x(t_1) = x_1$.

(B)$\Longrightarrow$(A). Since the function $x(t)$ is continuous on the segment $[t_1, t_1 + h]$, the function $t \mapsto \Phi(x(t))$ is also continuous on this segment as the composition of continuous functions $x(t)$ and $\Phi(x)$. Therefore, to the right-hand side of the integral equation (29.3) we can apply the theorem on the derivative of an integral with variable upper limit; we obtain

$$x(t) = \Phi(x(t)), \quad t \in [t_0, t_0 + h]; \tag{29.4}$$

substituting the value $t = t_1$ into the integral equation, we obtain the initial condition. Moreover, due to Eq. (29.4), the derivative $x'(t)$ is also continuous on the segment $[t_1, t_1 + h]$.

Therefore, to prove the existence and uniqueness of a continuously differentiable solution of the Cauchy problem (29.2), it suffices to prove the existence and uniqueness of a solution of the integral equation (29.3).[2]

To prove the existence and uniqueness of a solution of Eq. (29.3), introduce the Banach space $\mathcal{B} = C([t_1, t_1 + h]; \mathbb{B})$ with the norm

$$\|x\|_{\mathcal{B}} = \sup_{t \in [t_1, t_1 + h]} \|x(t)\|$$

and the operator (in general, nonlinear)

$$A : \mathcal{B} \to \mathcal{B}, \quad (Ax)(t) = x_1 + \int_{t_1}^{t} \Phi(x(\tau)) d\tau.$$

---

[2]Note that the existence and uniqueness of a solution of one equation implies not only the existence but also the uniqueness of a solution of the other equation!

Then Eq. (29.3) can be represented in the form

$$x = Ax. \tag{29.5}$$

Note that this operator (defined on the whole Banach space $\mathcal{B}$) is a contractive operator. Indeed, for any continuous functions $\tilde{x}(t)$ and $\tilde{\tilde{x}}(t)$ at any point $t \in [t_1, t_1 + h]$ we have

$$\left\| A\tilde{x}(t) - A\tilde{\tilde{x}}(t) \right\| = \left\| x_1 + \int_{t_1}^{t} \Phi(\tilde{x}(\tau))\, d\tau - x_1 - \int_{t_1}^{t} \Phi(\tilde{\tilde{x}}(\tau))\, d\tau \right\|$$

$$= \left\| \int_{t_1}^{t} \Big(\Phi(\tilde{x}(\tau)) - \Phi(\tilde{\tilde{x}}(\tau)) \Big) d\tau \right\| \leq \int_{t_1}^{t} \left\| \Phi(\tilde{x}(\tau)) - \Phi(\tilde{\tilde{x}}(\tau)) \right\| d\tau$$

$$\leq \int_{t_1}^{t} L \left\| \tilde{x}(\tau) - \tilde{\tilde{x}}(\tau) \right\| d\tau \leq \int_{t_1}^{t} L \sup_{t \in [t_1, t_1 + h]} \left\| \tilde{x} - \tilde{\tilde{x}} \right\| d\tau$$

$$= \int_{t_1}^{t} L \left\| \tilde{x} - \tilde{\tilde{x}} \right\|_{\mathbb{B}} d\tau \leq |t - t_1| L \left\| \tilde{x} - \tilde{\tilde{x}} \right\|_{\mathbb{B}}$$

$$\leq Lh \left\| \tilde{x} - \tilde{\tilde{x}} \right\|_{\mathbb{B}} \leq \frac{1}{2} \left\| \tilde{x} - \tilde{\tilde{x}} \right\|_{\mathbb{B}}.$$

Taking the supremum over all $t \in [t_1, t_1 + h]$, we obtain

$$\left\| A\tilde{x}(t) - A\tilde{\tilde{x}}(t) \right\|_{\mathbb{B}} \leq \frac{1}{2} \left\| \tilde{x} - \tilde{\tilde{x}} \right\|_{\mathbb{B}}. \tag{29.6}$$

Therefore, we can apply the fixed-point theorem, which proves the existence and uniqueness of a solution of the integral equation (29.3) and a similar result for the Cauchy problem (29.2). The lemma is proved. □

Lemma 29.2. *If $x_1(t)$ and $x_2(t)$ are respectively solutions of the Cauchy problem (29.1) on some intervals $\mathcal{T}_1$ and $\mathcal{T}_2$ starting at a point t_0 ($t_0 \in \mathcal{T}_1, \mathcal{T}_2$), then these functions coincide on $\mathcal{T}_1 \cap \mathcal{T}_2$.*

Proof. 1. For brevity, we introduce the notation $\mathcal{T}_3 = \mathcal{T}_1 \cap \mathcal{T}_2$. Consider the set

$$\mathcal{T}_4 = \big\{ t \in \mathcal{T}_3 \mid x_1(t) \neq x_2(t) \big\}.$$

If it is empty, then the lemma is proved. Assume that $\mathcal{T}_4 \neq \varnothing$. Note that this set is open in the metric space $\mathcal{T}_4$ as the preimage of the open set $(0, +\infty)$ under the continuous mapping $g(t) := \|x_1(t) - x_2(t)\|$.

2. We set $T = \inf \mathcal{T}_4$ and prove that $T \notin \mathcal{T}_4$, i.e., $x_1(T) = x_2(T)$. Indeed, if $T = 0$, then this follows from the definition of solutions (more precisely, from the initial condition of the Cauchy problem (29.1)). If $T > 0$, then any left semi-neighborhood contains points that do not belong to $\mathcal{T}_4$ and any right semi-neighborhood contains points that belong to $\mathcal{T}_4$ and hence do not belong to this open set.

3. The fact $T \notin \mathcal{T}_4$ and assumption on the nonemptiness of $\mathcal{T}_4$ imply, in particular, that $\mathcal{T}_3$ cannot have the point T as its endpoint and contains some segment $[T, T + h_1]$. Then we can choose a sufficiently small segment $[T, T + h]$ (where $h \leq \min(h_1, 1/(2L))$) and consider the Cauchy problem

$$\frac{d}{dt}x = \Phi(x), \ t \in [t_1, t_1 + h]; \quad x(T) = x_1(T). \tag{29.7}$$

Obviously. restrictions of the functions $x_1(t)$ and $x_2(t)$ are solutions of this problem but do not coincide identically on the segment $[T, T + h]$ since T is the *exact* lower bound of $\mathcal{T}_4$). This contradicts Lemma 29.1. Lemma 29.2 is proved. $\qquad\square$

Proof of Theorem 29.1. Theorem 29.1 follows from Lemmas 29.1 and 29.2. By Lemma 29.1, one can compose a solution of the Cauchy problem (29.1) from solutions of problems of the type (29.2) stepwise, using intervals of length h. Moreover, owing to the equality $x'(t) = \Phi(x(t))$, which is also valid at the boundary points of segments for the corresponding one-sided derivatives, and the continuity of the function $x(t)$ (and hence the continuity of the composite function $\Phi(x(t))$), we conclude that the derivative at matching points also exists and is continuous. Lemma 29.2 implies the uniqueness of the solution. The theorem is proved. $\qquad\square$

29.2   Example

Consider the generalized Kolmogorov–Petrovsky–Piskunov (GKPP) equation

$$\frac{\partial}{\partial t}\left(\varepsilon^4 \Delta u - \varepsilon^2 u\right) + k\varepsilon^2 \Delta u - f(u, x, \varepsilon) = 0$$

with the following initial and boundary conditions:

$$u(x, 0) = u_0(x), \ x \in \Omega, \quad u\big|_{\Gamma \times (0,T)} = 0, \ x \in \partial\Omega, \ t \in (0, T).$$

Consider the operator D acting on three times differentiable functions by the rule

$$Du = \left(\varepsilon^4 \Delta u - \varepsilon^2 u\right)_t + \varepsilon^2 k \Delta u - f(u, x), \tag{29.8}$$

where $f(u, x)$ is a continuous function satisfying the condition $f(0, x) = 0$ for all $x \in \Omega$ and the Lipschitz condition

$$\left|f(u_1, x) - f(u_2, x)\right| \leq C|u_1 - u_2| \tag{29.9}$$

for any $x \in \Omega$ and $u_{1,2} \in \mathbb{R}^1$, where $C > 0$ is a constant.

As an example, consider a function f such that

$$f(u, x) = \begin{cases} \gamma u\left(u^2 - U^2(x)\right), & |u| \leq U_0, \ x \in D, \\ \left(3\gamma U_0^2 - \gamma U^2(x)\right)u - 2\gamma U_0^3, & |u| \geq U_0, \end{cases} \tag{29.10}$$

where $U_0 > \max_D |U(x)|$.

Consider the following initial-boundary-value problem for the generalized Kolmogorov–Petrovsky–Piskunov equation:

$$Du = 0, \ x \in \Omega, \quad u(x, t) = 0, \ x \in \partial\Omega, \quad u(x, 0) = u_0(x), \ x \in \partial\Omega, \tag{29.11}$$

in a bounded domain $\Omega \subset \mathbb{R}^3$ with the boundary $\partial\Omega \in C^{(2,\delta)}$, $\delta \in (0,1]$.

We denote the scalar product and the norm in $L^2(\Omega)$ by $(u,v)_2$ and $\|u\|_2$, respectively, whereas the scalar product and the norm in $H_0^1(\Omega)$ by the same symbols without subscripts.

Introduce the operator

$$J : H_0^1(\Omega) \to H^{-1}(\Omega), \quad \langle Jv, w \rangle = \int_\Omega vw\, dx \quad \forall v, w \in H_0^1(\Omega).$$

Obviously, this operator is linear. We estimate its norm:

$$\left| \langle Jv, w \rangle \right| = \left| \int_\Omega vw\, dx \right| \le \|v\|_2 \|w\|_2 \le C_F^2 \|v\| \|w\|, \qquad (29.12)$$

where C_F is the constant from the Friedrichs inequality

$$\|w\|_2 \le C_F \|w\| \quad \forall w \in H_0^1(\Omega)$$

in the domain Ω. From (29.12) we have

$$\|J\| \le C_F^2. \qquad (29.13)$$

Next, we introduce the operator

$$\Delta : H_0^1(\Omega) \to H^{-1}(\Omega), \quad \langle \Delta v, w \rangle = -\int_\Omega (\nabla v, \nabla w)\, dx \quad \forall v, w \in H_0^1(\Omega).$$

Due to the estimate

$$\left| \int_\Omega (\nabla v, \nabla w)\, dx \right| \equiv |(\nabla v, \nabla w)| \le \|v\| \|w\|$$

we have $\|\Delta\| \le 1$ (in fact, setting $w = v$, we see that this norm is equal to 1, but it does not matter to us).

Finally, we introduce the nonlinear operator F by the rule $F(v) = f(v(x), x)$.

Lemma 29.3. *The operator $F(v)$ is Lipschitz continuous operator acting in $L^p(\Omega)$ (for any $p > 1$) with the Lipschitz constant equal to C from the formula (29.9).*

Proof. Since $|U(x)| < U_0 < +\infty$, from (29.10) we obtain the following estimate with some constants C_1 and C_2:

$$|f(v,x)| \le C_1 + C_2|u|,$$

which, due to the Krasnosel'sky theorem on Nemytsky operators, implies that the operator $v \mapsto F(v)$ transforms functions of the class $L^p(\Omega)$ to functions of the class $L^p(\Omega)$. Further, we have the estimate

$$\|F(v_1) - F(v_2)\|_p = \left(\int_\Omega |f(v_1,x) - f(v_2,x)|^p dx \right)^{1/p}$$

$$\le \{(29.9)\} \le \left(\int_\Omega C^p |v_1 - v_2|^p dx \right)^{1/p} = C\|v_1 - v_2\|_p.$$

The lemma is proved. $\qquad\qquad\qquad\qquad\qquad\qquad\qquad\qquad\qquad\qquad\qquad\square$

We fix $p = 2$ and assume that $F : L^2(\Omega) \to L^2(\Omega)$. Introduce also the embedding operators $J_1 : H_0^1(\Omega) \to L^2(\Omega)$ (acting naturally) and

$$J_2 : L^2(\Omega) \to H^{-1}(\Omega), \quad \langle J_2 v, w \rangle = \int_\Omega vw \, dx \quad \forall v, w \in H_0^1(\Omega).$$

We estimate their norms. Obviously, $\|J_1\| = C_F$ by the definition of the Friedrichs constant. Further,

$$|\langle J_2 v, w \rangle| = |(v, w)_2| \leq \|v\|_2 \|w\|_2 \leq C_F \|v\|_2 \|w\|,$$

which implies that $\|J_1\| \leq C_F$.

Remark 29.1. Obviously, $J = J_2 J_1$.

Now we can define the operator D. Namely, for any $v(x) \equiv v(x, t) \in C^1([0, T]; H_0^1(\Omega))$ or $v(x) \equiv v(x, t) \in C^1([0, T); H_0^1(\Omega))$ (where $T > 0$ is arbitrary and may be equal to $+\infty$ in the second case), we set

$$D(v) = \frac{d}{dt} \left(\varepsilon^4 \Delta v - \varepsilon^2 Jv \right) + \varepsilon^2 k \Delta v - J_2 F(J_1 v), \tag{29.14}$$

where d/dt means differentiation in the sense of the limit by the norm of $H^{-1}(\Omega)$:

$$\frac{d}{dt} v(t) = \lim_{\Delta t \to 0} \frac{1}{\Delta t} \left(v(t + \Delta t) - v(t) \right).$$

Obviously,

$$D : C^1([0, T]; H_0^1(\Omega)) \to C([0, T]; H^{-1}(\Omega)).$$

Definition 29.1. A *generalized solution* of the problem (29.11) is a function $u(x, t) \equiv u(x)(t)$ of the class $C^{(1)}([0, T]; H_0^1(\Omega))$, where $0 < T < +\infty$ (or of the class $C^{(1)}([0, T); H_0^1(\Omega))$, where $0 < T \leq +\infty$) satisfying the conditions

$$\begin{cases} D(u) = 0, & t \in [0, T] \quad \text{(respectively, } t \in [0, T)), \\ u(0) = u_0(x) \in H_0^1(\Omega). \end{cases} \tag{29.15}$$

Remark 29.2. Using integration by parts, one can prove that a classical solution of the problem (29.11) (if it exists) satisfies the definition of a generalized solution (Definition 29.1).

Remark 29.3. The problem (29.15) can be rewritten as follows:

$$\begin{cases} \forall w \in H_0^1(\Omega) \quad \langle D(u), w \rangle = 0, & t \in [0, T] \quad \text{(respectively, } t \in [0, T)), \\ u(0) = u_0(x) \in H_0^1(\Omega). \end{cases}$$

It turns out that the problem (29.15) can be reformulated in the form of an abstract Cauchy problem. For this, we will need some preparatory work.

First, we introduce the linear operator

$$A \equiv J - \varepsilon^2 \Delta : \ H_0^1(\Omega) \to H^{-1}(\Omega).$$

Due to the boundedness of the operators J and Δ proved above and estimates of their norms, we conclude that $\|A\| \leq C_F^2 + \varepsilon^2$. Thus, we have proved the following assertion.

Lemma 29.4. *The operator $A : H_0^1(\Omega) \to H^{-1}(\Omega)$ is a bounded linear operator with norm $\|A\| \leq C_F^2 + \varepsilon^2$.*

Corollary 29.1. *The operator A is radially continuous (see Definition 1.1(i)).*

Indeed,

$$\left|S(s_1) - S(s_2)\right| \leq \|A\| \, \|v_2\|^2 \, |s_1 - s_2|.$$

Lemma 29.5. *The operator A is strongly monotonic (see Definition 1.2(iii)).*

Proof. We have

$$\begin{aligned}
\langle Av_1 - Av_2, v_1 - v_2 \rangle &= \langle A(v_1 - v_2), v_1 - v_2 \rangle \\
&= \|v_1 - v_2\|_2^2 + \varepsilon^2 \|\nabla(v_1 - v_2)\|_2^2 \\
&\geq \varepsilon^2 \|\nabla(v_1 - v_2)\|_2^2 = \varepsilon^2 \|v_1 - v_2\|^2.
\end{aligned} \tag{29.16}$$

The lemma is proved. □

Lemma 29.6. *The operator A is coercive (see Definition 1.3).*

Proof. We have

$$\langle Av, v \rangle = \|v\|_2^2 + \varepsilon^2 \|\nabla v\|_2^2 \geq \varepsilon^2 \|\nabla v\|_2^2 = \varepsilon^2 \|v\|^2.$$

Therefore, the operator A is coercive with $\gamma(s) = \varepsilon^2 s$. □

Thus, the operator A is radially continuous, strictly monotonic, and coercive; by Theorem 20.2, we conclude that the following assertion holds.

Lemma 29.7. *The operator A has an inverse operator $A^{-1} : H^{-1}(\Omega) \to H_0^1(\Omega)$.*

Lemma 29.8. *The operator A^{-1} is Lipschitz continuous with the Lipschitz constant equal to $1/\varepsilon^2$.*

Proof. Note that, due to the definition of the norm on the dual space X^*,

$$\|f\|_{X^*} \geq \frac{|\langle f, w \rangle|}{\|w\|_X}. \tag{29.17}$$

Let $w_1, w_2 \in H^{-1}(\Omega)$ and $v_1 = A^{-1}w_1$ and $v_2 = A^{-1}w_2$. Then $w_1 = Av_1$ and $w_2 = Av_2$; taking into account (29.17) and the inequality (29.16) obtained in the proof of Lemma 29.4, we have

$$\begin{aligned}
\|w_1 - w_2\|_{X^*} &= \|Av_1 - Av_2\|_{X^*} \\
&\geq \frac{|\langle Av_1 - Av_2, v_1 - v_2 \rangle|}{\|v_1 - v_2\|} \geq \varepsilon^2 \|v_1 - v_2\| \\
&= \varepsilon^2 \|A^{-1}w_1 - a^{-1}w_2\|,
\end{aligned}$$

where, due to the invertibility of the operators A and A^{-1}, $v_1 \neq v_2$ if and only if $w_1 \neq w_2$. Thus, for all $v_1 \neq v_2$ we have

$$\|A^{-1}w_1 - A^{-1}w_2\| \leq \frac{1}{\varepsilon^2}\|w_1 - w_2\|_{X^*}.$$

The lemma is proved. $\square$

Now the generalized statement of the problem (see Definition 29.1) can be rewritten in the form

$$\begin{cases} \varepsilon^2 \dfrac{d}{dt}(Au) = -\varepsilon^2 k\Delta u - J_2 F(J_1 u), \\ u(0) = u_0 \in H_0^1(\Omega). \end{cases} \tag{29.18}$$

Owing to the smoothness of the solution with respect to t, the operators d/dt and A commute and we obtain (after division by ε^2)

$$\begin{cases} A\dfrac{d}{dt}(u) = -k\Delta u + \dfrac{1}{\varepsilon^2} J_2 F(J_1 u), \\ u(0) = u_0 \in H_0^1(\Omega). \end{cases} \tag{29.19}$$

Finally, using the continuous invertibility of the operator A proved above, we transform the problem to the form

$$\begin{cases} \dfrac{d}{dt}(u) = -A^{-1}\left(k\Delta u - \dfrac{1}{\varepsilon^2} J_2 F(J_1 u)\right), \\ u(0) = u_0 \in H_0^1(\Omega). \end{cases} \tag{29.20}$$

We denote the operator in the right-hand side by $\Phi(u)$. Thus, $\Phi : H_0^1(\Omega) \to H_0^1(\Omega)$ is an operator acting by the rule

$$\Phi(v) = -A^{-1}\left(k\Delta u - \frac{1}{\varepsilon^2} J_2 F(J_1 u)\right). \tag{29.21}$$

Obviously, this (nonlinear) operator is Lipschitz continuous. Indeed, it was proved above that the operator $A^{-1}\Delta : H_0^1(\Omega) \to H_0^1(\Omega)$ is a bounded linear operator; the operator $A^{-1} \circ J_2 \circ F \circ J_1$ is Lipschitz continuous as the composition of continuous linear operators A^{-1}, J_i, $i = 1, 2$, and the Lipschitz continuous operator F.

Thus, the original problem (in the generalized statement) is reduced to the abstract Cauchy problem

$$\begin{cases} \dfrac{d}{dt}u = \Phi(u), \\ u(0) = u_0 \in H_0^1(\Omega) \end{cases} \tag{29.22}$$

with the Lipschitz continuous right-hand side.

By Theorem 29.1, the abstract Cauchy problem (29.22) is globally solvable, i.e., there exists a unique solution $u(t) \in C([0, +\infty); H_0^1(\Omega))$ and any other its solution (on a finite interval $\mathcal{T}$) is its restriction from the half-line $[0, +\infty)$ to the interval $\mathcal{T}$.

29.3 Problems

Problem 29.1. Prove that the vector space $C([0,T]; \mathbb{B})$ is a Banach space.

Problem 29.2. Concretize the arguments on matching of solutions defined on segments, which were used in the proof of Theorem 29.1.

Problem 29.3. Formulate and prove a theorem on the global solvability of a homogeneous system of linear differential equations.

Lecture 30

Theorem on Nonextendable Solutions of Cauchy Problems

In this lecture, we prove the theorem on the existence, uniqueness, and extendability of solutions of abstract differential equations of the form $y' = A(t, y)$.

Let $\mathbb{B}$ be a Banach space with the norm $\| \cdot \|$. We also consider the metric space $\mathbb{R}_+ \times \mathbb{B}$ with the distance function

$$\rho\big((t_1, y_1), (t_2, y_2)\big) = \max(|t_1 - t_2|, \|y_1 - y_2\|). \tag{30.1}$$

Assume that a mapping $A(t, y) : \mathbb{R}_+ \times \mathbb{B} \to \mathbb{B}$ possesses the following properties:

(A_1) it is continuous with respect to the metric (30.1);

(A_2) there exists a function $\mu(t, s) : \mathbb{R}_+^2 \to \mathbb{R}_+$, which is bounded on any rectangle $[0; T] \times [0; S]$ $(T, S > 0)$, such that for all $t \geq 0$ and $z_1, z_2 \in \mathbb{B}$, the inequality

$$\big\|A(t, z_1) - A(t, z_2)\big\| \leq \mu\big(t, \max(\|z_1\|, \|z_2\|)\big)\|z_1 - z_2\|$$

holds.

Remark 30.1. Functions $A(t, y)$ satisfying the condition (A_2) (in particular, they can be independent of t) are said to be boundedly Lipschitz continuous since they are Lipschitz continuous in each bounded domain of the space $\mathbb{B}$ and, moreover, the Lipschitz constant depends on t and is bounded by a finite value if t varies on a bounded set.

Note that (A_1) implies the following property.

(A_3) the function $\nu(t) \equiv \|A(t, \theta)\|$ is bounded on each segment $[0; T]$.

Indeed, due to (A_1), the function $\|A(t, \theta)\|$ is continuous for all $t \geq 0$.

Further, (A_2) and (A_3) imply the property

(A_4) there exists a function $\lambda(t, s) : \mathbb{R}_+^2 \to \mathbb{R}_+$, which is bounded on each rectangle $[0; T] \times [0; S]$ $(T, S > 0)$ and satisfies the condition

$$\big\|A(t, u)\big\| \leq \lambda\big(t, \|u\|\big)$$

for all $t \geq 0$ and all $u \in \mathbb{B}$.

Indeed, we have

$$\|A(t,u)\| \leq \|A(t,\theta)\| + \|A(t,u) - A(t,\theta)\| \leq \nu(t) + \mu(t, \|u\|)\|u\| =: \lambda(t, \|u\|),$$

and

$$\sup_{\substack{t\in[0;T] \\ s\in[0;S]}} \lambda(t,s) \leq \sup_{t\in[0;T]} \nu(t) + S \sup_{\substack{t\in[0;T] \\ s\in[0;S]}} \mu(t,s).$$

and

$$\sup_{\substack{t\in[0;T] \\ s\in[0;S]}} \lambda(t,s) \leq \sup_{t\in[0;T]} \nu(t) + S \sup_{\substack{t\in[0;T] \\ s\in[0;S]}} \mu(t,s).$$

Lemma 30.1. *Let $y(t) \in C([a;b], \mathbb{B})$, where $[a;b] \subset \mathbb{R}_+$. Then the composite function $f(t) \equiv A(t, y(t))$, where A is the mapping introduced above, is continuous: $f(t) \in C([a;b], \mathbb{B})$.*

Proof. Note that the mapping $F : t \mapsto (t, y(t))$ acting from $[a;b]$ into $\mathbb{R}_+ \times \mathbb{B}$ with the metric (30.1) is continuous. Indeed, we have $\|y(t) - y(t_0)\| \to 0$ as $t \to t_0$ and, therefore,

$$\max\left(|t - t_0|, \|y(t) - y(t_0)\|\right) \to 0 \quad \text{as } t \to t_0.$$

Then the function $f(t)$ is continuous as the composition of continuous mappings

$$t \overset{F}{\mapsto} (t, y(t)) \overset{A}{\mapsto} A(t, y(t));$$

here we used the property (A_1). The lemma is proved.   □

Now we consider the following abstract Cauchy problem in the space $\mathbb{B}$:

$$\begin{cases} y'(t) = A(t, y), & t \geq 0; \\ y(0) = Y_0; & Y_0 \in \mathbb{B}. \end{cases} \tag{30.2}$$

We also consider the Cauchy problem with the initial condition at an arbitrary moment of time $t_0 \geq 0$:

$$\begin{cases} y'(t) = A(t, y), & t \geq t_0; \\ y(0) = y_0; & y_0 \in \mathbb{B}, \quad t_0 \geq 0. \end{cases} \tag{30.3}$$

Obviously, (30.2) is a particular case of (30.3).

Definition 30.1. Let $\mathcal{T} = [t_0; T)$, where $t_0 < T \leq +\infty$, or $\mathcal{T} = [0; T]$, where $t_0 < T < +\infty$. A *solution of the Cauchy problem* (30.2) *on the interval* $\mathcal{T}$ is an abstract function $y(t) \in C^1(\mathcal{T}, \mathbb{B})$ satisfying

  (i) the initial condition $y(t_0) = y_0$ and
 (ii) the equation $y'(t) = A(t, y(t))$ for each $t \in \mathcal{T}$, where differentiation is meant in the sense of the strong derivative in the space $\mathbb{B}$ and at boundary points of the interval $\mathcal{T}$ that belong to it, one-sided derivatives are taken.

Remark 30.2. Obviously, if $z(t)$ is a solution of the Cauchy problem (30.2) on an interval $\mathcal{T} = [0, T]$ (or $\mathcal{T} = [0, T)$), then the restriction of the function $z(t)$ to any interval $\mathcal{T}_1 = [t_0, t_1]$ (or $\mathcal{T}_1 = [t_0, t_1]$) is a solution of the Cauchy problem (30.3) with $y_0 = z(t_0)$ on the interval $\mathcal{T}_1$.

Definition 30.2. A solution $y_1(t) \in C^1(\mathcal{T}_1, \mathbb{B})$ of the Cauchy problem (30.2) on an interval $\mathcal{T}_1$ is said to be *nonextendable* if there is no solutions $y_2(t) \in C^1(\mathcal{T}_2, \mathbb{B})$ on an interval $\mathcal{T}_2$ of the same problem, which satisfy the conditions

(i) $\mathcal{T}_2 \supsetneq \mathcal{T}_1$;
(ii) $y_2(t) = y_1(t)$ for all $t \in \mathcal{T}_1$.

Preliminarily, we prove some auxiliary results.
Consider the Volterra integral equation

$$y(t) = y_0 + \int_{t_0}^{t} A(\tau, y(\tau)) \, d\tau. \tag{30.4}$$

Definition 30.3. A *solution of the integral equation* (30.4) *on a segment* $[t_0, t_0 + T]$ is a function $y(t) \in C([t_0, t_0 + T, \mathbb{B}])$ satisfying Eq. (30.4) for each $t \in [t_0, t_0 + T]$, where the integral is treated in the Riemann sense.

Remark 30.3. Lemma 30.1 implies that under the condition $y(t) \in C([t_0, t_0 + T, \mathbb{B}])$ we have $A(t, y(t)) \in C([t_0, t_0 + T], \mathbb{B})$, and hence the integral in (30.4) exist for all $t \in [t_0, t_0 + T]$.

Lemma 30.2. *For all $T > 0$, the following assertions are equivalent:*

(diff) $y(t) \in C^1([t_0, t_0 + T], \mathbb{B})$ *and* $y(t)$ *is a solution of the Cauchy problem* (30.3) *on the segment* $[t_0, t_0 + T]$;
(int) $y(t) \in C([t_0, t_0 + T], \mathbb{B})$ *and* $y(t)$ *is a solution of the integral equation* (30.4) *on the segment* $[t_0, t_0 + T]$.

Proof. (diff) $\Rightarrow$ (int). Obviously, $C^1([t_0, t_0 + T], \mathbb{B}) \subset C([t_0, t_0 + T], \mathbb{B})$. The right-hand side of the equation $y'(t) = A(t, y(t))$ is continuous (since $y'(t)$ is continuous) and hence for all $t \in [0; T]$, the integral

$$\int_{t_0}^{t} A(\tau, y(\tau)) \, d\tau \tag{30.5}$$

exists. Integrating both sides of the equation $y'(t) = A(t, y(t))$ from t_0 to t and using the initial condition $y(t_0) = y_0$, we see that

$$y(t) - y_0 = \int_{t_0}^{t} A(\tau, y(\tau)) \, d\tau$$

for all $t \in [t_0, t_0 + T]$, as required.

(int) $\Rightarrow$ (diff). Due to Lemma 30.1 and the continuity of the function $y(t)$, the integrand in (30.4) is a continuous function and hence for all $t \in [t_0, t_0 + T]]$ the integral (30.5) exist and admits differentiation by the upper limit. Then we conclude that $y(t) \in C^1([0; T], \mathbb{B})$ and $y'(t) = A(t, y(t))$ for all $t \in [t_0, t_0 + T]$ and $y(t_0) = y_0$, as required. The lemma is proved. □

The significance of Lemma 30.2 is that the question on the existence and uniqueness of a solution of the Cauchy problem (30.3) (or (30.2)) on a finite is reduced to a similar question for the integral equation (30.4).

It is easy to prove that the vector space $\mathcal{B}_{\mathcal{B}_{t_0}, T} \equiv C([t_0, t_0 + T], \mathbb{B})$ is a Banach space with respect to the norm

$$\|y\|_{\mathcal{B}_{t_0}, T} = \sup_{t \in [t_0, t_0 + T]} \|y(t)\|. \tag{30.6}$$

Therefore, for a fixed element $z_0 \in \mathbb{B}$, the "band"

$$\mathcal{B}^R_{t_0, z_0, T} \equiv \left\{ y(t) \in \mathcal{B}_{t_0, T} \,\Big|\, \sup_{t \in [t_0, t_0 + T]} \|y(t) - z_0\| \le R \right\}$$

is a closed ball in the Banach space $\mathcal{B}_{t_0, T}$ and hence a complete metric space with respect to the distance function generated by the norm $\| \cdot \|_{\mathcal{B}_{t_0}, T}$; here the parameters $t_0 \ge 0$, $z_0 \in \mathbb{B}$, and $R > 0$ are arbitrary.

Lemma 30.3. *Let $t_0 \ge 0$, $R > 0$, and $y_0 \in \mathbb{B}$ be arbitrary. Then there exists T' such that for all $T \in (0, T']$, a solution of the integral equation(30.4) on the segment $[t_0, t_0 + T]$ exists and is unique in the class $\mathcal{B}^R_{t_0, y_0, T}$. In other words, the integral equation (30.4) has a solution on the segment $[t_0, t_0 + T]$, which belongs to the set $\mathcal{B}^R_{t_0, y_0, T}$ and has no other solutions from this set.*

Proof. The proof is based on the fixed-point theorem. We introduce the operator

$$\mathbb{A}_{t_0, y_0, T} : \mathcal{B}_{t_0, T} \to \mathcal{B}_{t_0, T}, \quad \mathbb{A}_{t_0, y_0, T} z = y_0 + \int_{t_0}^{t} A(\tau, z(\tau)) \, d\tau,$$

depending on the parameters $t_0 \ge 0$, $y_0 \in \mathbb{B}$, and $T > 0$. Note that the fact that for each collection of values of the parameters, this operator maps the Banach space $\mathcal{B}_{t_0, T}$ into itself follows from Lemma 30.1 and the continuity of the integral with variable upper limit of a bounded function.

We choose T' so that for all $T \in (0, T']$, (a) the operator $\mathbb{A}_{t_0, y_0, T}$ maps the set into itself and (b) it is a contractive operator in this space. In what follows, for brevity we will omit the parameters t_0, y_0, and T of the operator A.

For (a), we perform the estimate

$$\|\mathbb{A}z(t) - y_0\|_{\mathcal{B}_{t_0}, T} = \sup_{t \in [t_0, t_0 + T]} \left\| \int_{t_0}^{t} A(\tau, z(\tau)) \, d\tau \right\| \le \sup_{t \in [t_0, t_0 + T]} \int_{t_0}^{t} \|A(\tau, z(\tau))\| \, d\tau$$

$$\le \int_{t_0}^{t_0+T} \|A(\tau, z(\tau))\| \, d\tau \le T \sup_{\substack{t \in [0; T] \\ s \in [0; S]}} \lambda(t, s).$$

For (b), we perform the estimate

$$\left\|\mathbb{A}z_1(t) - \mathbb{A}z_2(t)\right\|_{\mathbb{B}_{t_0,T}} = \sup_{t\in[t_0,t_0+T]} \left\|\int_{t_0}^{t} \mathbb{A}(\tau, z_1(\tau))\, d\tau - \int_{t_0}^{t} \mathbb{A}(\tau, z_2(\tau))\, d\tau\right\|$$

$$\leq \sup_{t\in[t_0,t_0+T]} \int_{t_0}^{t} \left\|A(\tau, z_1(\tau)) - A(\tau, z_2(\tau))\right\| d\tau$$

$$\leq \int_{t_)}^{t_0+T} \left\|A(\tau, z_1(\tau)) - A(\tau, z_2(\tau))\right\| d\tau$$

$$\leq \int_{t_0}^{t_0+T} \mu\left(\tau,\ \max\left(\|z_1(\tau)\|,\ \|z_2(\tau)\|\right)\right)\left\|z_1(\tau) - z_2(\tau)\right\| d\tau$$

$$\leq T \sup_{\substack{t\in[0;t_0+T]\\ s\in[0;\|z_0\|+R]}} \mu(t,s)\left\|z_1 - z_2\right\|_{\mathcal{B}_{t_0,T}}.$$

Thus, the conditions (a) and (b) are fulfilled for some T if this T satisfies the conditions

$$T \sup_{\substack{t\in[0;t_0+T]\\ s\in[0;\|z_0\|+R]}} \lambda(t,s) \leq R, \qquad T \sup_{\substack{t\in[0;t_0+T]\\ s\in[0;\|z_0\|+R]}} \mu(t,s) \leq \frac{1}{2}. \qquad (30.7)$$

We require that the conditions (30.7) hold for all $T \in (0,T']$, where T' is a certain number. First, we choose $\bar{T}$ arbitrarily and then take $T' \leq \bar{T}$ such that

$$T' \sup_{\substack{t\in[0;t_0+\bar{T}]\\ s\in[0;\|z_0\|+R]}} \lambda(t,s) \leq R, \qquad T' \sup_{\substack{t\in[0;t_0+\bar{T}]\\ s\in[0;\|z_0\|+R]}} \mu(t,s) \leq \frac{1}{2}; \qquad (30.8)$$

this can be done since for fixed $\bar{T}$, the number T' in the left-hand sides of both inequalities (30.8) is multiplied by a constant. Then

$$T' \sup_{\substack{t\in[0;t_0+T']\\ s\in[0;\|z_0\|+R]}} \lambda(t,s) \leq R, \qquad T' \sup_{\substack{t\in[0;t_0+T']\\ s\in[0;\|z_0\|+R]}} \mu(t,s) \leq \frac{1}{2}$$

and also

$$T \sup_{\substack{t\in[0;t_0+T]\\ s\in[0;\|z_0\|+R]}} \lambda(t,s) \leq R, \qquad T \sup_{\substack{t\in[0;t_0+T]\\ s\in[0;\|z_0\|+R]}} \mu(t,s) \leq \frac{1}{2}$$

for any $T \in (0,T']$. The lemma is proved. □

Due to Lemma 30.2, we can spread Lemma 30.3 to the Cauchy problem (30.3). Namely, the following assertion holds.

Lemma 30.4. *Let $t_0 \geq 0$, $R > 0$, and $y_0 \in \mathbb{B}$ be fixed arbitrarily. Then there exists T' such that for all $T \in (0,T']$, the solution of the Cauchy problem (30.3) on the segment $[t_0, t_0 + T]$ exists and is unique in the class $\mathcal{B}_{t_0,y_0,T}^{R}$.*

From this lemma, we obtain the result on the absence of "branching" of solutions of the problem (30.2).

Lemma 30.5. *Let $y_1(t)$ and $y_2(t)$ be solutions of the problem (30.2) on the intervals $\mathcal{T}_1$ and $\mathcal{T}_2$, respectively. Then one of these solutions is an extension of the other In particular, these solutions coincide if $\mathcal{T}_1 = \mathcal{T}_2$.*

Proof. Assume the contrary:

$$y_1(t) \not\equiv y_2(t) \quad \text{on } \mathcal{T}_1 \cap \mathcal{T}_2.$$

Consider the set

$$\mathcal{T}^{\neq} \equiv \{t \in \mathcal{T}_1 \cap \mathcal{T}_2 \mid y_1(t) \neq y_2(t)\}.$$

Note that $0 \notin \mathcal{T}^{\neq}$ due to the initial condition of the problem (30.2)). Further, the set $\mathcal{T}^{\neq}$ is open as a subset of $\mathcal{T}_1 \cap \mathcal{T}_2$ since it is the preimage of the set $(0, +\infty)$ under the continuous mapping $t \mapsto \|y_1(t) - y_2(2)\|$ defined on $\mathcal{T}_1 \cap \mathcal{T}_2$. We set $T^* = \inf \mathcal{T}^{\neq}$. Note that $T^* \notin \mathcal{T}^{\neq}$. Indeed, if $T^* = 0$, then this was proved above. If $T^* > 0$, then T^* is a boundary point of the set $\mathcal{T}^{\neq}$ and, therefore, $T^* \notin \mathcal{T}^{\neq}$ since this set is open in $\mathcal{T}_1 \cap \mathcal{T}_2$. Therefore, there exists $t_1 > T^*$ such that $t_1 \in \mathcal{T}^{\neq} \subset \mathcal{T}_1 \cap \mathcal{T}_2$ and, moreover, the intersection of any right semi-neighborhood of the point T^* with $\mathcal{T}^{\neq}$ is nonempty.

By Remark 30.2, the functions $y_1(t)$ and $y_2(t)$ are solutions of the Cauchy problem

$$\begin{cases} y'(t) = A(t, y), \quad t \geq T^*; \\ y(T^*) = y_1(T^*) \end{cases} \tag{30.9}$$

on the segment $[T^*, t_1]$. Due to the continuity of $y_1(t)$ and $y_2(t)$, we have

$$R_{12} = \max_{i=1,2} \sup_{t \in [T^*, t_1]} \|y_i(t) - y_1(T^*)\| < +\infty. \tag{30.10}$$

By Lemma 30.4, there exists T' such that for any $T \in (0, T']$, the solution of the Cauchy problem (30.9) on the segment $[T^*, T^* + T]$ satisfying the condition

$$\|y(t) - y_1(T^*)\| \leq R_{12} \tag{30.11}$$

is unique. Taking $T = \min(T', t_1 - T^*)$, we obtain a contradiction since, due to (30.10), the condition (30.11) is fulfilled and $y_1 \not\equiv y_2$ in any right semi-neighborhood of T^*. The lemma is proved. $\square$

Now we formulate and prove the mail theorem of this lecture.

Theorem 30.1 (theorem on nonextendable solutions). *There exists a unique nonextendable solution $\tilde{y}(t) \in C^1(\mathcal{T}_0; \mathbb{B})$ of the Cauchy problem (30.2) satisfying the following conditions:*

(1) $\mathcal{T}_0 = [0; T_0), 0 < T_0 \leq +\infty$;

(2) *in the case where $T_0 < +\infty$, the following limit relation holds:*

$$\lim_{t \to T_0 - 0} \|\tilde{y}(t)\| = +\infty; \qquad (30.12)$$

(3) *any other solution of the problem* (30.2) *is a restriction of the solution $\tilde{y}(t)$ to an interval $\mathcal{T} \subsetneq \mathcal{T}_0$.*

Remark 30.4. In the case where $T_0 = +\infty$, the relation $\lim\limits_{t \to +\infty} \|\tilde{y}(t)\| = +\infty$ may be valid or invalid.

Proof. By Lemma 30.4 a solution of the Cauchy problem (30.2) exists at least on a certain segment $[0, T]$. By Lemma 30.5, from any two solutions of the Cauchy problem (30.2), one of them is an extension of the other (in particular, they may coincide).

Now for any $T > 0$, we consider the set $C^1([0; T]; \mathbb{B})$, which either contains a solution of the problem (30.2) on the segment $[0; T]$ or not. We set

$$\mathbb{T} = \{T > 0 : \exists \text{a solution of the problem (30.2) from } C^1([0; T], \mathbb{B})\}, \quad T_0 = \sup \mathbb{T}.$$

If $T_0 = +\infty$, then there exists a solution $\tilde{y}(t) \in C^1([0; +\infty), \mathbb{B})$. Indeed, choosing a sequence $T_n \uparrow +\infty$ and the corresponding sequence of solutions $\{y_n(t)\}$, due to Lemma 30.5 we conclude that for all $n \in \mathbb{N}$, the solution y_{n+1} is an extension of the solution y_n. Therefore, the function

$$\tilde{y}(t) = \begin{cases} y_n(t), & t \in [T_{n-1}; T_n), \ n \geq 2, \\ y_1(t), & t \in [0; T_1), \end{cases}$$

is a nonextendable solution, defined on $[0; +\infty)$ and there are no other solution that are not restrictions of $\tilde{y}(t)$ to a narrower interval.

Now let $T_0 < +\infty$. Then the following two cases are hypothetically possible: (a) $T_0 \in \mathbb{T}$ and (b) $T_0 \notin \mathbb{T}$.

In the case (a), there exists a solution $y(t) \in C^1([0; T_0], \mathbb{B})$. Then, by Lemma 30.4 applied to the problem (30.3) with $t_0 = T_0$, the solution can be extended beyond the point T_0 so that both one-sided derivatives $y'_-(T_0)$ and $y'_+(T_0)$ exist and equal $A(T_0; y(T_0))$: left by the definition of a solution on $[0; T_0]$ and right by the definition of a solution of the Cauchy problem on the interval starting at the point T_0. Therefore, we obtain a solution on a wider interval and arrive at a contradiction with the definition of T_0. Thus, the case (a) is impossible.

In the case (b), arguing similarly to the case of $T_0 = +\infty$, we establish the existence and uniqueness of a solution $y(t)$ of the problem (30.2) on the semi-interval $[0; T_0)$. The case (b) splits into two subcases:

(b$_1$) $\varlimsup\limits_{t \to T - 0} \|y(t)\| = +\infty$, i.e., the solution is unbounded in any left semi-neighborhood of the point T_0;

(b$_2$) $\varlimsup\limits_{t \to T - 0} \|y(t)\| < +\infty$.

We prove that the subcase (b$_2$) is impossible. Indeed, let a function $y(t)$ be bounded in some semi-interval $(T_0 - \gamma; T_0)$:

$$\exists C \geq 0 \; \forall t \in (T_0 - \gamma; T_0) \quad \|y(t)\| \leq C.$$

Then due to (A_4) we have

$$\forall t \in (T_0 - \gamma; T_0) \quad \|A(t, y(t))\| \; \sup_{\substack{t \in [0; T_0] \\ s \in [0; C]}} \lambda(t, s) =: L.$$

Then the equation of the problem (30.2) implies that the derivative $y'(t)$ is bounded by the value $\mathcal{L}$ for $t \in (T_0 - \gamma; T_0)$. Therefore, the function $y(t)$ is Lipschitz continuous on $(T_0 - \gamma; T_0)$ and hence satisfies the Cauchy condition at the point T_0 from the left. Thus, there exists the limit

$$Y_0 = \lim_{t \to T_0 - 0} y(t).$$

We define the function $y(t)$ at the point T_0 by the value $y(T_0) = Y_0$. The function $Y(t)$ obtained is continuous at the point T_0 on the left. Then by Lemma 30.1, the function $A(t, Y(t))$ is also continuous at the point T_0 on the left and, therefore,

$$\lim_{t \to T_0 - 0} A(t, y(t)) = \lim_{t \to T_0 - 0} A(t, Y(t)) = A(T_0; Y_0). \tag{30.13}$$

Since for $t < T_0$ the equality $y' = A(t, y(t))$ holds, we obtain from (30.13) the relation

$$\lim_{t \to T_0 - 0} y'(t) = A(T_0; Y_0).$$

Due to the lemma on extension to a point, we conclude that the solution $y(t)$ is extendable from $[0; T_0)$ to $[0; T_0]$, which contradicts the condition of the case (b) (there is no solutions on $[0; T_0]$).

Thus, for $T_0 < +\infty$, only the case (b$_1$) is realized:

$$\overline{\lim_{t \uparrow T}} \|y(t)\| = +\infty. \tag{30.14}$$

Now we prove the following relation, which is stronger than (30.14):

$$\lim_{t \to T_0 - 0} \|y(t)\| = +\infty. \tag{30.15}$$

We must prove that for all $M > 0$ there exists $\delta > 0$ such that for all $t \in (T_0 - \delta; T_0) \cap [0; T_0)$, the inequality $\|y(t)\| > M$ holds. Assume the contrary, i.e., there exists $M > 0$ such that for all $\delta > 0$, the inequality $\|y(t)\| \leq M$ is valid for some $t \in (T_0 - \delta; T_0) \cap [0; T_0)$.

Fix such a number M. By the condition (A_4), we have

$$\forall t \in [0; T_0), \; \forall z \in \mathbb{B} \quad \left(\|z\| \leq 2M \Rightarrow \|A(t, z)\| \leq \sup_{\substack{t \in [0; T_0] \\ s \in [0; 2M]}} \lambda(t, s) =: E \right). \tag{30.16}$$

From (30.16) and the equation $y'(t) = A(t, y)$ we obtain

$$\|y'(t)\| \leq E \quad \text{for} \quad \|y(t)\| \leq 2M, \; t \in [0; T_0). \tag{30.17}$$

We choose $\delta \leq M/(2E)$ and take from (30.7) the corresponding value $t = t^*$: $\|y(t^*)\| \leq M$, $T_0 - \delta < t^*$. Due to (30.14), there exists t^{**} such that $T_0 < t^* < t^{**} < T_0$ and $\|y(t^{**})\| \geq 2M$. Then, due to the continuity of the function $y(t)$ ("the preimage of a closed set is closed"), there exists $t^{***} \in (t^*; t^{**}]$ such that

$$\|y(t^{***})\| = 2M \quad \forall t \in (t^*; t^{***}) : \|y(t)\| < 2M,$$

and hence, due to (30.17),

$$\|y'(t)\| \leq E \quad \text{for all } t \in [t^*; t^{***}].$$

However, in this case, we have simultaneously

$$\|y(t^{***}) - y(t^*)\| \leq |t^{***} - t^*| \cdot E < \delta \cdot E \leq \frac{1}{2} \cdot \frac{M}{E} \cdot E < M,$$
$$\|y(t^{***}) - y(t^*)\| \geq \|y(t^{***})\| - \|y(t^*)\| \geq M.$$

The contradiction obtained proves (30.15). The theorem is proved. $\square$

Remark 30.5. An operator $F : \mathbb{B}_1 \to \mathbb{B}_2$ is said to be *boundedly Lipschitz continuous* if there exists a function $\mu : [0, +\infty) \to [0, +\infty)$, which is bounded on each bounded subset of $[0, +\infty)$, satisfying the following condition:

$$\|F(u_1) - F(u_2)\|_2 \leq \mu(R)\|u_1 - u_2\|_1$$

for all $u_k \in \mathbb{B}_1$ such that $\|u_k\|_1 \leq R$, $k = 1, 2$. In other words, this function is Lipschitz continuous in any ball (perhaps, with different Lipschitz constants in different balls).

Remark 30.6. An operator $F : \mathbb{B}_1 \to \mathbb{B}_2$ is said to be *locally Lipschitz continuous* if for each point $u_0 \in \mathbb{B}_1$, there exists a neighborhood $B_\delta(u_0)$ in which this operator is Lipschitz continuous:

$$\|F(u_1) - F(u_2)\|_2 \leq \mu\Big(\max\big(\|u_1\|, \|u_2\|\big) \Big)\|u_1 - u_2\|_1 \quad \text{for all } u_1 \text{ and } u_2.$$

In Theorem 30.1, a boundedly Lipschitz continuous function is involved. The condition of local Lipschitz continuity is weaker.

Problem 30.1. Prove the impossibility of branching of solutions of Eq. (30.2).

Lecture 31

Benjamin–Bona–Mahony–Burgers Equation

31.1 Statement of the Problem and Its Equivalent Reformulations

Consider the initial-boundary-value problem

$$\frac{\partial}{\partial t}(u_{xx} - u) + u_{xx} + uu_x = 0, \quad (x,t) \in [0,l] \times (0,T_0), \tag{31.1}$$

$$u(x,0) = u_0(x), \qquad\qquad x \in [0,l], \tag{31.2}$$

$$u(0,t) = 0, \; lu_x(0,t) = u(l,t), \quad t \in [0,T_0), \tag{31.3}$$

where the initial function $u_0(x) \in C^2([0,l])$ satisfies the boundary condition (31.3). The value of T_0, which can be finite or infinite ($T_0 = +\infty$), will be specified below. We denote derivatives with respect to the variable x by subscripts (even for function depending only on x), whereas the stroke $'$ means derivatives with respect to time t.

Introduce the functional spaces

$$Z = C([0,l]), \quad \|z\|_Z = \|z\|_{C([0,T])},$$

$$Z_1 = \{z(x) \in C^1([0,l]) \mid z(0) = 0, \; lz_x(0) = z(l)\},$$

$$\|z\|_{Z_1} = \|z\|_{C([0,T])} + \|z_x\|_{C([0,T])},$$

$$Z_2 = \{z(x) \in C^2([0,l]) \mid z(0) = 0, \; lz_x(0) = z(l)\},$$

$$\|z\|_{Z_2} = \|z\|_{C([0,T])} + \|z_x\|_{C([0,T])} + \|z_{xx}\|_{C([0,T])}.$$

Obviously, the spaces Z_1 and Z_2 are complete with respect to the norms specified as closed subspaces of the spaces $C^1([0,l])$ and $C^2([0,l])$, respectively.

We also introduce the continuous operator

$$L : Z_2 \to Z, \quad Lz = z_{xx} - z.$$

The operator L possesses a continuous inverse operator $G : Z \to Z_2$, which can be presented explicitly:

$$(Gf)(x) = \int_0^l G(x,s)f(s)\,ds, \tag{31.4}$$

where $G(x,s)$ is the Green function of the problem

$$v_{xx} - v = f(x), \; x \in [0,l]; \qquad v(0) = 0; \qquad lv_x(0) = v(l).$$

One can easily verify that this Green function has the form
$$G(x,s) = G_0(x,s) + \frac{\sinh x (l \cosh s - \cosh l \sinh s)}{l - \sinh l},$$
where
$$G_0(x,s) = \begin{cases} -\sinh x \cosh s, & 0 \le x \le s \le l, \\ -\cosh x \sinh s, & 0 \le s \le x \le l \end{cases}$$
is the Green function of the first boundary-value problem for the equation $v_{xx} - v = f(x)$. The continuity of the operator $G : Z \to Z_2$ follows from the general properties of Green functions; it can also be verified immediately.

Definition 31.1. A *solution* of the problem (31.1)–(31.3) is a function
$$u(x)(t) \in C^1([0, T_0); Z_2), \tag{31.5}$$
where $T_0 < +\infty$ or $T_0 = +\infty$, satisfying Eq. (31.1) and the initial condition (31.2) (note that the boundary conditions are included in the definition of the spaces Z_1 and Z_2). In Eq. (31.1), the expression $u_{xx} - u$ is treated in the sense of the operator L, the second term—in the sense of the differentiation operator naturally acting from Z_2 into Z for each fixed t, and the third term is considered as the result of embedding into Z of the function uu_x obtained naturally for each fixed t.

Thus, the equality in (31.1) should be understood as the equality of two elements of the space Z, the second of which is the identical zero, i.e., Eq. (31.1) is interpreted as follows:
$$\frac{d}{dt}(Lu) + u_{xx} + uu_x = 0, \tag{31.6}$$
where the derivative with respect to time is considered as the ordinary (not partial) derivative in the sense of the space (31.5). The operators L and d/dt commute; therefore, Eq. (31.6) can be rewritten in the following form:
$$Lu' + u_{xx} + uu_x = 0. \tag{31.7}$$
Further, due to the invertibility of the operators $L = G^{-1}$, Eq. (31.7) is equivalent in the space (31.5) to the equation
$$u' + G(u_{xx}) + G(uu_x) = 0.$$
Since $G(u_{xx}) = G(u_{xx} - u + u) = u + Gu$, we obtain the equivalent form
$$u' + u + Gu + G(uu_x) = 0. \tag{31.8}$$
Performing in (31.8) the substitution
$$w(t) = e^t u(t), \tag{31.9}$$
we conclude that the function $u(x)(t) \in C^1([0, T_0); Z_2)$ is a solution of the problem (31.1)–(31.3) if and only if the function $w(x)(t) \in C^1([0, T_0); Z_2)$ (see (31.9)) is a solution of the Cauchy problem
$$w' = -\left(Gw + e^{-t}G(ww_x)\right), \quad w(x)(0) = u_0(x) \equiv w_0(x) \in Z_2. \tag{31.10}$$
By a standard way (see, e.g., the previous lecture), the problem (31.10) can be reduced to the integral equation
$$w(t) = w_0 - \int_0^t d\tau\, A(\tau, w(\tau)), \tag{31.11}$$
where $w_0 = u_0(x)$ and $A(t, z) = Gz + e^{-t}G(zz_x)$.

31.2 Integral Equation in the Space $C^1([0, T_0); Z_1)$

First, we search for a weakened solution of the integral equation (31.11): $w(x)(t) \in C^1([0, T_0); Z_1)$. Note that, as above, $w_0 \in Z_2 \subset Z_1$.

It is easy to see that the operator
$$A(t, z) : Z_1 \to Z_1, \quad z \mapsto Gz + e^{-t}G(zz_x)$$
is boundedly Lipschitz continuous due to the properties of the Green function (below, we prove a stronger assertion). Further, this operator is continuous by (t, z) due to the continuity of the product of continuous functions $e^{-t} : \mathbb{R} \to \mathbb{R}$ and $G(zz_x) : Z_1 \to Z_1$. Moreover, $A(t, 0) = 0$. Thus, the operator $A(t, z)$ satisfies all conditions of Theorem 30.1 and hence Eq. (31.11) has a unique nonextendable solution. Namely, the following theorem is valid.

Theorem 31.1. *A solution of the integral equation (31.11) (or, equivalently, of the Cauchy problem (31.10)) exists on a certain maximal interval $[0, T_0)$, where $0 < T_0 \leq +\infty$, and is unique on it. In the case where $T_0 < +\infty$, the following limit equality holds:*
$$\lim_{t \to T_0 - 0} \|Aw\|_{Z_1} = +\infty.$$

31.3 Improving Smoothness to $C^1([0, T_0); Z_2)$

Theorem 31.2. *Assume that on an interval $[0, T_0)$ (here T_0 may bi finite or infinite) there exists a solution $w(x)(t) \in C^1([0, T_0); Z_1)$ of the Cauchy problem (31.10) (or, equivalently, of the integral equation (31.11)). Then this solution belongs to the class $C^1([0, T_0); Z_2)$.*

Proof. Note that the operator
$$A(t, z) : Z_1 \to Z_2, \quad z \mapsto Gz + e^{-t}G(zz_x)$$
is boundedly Lipschitz continuous and its Lipschitz constant is independent of t. Indeed, for all $t \geq 0$ we have
$$\left\| A(t, \bar{z}) - A(t, \bar{\bar{z}}) \right\|_{Z_2} = \left\| G(\bar{z} - \bar{\bar{z}}) + e^{-t}G(\bar{z}\bar{z}_x) - e^{-t}G(\bar{\bar{z}}\bar{\bar{z}}_x) \right\|_{Z_2}$$
$$\leq \|G\|_{Z \to Z_2} \|\bar{z} - \bar{\bar{z}}\|_Z + e^{-t} \left\| G(\bar{z}\bar{z}_x) - G(\bar{\bar{z}}\bar{\bar{z}}_x) \right\|_{Z_2}$$
$$\leq \|G\|_{Z \to Z_2} \|\bar{z} - \bar{\bar{z}}\|_{Z_1} + \|G(\bar{z}\bar{z}_x - \bar{\bar{z}}\bar{\bar{z}}_x)\|_{Z_2}.$$
To estimate the second term on the right-hand side, we note that
$$\left\| \bar{z}\bar{z}_x - \bar{\bar{z}}\bar{\bar{z}}_x \right\|_Z = \left\| \bar{z}\bar{z}_x - \bar{\bar{z}}\bar{z}_x + \bar{\bar{z}}\bar{z}_x - \bar{\bar{z}}\bar{\bar{z}}_x \right\|_Z$$
$$\leq \left\| \bar{z}(\bar{z}_x - \bar{\bar{z}}_x) \right\|_Z + \left\| (\bar{z} - \bar{\bar{z}})\bar{\bar{z}}_x \right\|_Z$$
$$\leq \|\bar{z}\|_Z \|\bar{z} - \bar{\bar{z}}\|_{Z_1} + \|\bar{z} - \bar{\bar{z}}\|_Z \|\bar{\bar{z}}\|_{Z_1}$$
$$\leq \|\bar{z}\|_{Z_1} \|\bar{z} - \bar{\bar{z}}\|_{Z_1} + \|\bar{z} - \bar{\bar{z}}\|_{Z_1} \|\bar{\bar{z}}\|_{Z_1}$$
$$\leq 2\max\left(\|\bar{z}\|_{Z_1}, \|\bar{\bar{z}}\|_{Z_1} \right) \|\bar{z} - \bar{\bar{z}}\|_{Z_1}.$$

Therefore, due to the continuity of the linear operator $G : Z \to Z_2$, we obtain the required result with a Lipschitz constant depending on $\max\left(\|\bar{z}\|_{Z_1}, \|\bar{\bar{z}}\|_{Z_1}\right)$.

Thus, $A(t, z) : Z_1 \to Z_2$ is a boundedly Lipschitz continuous operator with the Lipschitz constant depending on $\max\left(\|\bar{z}\|_{Z_1}, \|\bar{\bar{z}}\|_{Z_1}\right)$ and independent of t. (Note that the local Lipschitz continuity of the operator $A(t, z) : Z_1 \to Z_1$ immediately follows from this fact due to the continuity of the embedding $Z_2 \to Z_1$ with the embedding constant ≤ 1 for the norms of the spaces Z_1 and Z_2 specified above.) Due to the theorem on the composition of continuous mapping, the bounded Lipschitz continuity implies that if $w(x, t) \in C^1([0, T_0); Z_1)) \subset C([0, T_0, Z_1))$, then $A(t, w(x)(t)) \in C([0, T_0); Z_2))$ and hence the integral in the right-hand side of Eq. (31.11) is the integral of a continuous function in the sense of the space Z_2. Therefore, the right-hand side of Eq. (31.11) belongs to Z_2 for each t (recall that $w_0 \in Z_2$) and is differentiable as a Z_2-valued function of t. Therefore, the same can be said about the left-hand side, and we conclude that $w(x)(t) \in C^1([0, T_0); Z_2)$. The theorem is proved. $\square$

31.4   Further Strengthening Results

However, one can proceed from solutions in the space Z (not Z_1). For this, we must extend the operator $A(t, z)$ to functions $z \in C([0, l])$. We analyze the operator G (see (31.4) and below). We set

$$g_1(x, s) = -\sinh x \cosh s + \frac{\sinh x(l \cosh s - \cosh l \sinh s)}{l - \sinh l},$$

$$g_2(x, s) = -\cosh x \sinh s + \frac{\sinh x(l \cosh s - \cosh l \sinh s)}{l - \sinh l}.$$

Then

$$(Gf)(x) = \int_0^x g_2(x, s)f(s)\,ds + \int_x^l g_1(x, s)f(s)\,ds. \tag{31.12}$$

Naturally, this definition is invalid for the function $f = zz_x$ if z is only continuous. Therefore, we formally write $zz_x = (z^2)_x/2$ and integrate by parts in (31.12):

$$G(zz_x)(x) = \frac{1}{2}\left[\int_0^x g_2(x, s)(z^2(s))_s ds + \int_x^l g_1(x, s)(z^2(s))_s ds\right]$$

$$= \frac{1}{2}\left[g_2(x, s)z^2(s)\Big|_{s=0}^{s=x} - \int_0^x \frac{\partial g_2}{\partial s}z^2(s)\,ds\right.$$

$$\left. + g_1(x, s)z^2(s)\Big|_{s=x}^{s=l} - \int_x^l \frac{\partial g_1}{\partial s}z^2(s)\,ds\right]$$

$$= \frac{1}{2} \left[g_1(x,l)z^2(l) - g_2(x,0)z^2(0) - \int\limits_0^x \frac{\partial g_2}{\partial s} z^2(s)\, ds - \int\limits_x^l \frac{\partial g_1}{\partial s} z^2(s)\, ds \right],$$

$$(31.13)$$

where in the last equality we used the continuity of the Green function. The formula (31.13) must be considered as the definition of the operator $G(zz_x)$ applicable for $z \in Z$. However, for $z \in Z_1$, the chain of equalities (31.13) can be read from the end; this shows that for such z the new definition is equivalent to the old definition. Now we assume that the second term of the expression for the operator $A(t,z)$ is obtained by the formula (31.13).

Now we note that the function $(G(zz_x))(x)$ is differentiable by x, as follows from properties of integrals depending on parameters and properties of Green functions. We have

$$\left(G(zz_x)\right)_x(x) = \frac{1}{2} \left[\frac{\partial g_1}{\partial x}(x,l)z^2(l) - \frac{\partial g_2}{\partial x}(x,0)z^2(0) - \frac{\partial g_2}{\partial s}(x,s)\Big|_{s=x} z^2(x) \right. \quad (31.14)$$

$$- \int\limits_0^x \frac{\partial^2 g_2}{\partial x \partial s}(x,s)z^2(s)\, ds + \frac{\partial g_1}{\partial s}(x,s)\Big|_{s=x} z^2(x) \quad (31.15)$$

$$\left. - \int\limits_x^l \frac{\partial^1 g_2}{\partial x \partial s}(x,s)z^2(s)\, ds \right]. \quad (31.16)$$

We also note that for $z \in Z$, the function $A(t,z)$ satisfies the boundary conditions (31.3) (this follows from properties of the Green function). Therefore, for $z \in Z$ we have $A(t,z) \in Z_1$.

Using the boundedness of the functions $g_1(x,s)$ and $g_2(x,s)$ and their first-order and second-order derivatives on the segment $[0,l]$ and the inequality

$$\left| z_1^2(x) - z_2^2(x) \right| = |z_1(x) - z_2(x)| \cdot |z_1(x) + z_2(x)|$$

$$\leq \|z_1 - z_2\|_{C([0,l])} \cdot \left(\|z_1\|_{C([0,l])} + \|z_2\|_{C([0,l])} \right),$$

we conclude that the operator $z \mapsto G(zz_x)$ acting from Z to Z_1 (and hence the operator $A(t,z)$, in which the second term is defined now by the formula (31.13)) is uniformly by t, boundedly Lipschitz continuous. The theorem on the continuity of the product implies, as above, that $A(t,z)$ is continuous by (t,z) with respect to the norms considered. Then this is also valid for the operator $A : Z \to Z$. Thus, for the operator $A(t,z)$ in the sense specified, all the condition of Theorem 30.1 are fulfilled and, therefore, Eq. (31.11) has a unique nonextendable solution of the class

$$w(x)(t) \in C^1([0,T_0); Z). \quad (31.17)$$

Further, using the uniform in t bounded Lipschitz continuity and the continuity of the operator $A(t,z) : Z \to Z_1$ by the variables (t,z), we obtain that the existence of a solution of the class (31.17) on the interval $[0,T_0)$ guarantees the existence of a solution of the class $w(x)(t) \in C^1([0, T_0); Z_1)$ on the same interval, which, due to Theorem 31.2, provides the existence of a classical solution $w(x)(t) \in C^1([0, T_0); Z_2)$ on this interval. Hence, we have proved the following theorem.

Theorem 31.3.

1. *A solution of the integral equation (31.11) (or, equivalently, of a Cauchy problem (31.10)) of the class $w(x)(t) \in C^1([0, T_0); Z)$ exists on a certain maximal interval $[0, T_0)$, where $0 < T_0 \leq +\infty$, and is unique on it. Moreover, in the case where $T_0 < +\infty$, the all limit relation holds:*

$$\lim_{t \to T_0 - 0} \|Aw\|_Z = +\infty.$$

2. *The existence of a solution of the integral equation (31.11) of the class $w(x)(t) \in C^1([0, T_0); Z)$ guarantees the existence of a solution in the class $w(x)(t) \in C^1([0, T_0); Z_2)$ with the same T_0.*

31.5   Blow-Up of Solutions

In the previous section, we have proved the existence of a unique maximal solution of the problem (31.1)–(31.3) in the "strengthened classical" sense

$$u(x)(t) \in C^1([0, T_0); C^2([0, l])).$$

Now we multiply both sides of Eq. (31.1) by the test function $l - x$ and integrate by $x \in (0, l)$. Taking into account the boundary conditions, we obtain the following formulas of integration by parts:

$$\int_0^l (l - x) u'_{xx}\, dx = -l u'_x(0, t) + u'(l, t) - u'(0, t) = 0,$$

$$\int_0^l (l - x) u_{xx}\, dx = -l u_x(0, t) + u(l, t) - u(0, t) = 0,$$

$$\int_0^l (l - x) u u_x\, dx = \frac{1}{2} \int_0^l u^2(x, t)\, dx.$$

Thus, using Eq. (31.1), we arrive at the following equality:

$$\frac{dJ}{dt} = \frac{1}{2} \int_0^l u^2\, dx, \quad J(t) = \int_0^l (l - x) u\, dx. \tag{31.18}$$

Applying the Cauchy inequality, we obtain the following estimate:

$$\left(J(t)\right)^2 = \left(\int_0^l (l - x) u\, dx\right)^2 \leq \int_0^l (l - x)^2 dx \int_0^l u^2 dx = \frac{l^3}{3} \int_0^l u^2 dx, \tag{31.19}$$

which implies (see (31.18))

$$\frac{dJ}{dt} \geq \frac{3}{2l^3} J^2(t). \tag{31.20}$$

Now we assume that the initial function $u_0(x)$ satisfies the condition

$$J(0) = \int_0^l (l-x) u_0(x) \, dx > 0. \tag{31.21}$$

Lemma 31.1. *Assume that the function $J(t)$ satisfies the following conditions:*

(i) $J(t) \in C[0,T) \cap C^1(0,T)$;
(ii) $J(0) > 0$;
(iii) *there exists $a > 0$ such that $dJ/dt \geq aJ^2$ for all $t \in (0,T)$,*

where $0 < T \leq +\infty$. Then

$$J(t) \geq \frac{J(0)}{1 - aJ(0)t} \quad \text{for all } t \in [0,T). \tag{31.22}$$

Proof. By the conditions (ii) and (iii), we have $J(t) \geq J(0)$ for all $t \in [0,T)$. Therefore,

$$\frac{J'}{J^2} \geq a \quad \Rightarrow \quad -\frac{1}{J}\Big|_{t=0}^{t=t} \geq at$$

$$\Rightarrow \quad \frac{1}{J(0)} - \frac{1}{J(t)} \geq at \quad \Rightarrow \quad \frac{1}{J(t)} \leq \frac{1}{J(0)} - at$$

$$\Rightarrow \quad J(t) \geq \frac{J(0)}{1 - aJ(0)t} \quad \text{for } t \in \left(0, \frac{1}{aJ(0)}\right).$$

The last inequality and the condition (i) imply that $T \leq 1/(aJ(0))$; therefore, the inequality (31.22) holds on the whole domain of existence of the function $J(t)$. The lemma is proved. $\qquad\square$

From Lemma 31.1 and (31.19) we obtain the following inequality:

$$J(t) \geq \frac{J(0)}{1 - aJ(0)t}, \tag{31.23}$$

where $a = 3/2l^3$, which implies that

$$T_0 \leq T_1 \equiv \frac{2l^3}{3} J(0)^{-1}, \tag{31.24}$$

i.e., the solution cannot exist on a wider interval of time.

Thus, we have prove the following theorem.

Theorem 31.4. *Under the condition (31.21), a solution of the problem (31.1)–(31.3) in the sense (31.5) cannot exist for all $t \in [0,+\infty)$. The interval $[0,T_0)$ on which a solution exists is limited by the condition (31.24).*

31.6   Main Result

Theorem 31.5. *Under the conditions of Theorems 31.1 and 31.4, a solution of the problem (31.1)–(31.3) exists in the classical sense*

$$u(x)(t) \in C^1([0, T_0); C^2([0, l]))$$

and blows up for a finite time, i.e., $T_0 \le T_1 \equiv l^3 J(0)^{-1}$ and

$$\lim_{t \to T_0 - 0} \|u(x)(t)\|_{C([0,l])} = +\infty.$$

Proof. Theorem 31.4 with the substitution (31.9) guarantees the existence and uniqueness of a classical solution of the class $C^1([0, T_0); C^2([0, l]))$ for values T_0 that provide the class $C^1([0, T_0); C([0, l]))$. Theorem 31.5 guarantees the blow-up of a classical solution for a finite time under the condition (31.21). Therefore, a solution of the class $C^1([0, T_0); C^2([0, l]))$ and hence a solution from $C^1([0, T_0); C([0, l]))$ exist only for $T_0 \le T_1$. Then, due to Theorem' 30.1, we have

$$\lim_{t \to T_0 - 0} \left\| A\big(t, w(x)(t)\big) \right\|_{C([0,l])} = +\infty. \tag{31.25}$$

Earlier we proved that the operator' $A(t) : Z \to Z$ is uniformly bounded with respect to t on each bounded set of the space $C([0, l])$. Therefore, from (31.25) we obtain

$$\lim_{t \to T_0 - 0} \left\| w(x)(t) \right\|_{C([0,l])} = +\infty,$$

and from the relation (31.9) and the inequalities $0 \le t < T_0 < +\infty$, we get

$$\lim_{t \to T_0 - 0} \left\| u(x)(t) \right\|_{C([0,l])} = +\infty.$$

The theorem is proved.   □

31.7   Problems

Problem 31.1. Verify that the function (31.13) satisfies the boundary conditions (31.3) if $z(x) \in C([0, l])$.

Problem 31.2. Verify that a solution of the class (31.5) is a classical solution, i.e., all derivatives involved in Eq. (31.1) exist and are continuous.

Lecture 32

Example of Global Solvability

32.1 Application of the Picard Theorem

Consider the initial-boundary-value problem

$$\frac{\partial}{\partial t}(\Delta u - u) + \Delta u - u^3 = 0, \quad u(x,0) = u_0(x), \quad u\big|_{\partial\Omega} = 0.$$

We analyze the generalized statement of this problem. Introduce the linear operator

$$Au = \Delta u - Iu, \quad A : H_0^1(\Omega) \to H^{-1}(\Omega);$$

the linear operators $\Delta : H_0^1(\Omega) \to H^{-1}(\Omega)$ and $I : H_0^1(\Omega) \to H^{-1}(\Omega)$ act by the rules

$$\langle \Delta v, w \rangle = - \int_\Omega (\nabla v, \nabla w)\, dx \quad \forall v, w \in H_0^1(\Omega), \tag{32.1}$$

$$\langle Iv, w \rangle = \int_\Omega vw\, dx \quad \forall v, w \in H_0^1(\Omega). \tag{32.2}$$

The boundedness of the operator Δ is verified by using the Cauchy inequality; for the operator I, the Friedrichs inequality is also required. Using the Browder–Minty theorem, we conclude that the operator A has a bounded inverse operator

$$A^{-1} : H^{-1}(\Omega) \to H_0^1(\Omega).$$

We search for a generalized solution of the problem (32.1) as a function

$$u(t) \in C^1([0,T), H_0^1(\Omega)), \quad 0 < T \le +\infty.$$

Then the equation from (32.1) can be formally rewritten in the form

$$\frac{d}{dt} Au + \Delta u - u^3 = 0 \tag{32.3}$$

or, using the identity $\frac{d}{dt}Au = Au'$ and the invertibility of the operator A, in the form

$$\frac{d}{dt} u = A^{-1}(u^3 - \Delta u). \tag{32.4}$$

The representation of the equation in the form (32.4) is formal since we do not know whether the function u^3 belongs to the domain $H^{-1}(\Omega)$ of the operator A^{-1}. However, now we prove this fact. Namely, the embedding theorem for Sobolev spaces implies that a bounded domain Ω, the continuous embedding

$$J_1 : H_0^1(\Omega) \to L^4(\Omega) \tag{32.5}$$

holds. Due to the Krasnosel'sky theorem on Nemytsky operators (see Theorem 6.2), the mapping $u \mapsto u^3$ transforms functions from the space $L^4(\Omega)$ into functions from the space $L^{4/3}(\Omega)$. (However, in the case considered, this is obvious from simpler arguments: $|u^3|^{4/3} = |u|^4$, so that if $u \in L^4(\Omega)$, then $u^3 \in L^{4/3}(\Omega)$.) On the other hand, $L^{4/3}(\Omega) = (L^4(\Omega))^*$, and hence by the well-known embedding theorem for Banach spaces and their dual spaces,[1] we obtain from (32.5) the continuous embedding

$$J_2 : L^{4/3}(\Omega) \to H^{-1}(\Omega)$$

treated in the natural sense:

$$\langle J_2 v, w \rangle = \int_\Omega vw \, dx \quad \forall v \in L^{4/3}(\Omega), \ \forall w \in H_0^1(\Omega). \tag{32.6}$$

Thus, we have

$$H_0^1(\Omega) \xrightarrow{J_1} L^4(\Omega) \xrightarrow{F_3} L^{4/3}(\Omega) \xrightarrow{J_1} H^{-1}(\Omega). \tag{32.7}$$

Therefore, we have the mapping $A^{-1} \circ J_2 \circ F_3 \circ J_1$. To prove that it is boundedly Lipschitz continuous, it suffices to prove this property for the mapping $F_3 : L^4(\Omega) \to L^{4/3}(\Omega)$ since the other mappings are bounded linear operators (see Problem 32.1). Thus, if we can prove that the mapping F_3 is boundedly Lipschitz continuous, then we can reduce the original problem to the abstract Cauchy problem

$$\frac{d}{dt}u = A^{-1}(J_2 F_3(J_1 u) - \Delta u), \quad u(0) = u_0. \tag{32.8}$$

We demonstrate the standard techniques based on the Hölder and Young inequalities on this example. We have

$$\left\| u_1^3 - u_2^3 \right\|_{4/3} = \left(\int_\Omega |u_1^3 - u_2^3|^{4/3} dx \right)^{3/4}$$

$$= \left(\int_\Omega |u_1 - u_2|^{4/3} \cdot |u_1^2 + u_1 u_2 + u_2^2|^{4/3} \right)^{3/4} = \{\text{Hölder inequality}\}$$

$$= \left(\left(\int_\Omega |u_1 - u_2|^{4/3 \cdot 3} dx \right)^{1/3} \cdot \left(\int_\Omega |u_1^2 + u_1 u_2 + u_2^2|^{4/3 \cdot 3/2} dx \right)^{2/3} \right)^{3/4}$$

---

[1] If X and Y are reflexive infinite-dimensional Banach spaces, then the condition "X is densely and continuously embedded into Y" is equivalent to the condition "Y^* is densely and continuously embedded into X^*."

$$= \left(\left(\int_\Omega |u_1 - u_2|^4 dx \right)^{1/3} \cdot \left(\int_\Omega |u_1^2 + u_1 u_2 + u_2^2|^2 dx \right)^{2/3} \right)^{3/4}$$

$$= \left(\int_\Omega |u_1 - u_2|^4 \right)^{1/4} \cdot \left(\int_\Omega |u_1^2 + u_1 u_2 + u_2^2|^2 \right)^{1/2}$$

$$= \|u_1 - u_2\|_4 \cdot \left(\int_\Omega |u_1^2 + u_1 u_2 + u_2^2|^2 \right)^{1/2}.$$

To estimate the second factor, we note that

$$|u_1^2 + u_1 u_2 + u_2^2|^2 = u_1^4 + u_2^4 + 3u_1^2 u_2^2 + 2(u_1^3 u_2 + u_1 u_2^3) \le C(u_1^4 + u_2^4),$$

where the last inequality is obtained by the Young inequality in the form

$$u_1^2 u_2^2 \le \frac{u_1^4 + u_2^4}{2}, \quad |u_1||u_2|^3 \le \frac{u_1^4}{4} + \frac{u_2^4}{4/3}, \quad |u_1|^3|u_2| \le \frac{u_1^4}{4/3} + \frac{u_2^4}{4}.$$

Therefore,

$$\left(\int_\Omega |u_1^2 + u_1 u_2 + u_2^2|^2 \right)^{1/2}$$

$$\le C_1 \left(\int_\Omega (u_1^4 + u_2^4) dx \right)^{1/2} = C_1 \left(\int_\Omega u_1^4 dx + \int_\Omega u_2^4 dx \right)^{1/2}$$

$$= C_1 \left(\|u_1\|_4^4 + \|u_2\|_4^4 \right)^{1/2} = C_1 \left(2 \max(\|u_1\|_4, \|u_2\|_4) \right)^2. \tag{32.9}$$

Thus, the bounded Lipschitz continuity of the mapping $A^{-1} \circ J_2 \circ F_3 \circ J_1$ is proved. Therefore, applying Theorem 30.1 to the Cauchy problem (32.8), we obtain the local (by t) solvability of the problem considered.

32.2 Global Solvability

We use the method of *a priori* estimates. Thus, a solution of the problem is a function $u(t) \in C^1([0, T); H_0^1(\Omega))$. Therefore, the left-hand side is an element of the space $C([0, T); H^{-1}(\Omega))$ and hence for each $t \ge 0$, we can apply the left-hand side of the equation to $u(t)$:

$$\left\langle \frac{d}{dt}(Au) + \Delta u - J_2 F_3(J_1 u) \right\rangle = 0, \quad t \in [0, T).$$

Using the commutativity of the operators d/dt and A, we obtain

$$\left\langle A\frac{d}{dt}u + \Delta u - J_2 F_3(J_1 u) \right\rangle = 0, \quad t \in [0, T),$$

or

$$\left\langle A\frac{d}{dt}u, u \right\rangle + \langle \Delta u, u \rangle - \langle J_2 F_3(J_1 u), u \rangle = 0, \quad t \in [0, T).$$

Representing the operator A and the duality bracket explicitly (see (32.1) and (32.2)), we have

$$-\int_\Omega (\nabla u', \nabla u)dx - \int_\Omega u'u\, dx - \int_\Omega (\nabla u, \nabla u)dx - \int_\Omega u^3 \cdot u\, dx = 0,$$

where the last term is obtained due to (32.6), or

$$-(u', u) - (u', u) - \int_\Omega (\nabla u, \nabla u)dx - \int_\Omega u^4 dx = 0. \tag{32.10}$$

Since in a real Hilbert space H,

$$\frac{d}{dt}\|v\|^2 = 2(v', v)$$

for any continuously differentiable function $v(t) \in C^1(\mathcal{T}, H)$, Eq. (32.10) can be reduced to the form

$$-\frac{1}{2}\frac{d}{dt}\left(\|u\|^2_{L^2(\Omega)} - \|u\|^2_{H^1_0(\Omega)}\right) - \int_\Omega u^4\, dx = 0. \tag{32.11}$$

Introducing the notation

$$E(t) := \|u\|^2_{L^2(\Omega)} + \|u\|^2_{H^1_0(\Omega)},$$

we have

$$\frac{1}{2}\frac{dE}{dt} = -\left(\|u\|^2_{H^1_0(\Omega)} + \int_\Omega u^4\, dx\right),$$

so that

$$\frac{1}{2}\frac{dE}{dt} \le 0.$$

Since

$$E(0) = \|u_0\|^2_{L^2(\Omega)} + \|u_0\|^2_{H^1_0(\Omega)} < +\infty,$$

we have the following inequality on the whole domain of the solution:

$$E(t) \le E(0). \tag{32.12}$$

Recall that we reduced the original problem to an abstract Cauchy problem in the space $H^1_0(\Omega)$. Therefore, exactly the norm of this space will appear in Theorem 30.1 in application to the problem considered. From (32.12) we obtain

$$\|u\|_{H^1_0(\Omega)} \le E(t) \le E(0)$$

and, in particular, for any finite $T_1 < +\infty$ we have

$$\lim_{t \to T_1} \|u\|_{H^1_0(\Omega)} < +\infty. \tag{32.13}$$

Theorem 30.1 implies that if a solution exists only on a finite interval, then its norm tends to infinity at the endpoint of this interval. Therefore, the solution exists on the whole half-line $[0, +\infty)$.

Remark 32.1. We see that the estimate (32.12) is not important here, but the following weaker condition, namely, the boundedness of the norm of a solution on each bounded interval.

32.3   Problems

Problem 32.1. Using the estimate (32.9), obtain an explicit expression of the function $\mu(t, s)$ from Theorem 30.1 for the mapping $A^{-1} \circ J_2 \circ F_3 \circ J_1$.

Problem 32.2. Prove the bounded Lipschitz continuity of the following operators:

(1) $u \mapsto u^5$, $L^6(\Omega) \to L^{6/5}(\Omega)$;
(2) $u \mapsto |u|^q$, $L^{q+1}(\Omega) \to L^{(q+1)/q}(\Omega)$, $q > 1$.

Problem 32.3. Prove the formula for differentiation of the scalar square of a differentiable function with values in a real Hilbert space:

$$\frac{d}{dt}\|u(t)\|^2 = 2(u', u).$$

Problem 32.4. Formulate generalized statements and prove the global solvability of a similar problem with the following right-hand sides:

(1) $f(x) \in L^2(\Omega)$;
(2) $f(x) \in H^{-1}(\Omega)$;
(3) $f(x, t) \in C([0, +\infty); L^2(\Omega))$;
(4) $f(x, t) \in C([0, +\infty); H^{-1}(\Omega))$;
(5) $f(x, u)$;
(6) $f(x, |\nabla u|)$,

where in the last two cases, $f(x, s)$ is a Carathéodory function satisfying the estimate

$$|f(x, t)| \le |s|^\gamma, \quad \gamma \in (0, 1),$$

uniformly in $x \in \Omega$

Lecture 33

Various Generalizations
and Limits of Applicability

33.1 Nonextendable Solutions of Volterra Integral Equations

In a Banach space $\mathbb{B}$ with norm $\|\cdot\|$, consider the integral equation

$$u(t) = \bar{u}(t) + \int_0^t K(t, \tau) A(\tau, u(\tau)) \, d\tau. \tag{33.1}$$

Conditions for the kernel $K(t, \tau) : \mathbb{R}_+^2 \to L(\mathbb{B}, \mathbb{B})$ and the functions $A(t, u) : \mathbb{R}_+ \times \mathbb{B} \to \mathbb{B}$ and $\bar{u}(t) : \mathbb{R}_+ \to \mathbb{B}$ will be stated below. All integrals are treated in the Riemann sense.

Definition 33.1. A function $u(t)$ is called a *solution* of Eq. (33.1) *on an interval* $\mathcal{T} \equiv [0; T]$[1] if $u(t) \in C(\mathcal{T}, \mathbb{B})$ and $u(t)$ satisfies Eq. (33.1) for all $t \in \mathcal{T}$. In what follows, we often omit the words "Eq. (33.1)."

Remark 33.1. We use the notion "solution on an interval" instead of the notion "solution." If $u_1(t)$ and $u_2(t)$ are solutions on intervals $\mathcal{T}_1$ and $\mathcal{T}_2$, respectively, and $\mathcal{T}_1 \neq \mathcal{T}_2$, then they are considered *different* solutions irrelevantly of the coincidence of the values of $u_1(t)$ and $u_2(t)$ on $\mathcal{T}_1 \cap \mathcal{T}_2$.

Definition 33.2. A solution u_2 on an interval $\mathcal{T}_2$ is called an *extension* of a solution $u_1(t)$ on an interval $\mathcal{T}_1$ if $\mathcal{T}_2 \supseteq \mathcal{T}_1$ and $u_2(t) = u_1(t)$ on $\mathcal{T}_1$.

Remark 33.2. We agree that each solution is considered as its own extension.

Definition 33.3. A solution u_2 on an interval $\mathcal{T}$ is said to be *nonextendable* if it has no extensions different on itself, i.e., if there is no solution $\tilde{u}(t)$ on an interval $\widetilde{\mathcal{T}}$ such that $\tilde{u}(t)$ is an extension of the solution $u(t)$, where $\widetilde{\mathcal{T}} \supsetneq \mathcal{T}$. If such a solution $\tilde{u}(t)$ exists, then the solution $u(t)$ is said to be *extendable*.

To state conditions for the function $A(t, u)$, we consider the metric space $\mathbb{R}_+ \times \mathbb{B}$ with the distance function

$$\rho\big((t_1, u_1), (t_2, u_2)\big) = \max(|t_1 - t_2|, \|u_1 - u_2\|). \tag{33.2}$$

---

[1]That is, $\mathcal{T} = [0; T]$ or $\mathcal{T} = [0; T)$, and in the latter case, the possibility $T = +\infty$ is allowed. Unless otherwise specified, the initial point of the interval $\mathcal{T}$ is 0 and $0 \in \mathcal{T}$.

Obviously, this space is complete. Assume that the mapping

$$A(t, u) : \mathbb{R}_+ \times \mathbb{B} \to \mathbb{B}$$

possesses the following properties:

(A_1) it is continuous in the sense of the metric (33.2);

(A_2) there exists a function $\mu(t, s) : \mathbb{R}_+^2 \to \mathbb{R}_+$, which is bounded on each rectangle $[0; T] \times [0; S]$ ($T, S > 0$) and satisfies the condition

$$\left\| A(t, u_1) - A(t, u_2) \right\| \le \mu\Big(t, \ \max\big(\|u_1\|, \|u_2\| \big)\Big) \|u_1 - u_2\|$$

for all $t \ge 0$ and all $u_1, u_2 \in \mathbb{B}$.

Note that (A_1) implies the following property:

(A_3) the function $\nu(t) \equiv \|A(t, \theta)\|$ is bounded on each segment $[0; T]$.

Indeed, due to (A_1), the function $\|A(t, \theta)\|$ is continuous for all $t \ge 0$.

Further, (A_2) and (A_3) imply the property

(A_4) there exists a function $\lambda(t, s) : \mathbb{R}_+^2 \to \mathbb{R}_+$, which is bounded on each rectangle $[0; T] \times [0; S]$ ($T, S > 0$) and satisfies the condition

$$\left\| A(t, u) \right\| \le \lambda\big(t, \|u\|\big)$$

for all $t \ge 0$ and all $u \in \mathbb{B}$.

Indeed, we have

$$\left\| A(t, u) \right\| \le \left\| A(t, \theta) \right\| + \left\| A(t, u) - A(t, \theta) \right\| \le \nu(t) + \mu\big(t, \|u\|\big)\|u\| =: \lambda\big(t, \|u\|\big),$$

and

$$\sup_{\substack{t \in [0;T] \\ s \in [0;S]}} \lambda(t, s) \le \sup_{t \in [0;T]} \nu(t) + S \sup_{\substack{t \in [0;T] \\ s \in [0;S]}} \mu(t, s).$$

Now we formulate and prove the main theorem of this lecture.

Theorem 33.1. *Assume that the following conditions hold:*

(i) *$\bar{u}(t) \in C(\mathbb{R}_+, \mathbb{B})$;*

(ii) *the kernel $K(t, \tau)$ is continuous on $\mathbb{R}_+^2$ in the uniform operator topology, i.e., with respect to the norm of the Banach algebra $L(\mathbb{B}, \mathbb{B})$;*

(iii) *the function $A(t, u)$ possesses the properties (A_1) and (A_2).*

Then the following assertions are valid.

1. *There exists at least one solution $u(t)$ on an interval $\mathcal{T}$, $\mathcal{T} \ne \varnothing$, $\mathcal{T} \ne \{0\}$.*

2. *For each two solutions u_1 and u_2, one of them is an extension of the other, (In particular, two coinciding solutions are extensions of each other.)*

3. *If $u(t)$ is a solution on a segment $[0; T]$, then it is extendable. (In particular, the "solution" $\bar{u}(0)$ is extendable from the "segment" $\{0\}$, as follows from (1).)*

4. *There exists $T_0 > 0$ and a solution $u_0(t)$ on the interval $\mathcal{T}_0 = [0; T_0)$ such that $u_0(t)$ is a nonextendable solution.*
5. *A nonextendable solution of unique.*
6. *For the nonextendable solution in the case $T_0 < +\infty$, the relation*

$$\limsup_{t \to T_0 - 0} \|u(t)\| = +\infty \tag{33.3}$$

holds. Moreover, if $K(t, \tau) \equiv I$ (the identity operator), then the nonextendable solution is not only unbounded, but it is infinitely large:

$$\lim_{t \to T_0 - 0} \|u(t)\| = +\infty. \tag{33.4}$$

In the case $T_0 = +\infty$, the relation (33.3) (respectively, (33.4)) may be valid or invalid.

Remark 33.3. In particular, one can consider kernels $K(t, \tau) : \mathbb{R}_+^2 \to \mathbb{R}$: the Banach algebra $\mathbb{R}$ is isometrically isomorphic to the subalgebra of scalar operators in $L(\mathbb{B}, \mathbb{B})$.

Proof. 1. For each $T > 0$, consider the Banach space

$$\mathbb{B}_T := C([0; T], B), \quad \|u\|_{\mathbb{B}_T} \equiv \sup_{t \in [0; T]} \|u(t)\|,$$

and the operator

$$A_T : \mathbb{B}_T \to \mathbb{B}_T, \quad A_T(u) := \bar{u}(t) + \int_0^t K(t, \tau) A(\tau, u(\tau)) \, d\tau.$$

Note that for a continuous function $u(t)$, the integral on the right-hand side of the last formula is continuous; this follows from Lemma 30.1 and standard estimates based in the uniform continuity of the kernel $K(t, \tau)$ on any rectangle $[0; T_1] \times [0; T_2])$. Therefore, the function $u(t) \in C([0; T], \mathbb{B})$ is a solution of Eq. (33.1) if and only if it is a solution of the equation

$$u = A_T(u) \tag{33.5}$$

in the Banach space $\mathbb{B}_T$.

We specify the choice of $T > 0$ necessary for the unique solvability of Eq. (33.5) by the method of contractive mappings. Fix arbitrary $R > 0$ and consider the closed subset

$$\mathbb{B}_T^R = \left\{ u(t) \in \mathbb{B}_T \ \middle| \ \sup_{t \in [0; T]} \|u(t) - \bar{u}(t)\| \equiv \|u - \bar{u}\|_{\mathbb{B}_T} \leq R \right\}.$$

Due to general properties of metric spaces, the set $\mathbb{B}_T^R$ is also a complete metric space with respect to the distance function generated by the norm of the space $\mathbb{B}_T$. Thus, we need that the operator A_T (a) has the image lying in the set $\mathbb{B}_T^R$ and (b) possesses the contraction property in it.

To prove (a), we consider the estimate

$$\left\| \int_0^t K(t,\tau) A(\tau, u(\tau))\, d\tau \right\|_{\mathbb{B}_T} \equiv \sup_{t\in[0;T]} \left\| \int_0^t K(t,\tau) A(\tau, u(\tau))\, d\tau \right\|$$

$$\leq \int_0^T \|K(t,\tau)\| \big\| A(\tau, u(\tau)) \big\|\, d\tau$$

$$\leq T \sup_{t,\tau\in[0;T]} \|K(t,\tau)\| \sup_{\substack{t\in[0;T]\\ s\in\left[0;\|\bar{u}(t)\|_{\mathbb{B}_T}+R\right]}} \lambda(t,s). \qquad (33.6)$$

To prove (b), we consider the estimate

$$\left\| \int_0^t K(t,\tau) A(\tau, u_1(\tau))\, d\tau - \int_0^t K(t,\tau) A(\tau, u_2(\tau))\, d\tau \right\|_{\mathbb{B}_T}$$

$$\leq \int_0^T \|K(t,\tau)\| \big\| A\big(\tau, u_1(\tau)\big) - A\big(\tau, u_2(\tau)\big) \big\|\, d\tau$$

$$\leq \sup_{t,\tau\in[0;T]} \|K(t,\tau)\| \sup_{\substack{t\in[0;T]\\ s\in\left[0;\|\bar{u}(t)\|_{\mathbb{B}_T}+R\right]}} \mu(t,s) \int_0^T \big\| u_1(\tau) - u_2(\tau) \big\|\, d\tau$$

$$\leq T \sup_{t,\tau\in[0;T]} \|K(t,\tau) \sup_{\substack{t\in[0;T]\\ s\in\left[0;\|\bar{u}(t)\|_{\mathbb{B}_T}+R\right]}} \mu(t,s) \|u_1 - u_2\|_{\mathbb{B}_T}. \qquad (33.7)$$

Note that, due to the properties of the functions μ and λ and the continuity of the functions $\bar{u}(t)$ and $K(t,\tau)$, the suprema in (33.6) and (33.7) are finite (and, obviously, do not increase when T decreases). Therefore, there exists $T > 0$ such that the following conditions hold:

$$\begin{cases} T \sup\limits_{t,\tau\in[0;T]} \|K(t,\tau)\| \sup\limits_{\substack{t\in[0;T]\\ s\in\left[0;\|\bar{u}(t)\|_{\mathbb{B}_T}+R\right]}} \lambda(t,s) \leq R, \\[3em] T \sup\limits_{t,\tau\in[0;T]} \|K(t,\tau)\| \sup\limits_{\substack{t\in[0;T]\\ s\in\left[0;\|\bar{u}(t)\|_{\mathbb{B}_T}+R\right]}} \mu(t,s) \leq \dfrac{1}{2}. \end{cases}$$

In this case, due to the contractive mapping theorem, Eq. (33.5) is uniquely solvable and hence the initial equation (33.1) has a unique solution on the interval $[0;T]$.

2. First, we consider the case where $\mathcal{T}_1 = \mathcal{T}_2 =: \mathcal{T}$. Assume that $u_1(t) \not\equiv u_2(t)$ on $\mathcal{T}$. Note that the set of points t at which $u_1(t) = u_2(t)$ is a closed subset of the interval $\mathcal{T}$ as the preimage of the closed set $\{\theta\}$ under the continuous mapping $u_2 - u_1$; therefore, the set $\mathfrak{T}$ on which the equality of solutions is violated is open in $\mathcal{T}$. Therefore, there exists the point $T^* = \inf \mathfrak{T}$ (note that $u_1(T^*) = u_2(T^*) =: u^*$) and a number T^{**} such that $(T^*; T^{**}] \subset \mathfrak{T}$. Then both functions $u_1(t)$ and $u_2(t)$ are solutions of the equation

$$u(t) = \bar{u}(t) + u^* - \bar{u}(T^*) + \int_{T^*}^{t} K(t, \tau) A(\tau, u(\tau)) \, d\tau$$

on the segment $[T^*; T^{**}]$. Decreasing T^{**} if necessary, and using arguments similar to those used above, we can prove the uniqueness of a solution of this equation on the segment $[T^*; T^{**}]$ and obtain a contradiction.

Now we consider the case where the intervals $\mathcal{T}_1$ and $\mathcal{T}_2$ do not coincide; for definiteness, let $\mathcal{T}_2 \supsetneq \mathcal{T}_1$. Then, passing to the functions $u_1(t)$ and $u_2|_{\mathcal{T}_1}(t)$, we reduce the problem to the previous case whose impossibility has been proved.

3. We represent Eq. (33.1) in the form

$$u(t) = \bar{u}(t) + \int_{T}^{0} K(t, \tau) A(\tau, u(\tau)) \, d\tau + \int_{T}^{t} K(t, \tau) A(\tau, u(\tau)) \, d\tau.$$

Note that $\bar{u}(t)$ is a given function and the function $\int_{T}^{0} K(t, \tau) A(\tau, u(\tau)) \, d\tau$ of the argument t is also known for all t since integration in this integral, depending on u, is performed over the segment $[0, T]$, where the u is known. Therefore, we can set

$$\bar{\bar{u}}(t) := \bar{u}(t) + \int_{T}^{0} K(t, \tau) A(\tau, u(\tau)) \, d\tau$$

and consider for $t \geq T$ the equation

$$u(t) = \bar{\bar{u}}(t) + \int_{T}^{t} K(t, \tau) A(\tau, u(\tau)) \, d\tau,$$

which is similar to the original equation. It remains to apply arguments similar to those used in item 1. Obviously, the solutions obtained can be continuously matched and generate an extension of the solution $u(t)$ from the segment $[0, T]$ to a wider segment.

4. Consider the equation

$$u(t) = \bar{u}(t) + u(T) - \bar{u}(T) + \int_{T}^{t} K(t, \tau) A(\tau, u(\tau)) \, d\tau$$

for $t \geq T$ and apply arguments used above. Obviously, solutions obtained can be continuously matched, and we get an extension of the solution $u(t)$.

5. Consider the set

$$\mathfrak{T} = \Big\{ T > 0 : \exists \text{ a solution of (33.1) on } [0; T] \Big\},$$

$$T_0 = \sup \mathfrak{T} \leq +\infty.$$

Due to item 1, the set $\mathfrak{T}$ is nonempty and contains a certain segment of nonzero length. Therefore, $T_0 > 0$ and there exists a sequence of solution $\{u_n(t)\}_{n=1}^\infty$ on the segments $[0; T_n]$ such that $T_n > 0$ and $T_n \uparrow T_0$. Due to item 2, any two solutions of Eq. (33.1) coincide in their common domain. Therefore, for all $n \in \mathbb{N}$, the solution $u_{n+1}(t)$ is an extension of the solution $u_n(t)$. Then we can construct the function

$$u(t) = \begin{cases} u_1(t), & t \in [0; T_1], \\ u_{n+1}(t), & t \in (T_n; T_{n+1}], \ n \in \mathbb{N}. \end{cases}$$

This function is a solution of Eq. (33.1) on the interval $[0; T_0)$. If $T_0 = +\infty$, then, obviously, $u(t)$ is nonextendable. If $T_0 < +\infty$, then, by the definition of T_0, the maximal domain of the solution is either semi-interval $[0; T_0)$ or the segment $[0; T_0]$. However, the latter case is impossible due to item 3 since in the opposite case, there exists a solution on an interval wider than the segment $[0; T_0]$ and hence of a certain segment $[0; T_1]$, where $T_1 > T_0$. Therefore, the solution $u(t)$, $t \in [0; T_0)$, is nonextendable.

Thus, there exists a nonextendable solution defined on a semi-open interval $[0; T_0)$, where $T_0 \leq +\infty$.

6. Let $u_1(t)$ and $u_2(t)$ be two nonextendable solutions. Then, due to item 2, one of them is an extension of the other. Therefore, either they coincide or one of them is extendable.

7. Let $u(t)$ be a solution on $[0; T_0)$, $T_0 < +\infty$, and $u(t)$ be a nonextendable solution. Assume that the solution $u(t)$ is bounded, i.e., there exists a number C_0 such that

$$\|u(t)\| \leq C_0, \quad t \in [0; T_0).$$

The integral on the right-hand side of Eq. (33.1) satisfies the Cauchy condition in a left semi-neighborhood of the point T_0; this follows from the inequality

$$\left\| \int_0^{t_2} K(t_2, \tau) A(\tau, u(\tau)) \, d\tau - \int_0^{t_1} K(t_1, \tau) A(\tau, u(\tau)) \, d\tau \right\|$$

$$\leq \int_0^{t_1} \left\| K(t_2, \tau) - K(t_1, \tau) \right\| \left\| A(\tau, u(\tau)) \right\| d\tau$$

$$+ \int_{t_1}^{t_2} \left\| K(t_2, \tau) \right\| \left\| A(\tau, u(\tau)) \right\| d\tau, \tag{33.8}$$

where $0 < t_1 < t_2 < T_0$, the uniform continuity of the kernel $K(t, \tau)$ on any rectangle, and the property (A_4). On the other hand, the function $\bar{u}(t)$ is continuous everywhere by assumption. Therefore, the function $u(t)$ can be continuously extended to the point T_0. We denoted the extended function by $\tilde{u}(t)$. Then the function

$$\bar{u}(t) + \int_0^t K(t,\tau) A(\tau, \tilde{u}(\tau)) \, d\tau,$$

which obviously coincides with

$$u(t) = \bar{u}(t) + \int_0^t K(t,\tau) A(\tau, u(\tau)) \, d\tau$$

on $[0; T_0)$, exists and is continuous on $[0; T_0]$. Therefore, its value at the point T_0 is equal to $\tilde{u}(T_0)$ due to the uniqueness of the continuous extension to the closure. Therefore, the function $\tilde{u}(t)$ is a solution of Eq. (33.1) on the segment $[0; T_0]$, which contradicts the assumption that the solution $u(t)$ on the interval $[0; T_0)$ is nonextendable. Thus, the relation (33.3) is proved.

Remark 33.4. Further arguments are similar to those performed in Lecture 30 for differential equations.

Now we verify that in the case $K(t,\tau) \equiv I$, the limit relation (33.4) holds. We must prove that for all $M > 0$, there exists $\delta > 0$ such that the inequality $\|u(t)\| > M$ holds for all $t \in (T_0 - \delta; T_0) \cap [0; T_0)$. Assume the contrary, i.e.,

$$\exists M > 0 \; \forall \delta > 0 \; \exists t \in (T_0 - \delta; T_0) \cap [0; T_0) \quad \|u(t)\| \le M. \tag{33.9}$$

Fix M from (33.9). Due to the property (A_4) we have

$$\forall t \in [0; T_0), \; \forall z \in \mathbb{B} \quad \left(\|z\| \le 2M \; \Rightarrow \; \|A(t,z)\| \le \sup_{\substack{t \in [0;T_0] \\ s \in [0;2M]}} \lambda(t,s) =: E \right). \tag{33.10}$$

Choose $\delta \le M/(4E)$ from the condition

$$\left\| \bar{u}(t'') - \bar{u}(t') \right\| < \frac{M}{4} \quad \text{for } |t'' - t'| < \delta.$$

This is possible since the function $\bar{u}(t)$ is continuous on $\mathbb{R}_+$ and hence uniformly continuous on the segment $[0; T_0]$.) From (33.9), we take $t = t^*$ such that $T_0 - \delta < t^* < T_0$, $\|u(t^*)\| \le M$. By (33.3), there exists t^{**} such that $T_0 - \delta < t^* < t^{**} < T_0$ and $\|u(t^{**})\| \ge 2M$. Then, due to the continuity of the function $u(t)$, there exists $t^{***} \in (t^*; t^{**})$ such that

$$\left\| u(t^{***}) \right\| = 2M, \quad \|u(t)\| < 2M \quad \text{for all } t \in (t^*; t^{***}). \tag{33.11}$$

Then, on the one hand, we have

$$\left\| u(t^{***}) - u(t^*) \right\| \ge \left\| u(t^{***}) \right\| - \left\| u(t^*) \right\| = M.$$

On the other hand, due to Eq. (33.1), the assertions (33.10) and (33.11), and the choice of δ, we get

$$\left\| u(t^{***}) - u(t^*) \right\| \le \left\| \bar{u}(t^{***}) - \bar{u}(t^*) \right\| + \int_{t^*}^{t^{***}} \left\| A(\tau, u(\tau)) \right\| \, d\tau$$

$$< \frac{M}{4} + |t^{***} - t^*| E \le \frac{M}{4} + \delta E \le \frac{M}{2},$$

a contradiction. The theorem is proved. $\qquad \square$

Remark 33.5. It is easy to see that in the case where the kernel K is independent of t and continuously depends on τ, it can be "included" into $A(t, u)$ and hence the relation (33.4) is also valid in this case.

33.2   Example of Nonextendable Solution, which Does not Have a Limit

In the case of a kernel depending on t and satisfying the conditions of Theorem 33.1, the relation (33.4) may be violated. We present an example. Consider the function

$$u(t) = 1 + \frac{1}{T_0 - t} \cos^2 \frac{1}{T_0 - t} \in C[0; T_0), \quad T_0 = \frac{2}{\pi}, \tag{33.12}$$

and construct an integral equation of the form (33.1), whose solution is the function (33.12), such that its kernel depends only on the variable t. It is easy to see that as $t \to T_0 - 0$, the function (33.12) has no limit (since its attains the value 1 in an arbitrarily small neighborhood of the point T_0) but

$$\limsup_{t \to T_0 - 0} u(t) = +\infty.$$

Thus, the function (33.12) satisfies the relation (33.3) but does not satisfy the relation (33.4). We must find a function $K(t) \in C[0; +\infty)$ such that

$$u(t) = u(0) + K(t) \int_0^t (u(s))^k ds \quad \text{for } t \in [0; T_0)$$

a natural number k will be chosen below. Since $u(0) = 1$, we have

$$1 + \frac{1}{T_0 - t} \cos^2 \frac{1}{T_0 - t} = 1 + K(t) \int_0^t \left(1 + \frac{1}{T_0 - s} \cos^2 \frac{1}{T_0 - s} \right)^k ds$$

or

$$K(t) = \frac{\dfrac{1}{T_0 - t} \cos^2 \dfrac{1}{T_0 - t}}{\displaystyle\int_0^t \left(1 + \frac{1}{T_0 - s} \cos^2 \frac{1}{T_0 - s} \right)^k ds}, \quad T_0 = \frac{2}{\pi}. \tag{33.13}$$

For all natural k, the fraction in the right-hand side is a function continuous on the interval $t \in (0; T_0)$. Applying L'Hôpital's rule, we easily conclude that $K(t) \to 0$ as $t \to +0$. If, moreover,

$$K(t) \to 0 \quad \text{as } t \to T_0 - 0, \tag{33.14}$$

then the function $K(t)$ can be extended to a continuous function of the argument $t \in [0; +\infty)$, which satisfies the condition of Theorem 33.1. We will seek to fulfil the condition (33.14).

By the binomial theorem, taking into account the nonnegativity of the second term, for all $k \geq 3$, $k \in \mathbb{N}$, $s \in [0; T_0)$, we have

$$\left(1 + \frac{1}{T_0 - s} \cos^2 \frac{1}{T_0 - s}\right)^k \geq \frac{k(k-1)(k-2)}{6} \frac{1}{(T_0 - s)^3} \cos^6 \frac{1}{T_0 - s}; \quad (33.15)$$

therefore,

$$\int_0^t \left(1 + \frac{1}{T_0 - s} \cos^2 \frac{1}{T_0 - s}\right)^k ds \geq \frac{k(k-1)(k-2)}{6} \int_0^t \frac{1}{(T_0 - s)^3} \cos^6 \frac{1}{T_0 - s} ds$$

$$(33.16)$$

for all $t \in [0; T_0)$. Calculate the integral in the right-hand side of the last formula:

$$\int_0^t \frac{1}{(T_0 - s)^3} \cos^6 \frac{1}{T_0 - s} ds = \left\{ y = \frac{1}{T_0 - s}\right\} = \int_{1/T_0}^{1/(T_0 - t)} y \cos^6 y \, dy$$

$$= \frac{5}{32} y^2 + \frac{15}{32} \left(\frac{y \sin 2y}{2} + \frac{\cos 2y}{4}\right)$$

$$+ \frac{6}{32} \left(\frac{y \sin 4y}{4} + \frac{\cos 4y}{16}\right) + \frac{1}{32} \left(\frac{y \sin 6y}{6} + \frac{\cos 6y}{36}\right) \Bigg|_{1/T_0}^{1/(T_0 - t)}$$

$$= \frac{5}{32} \frac{1}{(T_0 - t)^2} + O\left(\frac{1}{T_0 - t}\right) \quad \text{as } t \to T_0 - 0. \quad (33.17)$$

From (33.13) and (33.15)–(33.17) we obtain

$$0 \leq K(t) \leq \frac{6}{k(k-1)(k-2)} \frac{\dfrac{1}{T_0 - t} \cos^2 \dfrac{1}{T_0 - t}}{\dfrac{5}{32} \dfrac{1}{(T_0 - t)^2} + O\left(\dfrac{1}{T_0 - t}\right)}$$

$$= \frac{32 \cdot 6}{5k(k-1)(k-2)} \frac{\cos^2 \dfrac{1}{T_0 - t}}{\dfrac{1}{T_0 - t} + O(1)} \to +0 \quad \text{as } t \to T_0 - 0.$$

This proves the limit relation (33.14), for example, for $k = 3$.

Thus, the equation has the form

$$u(t) = 1 + \int_0^t K(t)(u(s))^3 ds,$$

where

$$K(t) = \begin{cases} \dfrac{\dfrac{1}{T_0 - t} \cos^2 \dfrac{1}{T_0 - t}}{\displaystyle\int_0^t \left(1 + \dfrac{1}{T_0 - s} \cos^2 \dfrac{1}{T_0 - s}\right)^3 ds}, & t \in (0; T_0), \\[4mm] 0, & t \in \{0\} \cup [T_0; +\infty), \end{cases}$$

$T_0 = 2/\pi$, and the corresponding solution is

$$u(t) = 1 + \frac{1}{T_0 - t} \cos^2 \frac{1}{T_0 - t}.$$

Remark 33.6. This result is related to the result obtained in [33], which states that a nonextendable solution of the Cauchy problem for an autonomous abstract differential equation $u' = f(u)$ with locally Lipschitzian right-hand side $f(u)$ in an arbitrary infinite-dimensional Banach space $\mathbb{B}$ can be (in contrast to the cases (33.3) and (33.4)) even bounded if $f(u)$, being a locally Lipschitz continuous function, is not bounded on each bounded subset of the space $\mathbb{B}$.

Remark 33.7. Ii is important to distinguish between *boundedly Lipschitz continuous* and *locally Lipschitz continuous* functions. In the latter case, for any point of the domain, there exists a neighborhood in which a function is Lipschitz continuous. In infinite-dimensional Banach spaces, these condition are not equivalent.

33.3 Peano Theorem

Theorem 33.2 (Peano theorem). *Consider the following differential equation for a scalar function $u(t)$:*

$$u' = f(t, u). \tag{33.18}$$

If the right-hand side $f(t, u)$ is continuous in a bounded closed domain G, then through each interior point (t_0, u_0) of this domain, at least one integral curve of this equation passes.

It is easy to see that the uniqueness is not guaranteed by this theorem. Indeed, consider the Cauchy problem

$$u' = 3u^{2/3}, \quad u(0) = 0. \tag{33.19}$$

The problem (33.19) has the trivial solution $u = 0$ and the solution $u = t^3$. Moreover, for any t_0, the function $(t - t_0)^3$ is also a solution of the equation in (33.19) and, moreover,

$$\frac{d}{dt}(t - t_0)^3 \bigg|_{t=t_0} = 0.$$

Therefore, for arbitrary $t_0 > 0$, the smoothly matching of the solutions $u = 0$ and $u = (t - t_0)^3$ leads to the solution

$$u(t) = \begin{cases} 0, & t \in [0, t_0), \\ (t - t_0)^3, & t \in [t_0, +\infty), \end{cases} \tag{33.20}$$

of the problem (33.19). Thus, the problem (33.19) has infinitely many solutions defined for $t \geq 0$.

There exist even more "pathological" examples. We do not present them here because of their bulkiness, but note that there exists a continuous function $f(t, u)$ defined on the whole plane such that for any pair (t_0, u_0), the Cauchy problem

$$u' = f(t, u), \quad u(0) = u_0$$

has more than one solution *on each segment* $[t_0, t_0 + \varepsilon]$ (the problem (33.19) possesses this property only for $u_0 = 0$).

We formulated the Peano theorem above only for scalar functions. The fact is that the Peano theorem is valid only for finite-dimensional vector space; in any infinite-dimensional Banach space, the problem (33.18) may have no solutions (even local in time). This result was obtained in [26].

33.4 Problems

Problem 33.1. Based on the problem (33.19) (or similar problems), construct a Cauchy problem with the following properties:

(1) its trivial solution $u = 0$ exists on a half-line;
(2) for any $T > 0$, there exists a nontrivial solution (possibly, extendable) on $[0, T)$;
(3) none of its nontrivial solutions can be extended to the entire half-line.

Problem 33.2. Give an example of a locally Lipschitz continuous function, which is not boundedly Lipschitz continuous.

Bibliography

[1] R. Adams, *Sobolev Spaces*, Academic Press, New York–San Francisco–London (1975).

[2] A. Ambrosetti, Critical points and nonlinear variational problems, *Mem. Soc. Math. Fr.* **49**, 1–139 (1992).

[3] C. Baiocchi and A. Capelo, *Variational and Quasivariational Inequalities. Applications to Free Boundary Problems*, Wiley, Chichester (1984).

[4] M. S. Berger, A Sturm–Liouville theorem for nonlinear elliptic partial differential equations, *Ann. Scu. Norm. Super. Pisa* **20**, No. 3, 543–582 (1966).

[5] M. S. Berger, On von Kármán's equation and the buckling of a thin elastic plate. I. The clamped plate, *Commun. Pure Appl. Math.* **20**, 687–719 (1967).

[6] V. I. Bogachev and O. G. Smolyanov, *Real and Functional Analysis*, Springer, Cham (2020).

[7] H. Cartan, *Cours de Calcul Différentiel*, Hermann, Paris (1997).

[8] C. D. Clark, A variant of the Lusternik–Schnirelman theory, *Indiana Univ. Math. J.* **22**, No. 1, 65–74 (1972).

[9] P. Clément, H. J. A. M. Heijmans, S. Angenent, C. J. van Duijn, and B. de Pagter, *One-Parameter Semigroups*, North-Holland, Amsterdam (1987).

[10] M. G. Crandall and T. M. Liggett, Generation of semigroups of nonlinear transformations on general Banach spaces, *Am. J. Math.* **93**, 265–298 (1971).

[11] B. P. Demidovich, *Lectures on Mathematical Stability Theory* [in Russian], Nauka, Moscow (1967).

[12] Z. Denkowski, S. Migorski, and N. S. Papageorgiou, *An Introduction to Nonlinear Analysis*, Kluwer Academic/Plenum Publishers, Boston–Dordrecht–London–New York (2003).

[13] G. Dinca, P. Jebelean, and J. Mawhin, Variational and topological methods for Dirichlet problems with p-Laplacian, *Portug. Math.* **58**, No. 3, 339–378 (2001).

[14] P. Drabek and J. Milota, *Methods of Nonlinear Analisys. Applications to Differential Equations*, Birkhäuser, Basel (2007).

[15] Yu. A. Dubinskii, Nonlinear elliptic and parabolic equations, *J. Sov. Math.* **12**, No. 5, 475–554 (1979).

[16] Yu. A. Dubinskii, Weak convergence for nonlinear elliptic and parabolic equations, *Mat. Sb. (N.S.)* **67 (109)**, No. 4, 609–642 (1965).

[17] N. Dunford and J. T. Schwartz, *Linear Operators. Part* I. *General Theory*, Wiley, New York (1988).

[18] R. E. Edwards, *Functional Analysis. Theory and Applications.* Holt Rinehart and Winston, New York (1965).

[19] L. C. Evans, *Partial Differential Equations*, Am. Math. Soc., Providence, Rhode Island (1998).

[20] S. Fučík and A. Kufner, *Nonlinear Differential Equations*, Elsevier, Amsterdam–Oxford–New York (1980).

[21] H. Fujita, On the blowing up solutions of the Cauchy problem for $u_t = \Delta u + u^{1+\alpha}$, *J. Fac. Sci. Univ. Tokyo*, Sect. IA **13**, 109–124 (1966).

[22] H. Gajewski, K. Gröger, and K. Zacharias, *Nichtlineare Operatorgleichungen und Operatordifferentialgleichungen*, Akademie-Verlag, Berlin (1974).

[23] L. Gasiński and N. S. Papageorgiou, *Nonlinear Analisys*, Chapman and Hall, Boca Raton, FL (2005).

[24] L. Gasiński and N. S. Papageorgiou, *Nonsmooth Critical Point Theory and Nonlinear Boundary-Value Problems*, Chapman and Hall, Boca Raton, FL (2005).

[25] D. Gilbarg and N. S. Trudinger, *Elliptic Partial Differential Equations of Second Order*, Springer, Berlin (2001).

[26] A. N. Godunov, Peano's theorem in Banach spaces, *Funct. Anal. Appl.* **9**, No. 1, 53–55 (1975).

[27] E. Hille and R. S. Phillips, *Functional Analysis and Semi-Groups*, Am. Math. Soc., Providence, Rhode Island (1974).

[28] Y. X. Huang, Eigenvalues of the p-Laplacian in $\mathbb{R}^N$ with indefinite weight, *Comment. Math. Univ. Carol.* **36**, No. 3, 519–527 (1995).

[29] V. Hutson and J. S. Pym, *Applications of Functional Analysis and Operator Theory*, Academic Press, London (1980).

[30] L. V. Kantorovich and G. P. Akilov, *Functional Analysis*, Pergamon Press, Oxford (1982).

[31] T. Kato, Nonlinear semigroups and evolution equations, *J. Math. Soc. Jpn.* **19**, No. 4, 509–520 (1967).

[32] V. S. Klimov, On functionals with an infinite number of critical values, *Math. USSR-Sb.* **29**, No. 1, 91–104 (1976).

[33] V. Komornik, P. Martinez, M. Pierre, and J. Vanconsenoble, Blow-up of bounded solutions of differential equations, *Acta Sci. Math. (Szeged)* **69**, 651–657 (2003).

[34] Y. Komura, Nonlinear semi-groups in Hilbert space, *J. Math. Soc. Jpn.* **19**, No. 4, 493–507 (1967).

[35] M. A. Krasnosel'skii, *Topological Methods in the Theory of Nonlinear Integral Equations*, Pergamon Press, Oxford (1964).

[36] S. G. Krein, *Linear Equations in Banach Spaces*, Birkhäuser, Boston–Basel–Stuttgart (1982).

[37] N. V. Krylov, *Lectures on Elliptic and Parabolic Equations in Hölder spaces*, Am. Math. Soc., Providence, Rhode Island (1996).

[38] I. A. Kuzin, Solvability of some elliptic problems with critical exponent of nonlinearity, *Math. USSR-Sb.* **68**, No. 2, 339–349 (1991).

[39] I. A. Kuzin, Multiple solvability of certain elliptic problems with critical nonlinearity exponents, *Math. Notes* **52**, No. 1, 668–672 (1992).

[40] I. A. Kuzin, Comparison theorems for variational problems and their application to elliptic equations in $\mathbb{R}^N$, *Russian Acad. Sci. Izv. Math.* **43**, No. 2, 331–346 (1994).

[41] I. Kuzin and S. Pohozaev, *Entire Solutions of Semilinear Elliptic Equations*, Birkhäuser, Basel (1997).

[42] L. D. Landau and E. M. Lifshitz, *Electrodynamics of Continuous Media*, Pergamon Press, Oxford–London–New York–Paris (1960).

[43] E. M. Landis, *Second Order Equations of Elliptic and Parabolic Type*, Am. Math. Soc.. Providence, Rhode Island (1998).

[44] P. Lindqvist, On the equation $\mathrm{div}(|\nabla u|^{p-2}\nabla u) + \lambda |u|^{p-2}u = 0$, *Proc. Am. Math. Soc.* **109**, 157–164 (1990).

[45] P. Lindqvist, Notes on the p-Laplace equation, *electronic document*; available at the URL http://www.math.ntnu.no/ lqvist/p-laplace.pdf

[46] J. L. Lions, *Quelques Méthodes de Résolution des Problèmes aux Limites Non Linéaires*, Dunod, Paris; Gauthier-Villars, Paris (1969).

[47] J. L. Lions, *Optimal Control of Systems Governed by Partial Differential Equations*, Springer-Verlag, Berlin–Heidelberg–New York (1971).

[48] J. L. Lions and E. Magenes, *Nonhomogeneous Boundary Value Problems and Applications*, Vol. I, Springer-Verlag, Berlin–Heidelberg–New York (1972).

[49] L. Lusternik and L. Schnirelmann, *Méthodes Topologiques dans les Problèmes Variationnels*, Hermann, Paris (1934).

[50] L. A. Lyusternik, Topology and the calculus of variations, *Usp. Mat. Nauk* **1**, No. 1 (11), 30–56 (1946).

[51] K. Maurin, *Methods of Hilbert Spaces*, PWN, Warszawa (1967).

[52] E. Mitidieri and S. I. Pokhozhaev, A priori estimates and blow-up of solutions to nonlinear partial differential equations and inequalities, *Proc. Steklov Inst. Math.* **234**, 1–362 (2001).

[53] I. Miyadera, *Nonlinear Semigroups*, Am. Math. Soc., Providence, Rhode Island (1991).

[54] S. Mizohata, *The Theory of Partial Differential Equations*, Cambridge Univ. Press, London (1973).

[55] L. Nirenberg, Variational and topological methods in nonlinear problems, *Bull. Am. Math. Soc.* **4**, No. 3, 267–302 (1981).

[56] V. G. Osmolovskii, *Nonlinear Sturm–Liouville Problem* [in Russian], Saint-Petersburg Univ., Saint-Petersburg (2003).

[57] L. V. Ovsyannikov, *Investigation of gas flows with a straight sonic line*, thesis, Leningrad (1948).

[58] S. I. Pokhozhaev, On the method of fibering a solution in nonlinear boundary-value problems, *Proc. Steklov Inst. Math.* **192**, 157–173 (1992).

[59] S. I. Pokhozhaev, On a problem of L. V. Ovsyannikov, *J. Appl. Mech. Tech. Phys.* **30**, No. 2, 169–174 (1989).

[60] P. H. Rabinowitz, Variational methods for nonlinear elliptic eigenvalue problems, *Indiana Univ. Math. J.* **23**, No. 8, 729–754 (1974).

[61] A. P. Robertson and W. Robertson, *Topological Vector Spaces*, Cambridge Univ. Press, London (1973).

[62] W. Rudin, *Functional Analysis*, McGraw-Hill, New York (1973).

[63] J. Schwartz, *Nonlinear Functional Analysis*, Gordon and Breach, New York (1969).

[64] I. V. Skrypnik, *Methods for Analysis of Nonlinear Elliptic Boundary-Value Problems*, Am. Math. Soc. Providence, Rhode Island (1994).

[65] M. Struwe, *Variational Methods. Applications to Nonlinear Partial Differential Equations and Hamiltonian Systems*, Springer-Verlag, Berlin–Heidelberg (2008).

[66] V. A. Trenogin, *Functional Analysis* [in Russian], Nauka, Moscow (1993).

[67] M. M. Vainberg, *Variational Methods for the Study of Nonlinear Operators*, Holden-Day, San Francisco–London–Amsterdam (1964).

[68] V. S. Vladimirov, *Equations of Mathematical Physics*, Marcel Dekker, New York (1971).

[69] Zhuoqun Wu, Jingxue Yin, and Chunpeng Wang, *Elliptic and Parabolic Equations*, World Scientific, New Jersey (2006).

[70] K. Yosida, *Functional Analysis*, Springer-Verlag, Berlin (1994).

[71] V. A. Zorich, *Mathematical Analysis*, Springer, Berlin (2015).

Index